Rab GTPases and Membrane Trafficking

Edited by

Guangpu Li

Department of Biochemistry and Molecular Biology
The University of Oklahoma Health Sciences Center
USA

&

Nava Segev

Department of Biochemistry and Molecular Genetics
The University of Illinois at Chicago
USA

CONTENTS

FOREWORD

Since the 19th century, when Metchnikoff fed fragments of blue litmus paper to phagocytes and observed that they turned red, indicating the presence of an "acid compartment", biologists and physicians have been intrigued with the mechanisms of membrane internalization and trafficking and their implications for understanding physiology. In the 20th century, when much was already known at the morphological level about cell secretion and endocytosis, a major advance, the discovery of the Ypt/Rab family of GTPases, heralded the entry of an era that has propelled research in membrane trafficking into the 21st century. The discovery of Ypt1 in yeast and the subsequent discovery of the Rab GTPases in mammalian cells has revealed a vast network of interacting transport/trafficking pathways in cells not unlike, metaphorically speaking, a national highway system for packaging and distributing cargo and products from one location to another.

The course of evolution has been characterized by several large expansions in the signaling repertoire of cells and tissues accompanied by the five-fold expansion in the Rab GTPase catalog from yeast to man. This expansion, along with the corresponding need for additional Rab-specific effectors, motors, exchange factors and GAPs, has broadened our understanding of the immense impact of the Ypt/Rab family on human physiology and health.

A new offering by Li and Segev entitled "Rab GTPases and Membrane Trafficking" brings us up to date on many of the recent developments in Ypt/Rab biology and physiology. Following an excellent introduction by the Li and Segev that touches on many of the recent advances in this rapidly developing field, a collection of a dozen chapters focus on selected Ypt/Rabs and the transport and signaling functions that they oversee. A superb collection of contributors brings both diversity and substance to the volume while highlighting the novelty of individual Ypt/Rab GTPases. The last selection on the evolution of the Ypt/Rab family provides an excellent finale. The eBook makes good use of cartoons, is amply referenced and should be of great value to graduate students and aficionados alike.

Philip D. Stahl
Washington University
St. Louis, MO
USA

PREFACE

Ypt/Rab GTPases, which form the largest branch of the Ras-related small GTPase superfamily, regulate intracellular membrane trafficking in all eukaryotes. Like other GTPases, Ypt/Rab proteins act as molecular switches that alternate between the GTP-bound active and the GDP-bound inactive conformations. In their active form Ypt/Rabs interact with effectors, which in turn promote specialized functions in intracellular trafficking through the exocytic and endocytic pathways. Since their discovery over two decades ago, a wealth of knowledge has accumulated about the roles that Ypt/Rab proteins play in vesicular transport steps including vesicle budding, movement, and fusion, but a united mechanism remains elusive and the subject remains a very active area of research. In recent years, Ypt/Rabs and trafficking have emerged as important players in other cellular processes, such as signal transduction, cell growth and differentiation. Importantly, Rabs have been implicated in various human diseases ranging from diabetes to cancer, justifying a timely eBook on the subject.

This eBook covers well-characterized Ypt/Rabs involved in both exocytic and endocytic pathways as well as newly identified and uncharacterized Rabs. Chapter 1 provides an overview of the general principles of the Ypt/Rab GTPase cycle and the vesicular trafficking steps regulated by them. Chapters 2-6 describe the regulation and functions of exocytic Rabs, including Ypt1-Rab1, Ypt31/Ypt32-Rab11, Rab6, Sec4, Rab3, and Rab27. In yeast, Ypt1 and Sec4, the founding members of the Ypt/Rab family, control the first and final steps of the exocytic pathway, respectively. Ypt31/Ypt32-Rab11 and Ypt6-Rab6 regulate traffic through the Golgi complex. On the other hand, Rab3 and Rab27 are found only in higher eukaryotes, are specifically expressed in secretory cells and play a role in regulated secretion. Chapters 7-11 describe endocytic and recycling Rabs, including Rab5, Rab21, Rab22, Rab4, Rab8, Rab11, Rab7, and Rab9. Members of the Rab5 sub-family, Rab5, Rab21, and Rab22 regulate early endosomal sorting, fusion and movement, whereas Rab4, Rab8 and Rab11 regulate recycling pathways. Rab7 and Rab9 regulate transport from late endosomes to lysosomes and the Golgi, respectively. The final chapter, Chapter 12, describes newly identified and uncharacterized Rabs, which represent a great majority of the 7,500 Rabs identified in 250 sequenced genomes. Rabs with unknown function have usually emerged late in evolution and are expressed in specific tissues at low levels.

By covering Ypt/Rab GTPases in one eBook, a comprehensive picture emerges in which each Ypt/Rab controls multiple vesicular trafficking steps *via* interactions with multiple effectors. This eBook should provide a useful resource for researchers, teachers and students interested in the field.

Guangpu Li
Department of Biochemistry and Molecular Biology
The University of Oklahoma Health Sciences Center
USA

Nava Segev
Department of Biochemistry and Molecular Genetics
The University of Illinois at Chicago
USA

List of Contributors

Anna Akhmanova

Department of Biology
Utrecht University
The Netherlands
a.akhmanova@uu.nl

Margarita Cabrera

University of Osnabrück
Department of Biology/Chemistry
Barbarastrasse 13
49076 Osnabrück
Germany
margarita.cabrera@biologie.uni-osnabrueck.de

Shu H. Chen

Department of Biological Sciences
University of Illinois at Chicago
Chicago, IL 60607
USA
shuhui.chen@nih.gov

François Darchen

CNRS/Université Paris Descartes UMR 8192
Centre Universitaire des Saints-Pères
47 rue des Saints-Pères
75006 Paris
France
francois.darchen@parisdescartes.fr

Magda Deneka

Department of Cell Biology
University Medical Center Utrecht
3508 GA Utrecht
The Netherlands
m.deneka@umcutrecht.nl

Claire Desnos

CNRS/Université Paris Descartes UMR 8192
Centre Universitaire des Saints-Pères
47 rue des Saints-Pères
75006 Paris
France
claire.desnos@parisdescartes.fr

Eric J. Espinosa

Department of Biochemistry
Stanford University School of Medicine
Stanford, CA 94305-5307
USA
eespinos@stanford.edu

Vishnu Ganesan

Department of Biology
University of Pennsylvania
Philadelphia, PA 19104-6018
USA
vishnug@seas.upenn.edu

James R. Goldenring

Departments of Surgery and Cell and Developmental Biology
Epithelial Biology Center
Vanderbilt University School of Medicine
2213 Garland Ave.
Nashville, TN 37232-2733
USA
jim.goldenring@vanderbilt.edu

Bruno Goud

Institut Curie
CNRS UMR144
26 rue d'Ulm
75248 Paris cedex 05
France
bruno.goud@curie.fr

Wei Guo

Department of Biology
University of Pennsylvania
Philadelphia, PA 19104-6018
USA
guowei@sas.upenn.edu

John A. Hammer III

Laboratory of Cell Biology
National Heart, Lung and Blood Institute
National Institutes of Health
Bethesda, MD 20824
USA
hammer@nhlbi.nih.gov

Andreas Knödler

Department of Biology
University of Pennsylvania
Philadelphia, PA 19104-6018
USA
knoedler@sas.upenn.edu

Lynne A. Lapierre

Departments of Surgery and Cell and Developmental Biology
Epithelial Biology Center
Vanderbilt University School of Medicine
2213 Garland Ave.
Nashville, TN 37232-2733

USA
lynne.lapierre@vanderbilt.edu

Guangpu Li

Department of Biochemistry and Molecular Biology
University of Oklahoma Health Sciences Center
Oklahoma City, OK 73104
USA
guangpu-li@ouhsc.edu

Mirjana Nordmann

University of Osnabrück
Department of Biology/Chemistry
Barbarastrasse 13
49076 Osnabrück
Germany
mirjana.nordmann@biologie.uni-osnabrueck.de

José B. Pereira-Leal

Instituto Gulbenkian de Ciência
Rua da Quinta Grande 6
P-2781-901
Portugal
jleal@igc.gulbenkian.pt

Suzanne R. Pfeffer

Department of Biochemistry
Stanford University School of Medicine
Stanford, CA 94305-5307
USA
pfeffer@stanford.edu

Ioana Popa

Department of Cell Biology
University Medical Center Utrecht
3508 GA Utrecht
The Netherlands
i.popa@umcutrecht.nl

Maria Luisa Rodrigues

Instituto Gulbenkian de Ciência
Rua da Quinta Grande 6
P-2781-901
Portugal
mrodrigues@igc.gulbenkian.pt

Joseph T. Roland

Departments of Surgery and Cell and Developmental Biology
Epithelial Biology Center
Vanderbilt University School of Medicine
2213 Garland Ave.
Nashville, TN 37232-2733

USA
joseph.t.roland@vanderbilt.edu

Emma Martinez Sanchez

Department of Cell Biology
University Medical Center Utrecht
3508 GA Utrecht
The Netherlands
e.sanchez@umcutrecht.nl

Nava Segev

Department of Biochemistry and Molecular Genetics
University of Illinois at Chicago
Chicago, IL 60607
USA
nava@uic.edu

Peter van der Sluijs

Department of Cell Biology
University Medical Center Utrecht
3508 GA Utrecht
The Netherlands
p.vandersluijs@umcutrecht.nl

David Taussig

Department of Biological Sciences
University of Illinois at Chicago
Chicago, IL 60607
USA
dtauss2@uic.edu

Christian Ungermann

University of Osnabrück
Department of Biology/Chemistry
Barbarastrasse 13
49076 Osnabrück
Germany
christian.ungermann@biologie.uni-osnabrueck.de

Xufeng Wu

Laboratory of Cell Biology
National Heart, Lung and Blood Institute
National Institutes of Health
Bethesda, MD 20824
USA
wux@nhlbi.nih.gov

2

CHAPTER 1

Ypt/Rab GTPases and Intracellular Membrane Trafficking: An Overview

Guangpu Li[1,*] and Nava Segev[2]

[1]*The University of Oklahoma Health Sciences Center, USA and* [2]*The University of Illinois at Chicago, USA*

Abstract: Ypt/Rab GTPases form the broadest group of GTPases in eukaryotic cells. These conserved GTPases are key regulators of intracellular trafficking. Like other monomeric GTPases, Ypt/Rabs switch between GDP- and GTP-bound forms with the help of their upstream regulators. When in the GTP-bound form, they interact with downstream effectors that mediate all vesicular transport steps, from vesicle formation to fusion. In addition, Ypt/Rabs are considered as candidates to coordinate intermediate/individual steps during vesicular transport. Due to their central role in intracellular traffic regulation, Ypt/Rabs have been implicated in human diseases ranging from diabetes and cancer to neurological and immunological disorders.

Keywords: Ypt, Rab, GTPase, GTP-binding Protein, Membrane Trafficking, Vesicular Transport.

1. HISTORY

The Ypt/Rab GTPase field emerged as its own discipline more than two decades ago from studies done in yeast. Until then, large and small GTPases were thought to function in transduction of signals across the outer-cell membrane [1, 2]. Ypt1 was the first GTPase that was shown to function inside cells and to regulate intracellular trafficking [3]. Since then, the field has evolved through multiple transitions, often breaking dogmas widely accepted but poorly grounded.

In the multiple intracellular trafficking pathways, membranes and proteins move inside cells in an organized manner. Outward transport, termed exocytosis, involves translocation of newly synthesized proteins into the endoplasmic reticulum (ER), followed by their movement through the Golgi apparatus towards the plasma membrane (PM). Branches of this pathway carry enzymes from the Golgi to endosomes or lysosomes. Inward transport, termed endocytosis, involves uptake of proteins from the PM and moving them through a set of endosomes to the lysosome for degradation. In each organelle, proteins are sorted to vesicles heading forward to the next compartment or recycling back to the original compartment, *e.g.*, endosome to PM recycling. Ypt/Rabs orchestrate these highly organized trafficking pathways (see Fig. **4**).

1.1. Ypt/Rabs and Trafficking: From Yeast to Mammals

The first two Ypt/Rab GTPases were discovered in yeast. At that period, GTP-binding proteins were supposed to act in transduction of signals through the PM [1, 2]. Sec4 was cloned by complementation of the *sec4* mutation, which conferred a late secretory defect [4]. This suggested a role for Sec4 in intracellular trafficking; however, this function was still associated with the PM. The real surprise came with Ypt1. Ypt1 was initially identified as a yeast Ras homolog with a very different function from that of Ras1 and Ras2 [5]. A role of Ypt1 in intracellular transport was shown using reverse genetics. Specifically, an ER-to-Golgi defect was discovered in *ypt1* mutant cells, and the Ypt1 protein was localized to the yeast Golgi. This was the first GTPase shown to act not at the PM, but on an intracellular organelle [3].

The conservation of Ypts and their intracellular localization in evolution was first shown for Ypt1. Importantly, the closest human homolog of Ypt1, Rab1, not only shares 70% amino-acid sequence identity with Ypt1, but like Ypt1, it was shown to localize to the Golgi [3]. More Ypts and Rabs were discovered soon after by searching for homologs of Ypt1 and Sec4 [6]. The full sets of Ypts and Rabs were identified when the yeast and human genome-sequencing projects were completed. Currently, there are known to be eleven yeast Ypts and more than 60 human Rabs [7, 8].

*****Address correspondence to Guangpu Li:** Department of Biochemistry and Molecular Biology, University of Oklahoma Health Sciences Center, Oklahoma City, OK 73104, USA; Tel: 405-271-2227; Fax: 405-271-3910; E-mail: guangpu-li@ouhsc.edu

Multiple alignment analyses of the yeast Ypts with the human Rabs, which were studied extensively, show that they fall into two classes: exocytic and endocytic Ypt/Rabs (Fig. **1**). The nine Ypts shown to play a role in intracellular trafficking, excluding Ypt10 and Ypt11, fall into six groups, with Ypt31/32 and Ypt51(Vps21)/Ypt52/Ypt53 having higher sequence similarity to each other than to any Rab: Ypt1, Ypt31/32, Sec4, Ypt6, Ypt51/52/53, and Ypt7. Each of these six Ypt groups has a ubiquitous Rab group to which it is more similar than to Ypts from other groups. There are at least two examples of Rabs acting as functional homologs of Ypts: Ypt1-Rab1A and Ypt6-Rab6A [9, 10]. The ability of mammalian Rabs to function in the yeast cell suggests that the Rabs can make all the necessary interactions that the Ypts make with their accessory factors.

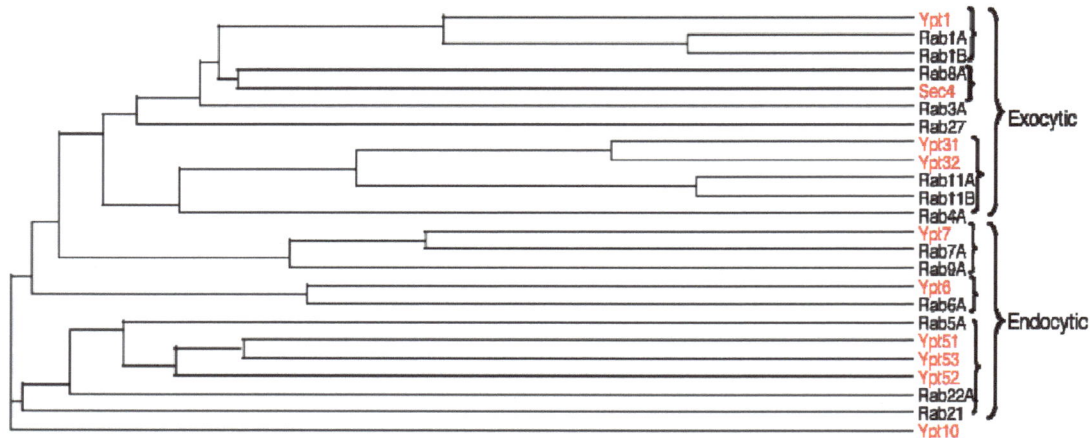

Figure 1: Multiple alignment analyses of the Ypts and Rabs discussed in this eBook. Sequences of ten *S. cerevisiae* Ypts were used for multiple alignment analyses with the human Rabs. The Ypt/Rabs fall into two groups of exocytic and endocytic GTPases. Each of the seven Ypt groups has a close human Rab homolog group (in brackets). Sequences were retrieved from UniProt and aligned using Clustalw. Ypts are shown in red.

1.2. Upstream Regulation: From Hydrolysis-Driven Action to Activation by Exchange

Like all GTPases, Ypt/Rabs cycle between the GTP- and the GDP-bound forms with the help of their upstream regulators: guanine-nucleotide exchange factors (GEFs) and GTPase-activating proteins (GAPs).

The first analogy of Ypt GTPase function was made to EF-Tu. In a review that made its way to textbooks, Bourne compared the role of GTP hydrolysis in EF-Tu function to that of the Ypts. This early model entailed that GTP hydrolysis provides a mechanism for vectorial transport due to free energy loss upon hydrolysis [11]. GAP was assumed to be on the acceptor membrane and stimulate this hydrolysis. The observation that a Sec4 mutant defective in GTP hydrolysis exhibits a secretory defect further supported this model [12].

Work done in yeast and mammalian cells overturned this dogma. In yeast, a number of papers on mutations and regulators of Ypt1 challenged the then-prevailing model. First, it was shown that cells expressing a Ypt1 mutant protein defective in GTP hydrolysis does not have a serious defect in ER-to-Golgi transport. In addition, the GAP activity for Ypt1 did not seem to localize to the Golgi, contrary to the model accepted at the time [13, 14]. More importantly, inhibition of nucleotide exchange was found to have a major effect on ER-to-Golgi transport, and the GEF for Ypt1 was localized to the Golgi [15]. In mammalian cells, mutations that lock Rab5 in its GTP-bound or nucleotide-free forms were used to show that Rab5 has to be in the GTP-bound form to mediate endosome fusion and endocytosis [16-18]. In contrast, GTP hydrolysis is required after membrane fusion to inactivate Rab5. Based on these findings, it was suggested that nucleotide exchange and GEFs are crucial for Ypt/Rab function whereas GTP hydrolysis and GAP play a role in the recycling of Ypt/Rabs for multiple rounds of action [19, 20]. Currently it is the prevailing view concerning the upstream regulation of Ypt/Rabs [21].

Ypt/Rabs attach to membranes *via* their geranylgeranyl lipid tail. Two accessory factors help Ypt/Rab recycling between membranes. First, after they complete their function and hydrolyzed their GTP, the GDP-dissociation inhibitor, GDI, extracts GDP-bound Ypt/Rab proteins from membranes and keeps them soluble in the cytoplasm. Second, the GDI displacement factor, GDF, helps to recruit GDI-attached GDP-bound Ypt/Rabs from the cytoplasm to the right membranes [22, 23].

1.3. Mechanisms: From Vesicle Fusion to Every Vesicle-Transport Step

Initially, both Ypt1 and Sec4 were shown to be required for vesicle fusion: Ypt1 for fusion of ER vesicles with the *cis*-Golgi, and Sec4 for fusion of *trans*-Golgi vesicles with the PM [24, 25]. Based on these findings, it was extrapolated that Ypt/Rabs mediate vesicle targeting and fusion [26, 27]. The first outliers from this axiom were the Ypt31/32 GTPases, which were shown to act in the formation of *trans*-Golgi vesicles [28]. The question that emerged was: How can such small proteins, ~200 amino acids long, mediate multiple vesicular transport events? Through work done on multiple Ypt/Rabs in many labs it has become clear that Ypt/Rabs do that by interacting with multiple effectors that themselves mediate vesicle formation, motility, tethering and fusion [20, 21].

A role for GTPases in coordination of the exocytic pathway steps was first suggested based on genetic interactions between exocytic Ypts and GEFs for Arfs, which form another family of GTPases that regulates trafficking [29]. Currently, Ypt/Rabs are implicated in integration of intermediate/individual steps of vesicular transport, coordination of multiple transport steps of the same pathway, and coordination of different trafficking pathways [30-32].

1.4. Function: from Transport Steps to Compartment Specificity and Disease

Because originally Ypt1 and Sec4 were each assigned to a specific transport step, it was assumed that Ypt/Rabs regulate distinct transport steps [27]. However, it was then shown that Ypt1 mediates at least two steps in the *cis* Golgi, and Ypt31/32 do the same in the *trans*-Golgi [28, 33, 34]. These findings suggested that Ypt/Rabs are not specific to individual transport steps, but to individual compartments [19]. It turns out that in mammalian cells there are more Rabs than cellular compartments. While some Rabs are cell- or developmental stage-specific, it is clear that there can be more than one Rab per compartment. Currently it is thought that different Rabs present on the same compartment define separate membrane domains [35, 36].

Since Ypt/Rabs regulate a basic cellular process required for the metabolism and life cycle of cells, it is clear that Rabs are important for human health and disease. Indeed, Rabs were implicated in processes that require secretion, *e.g.*, transport of insulin or antibodies. It is therefore not surprising that defects in Rabs have been found to be underlying causes in diabetes and immunological disorders [37, 38]. In addition, Rabs have been connected to other diseases like cancer and neurological disorders [39, 40]. Finally, there are now examples of inherited disorders caused by mutations in Rabs or their regulators [41, 42]. As a result, Rabs are currently considered as therapeutic targets for multiple diseases [43, 44].

2. UPSTREAM REGULATION

Ypt/Rabs are regulated at two levels: nucleotide switching and cycling between membranes (Fig. **2**). The two types of nucleotide-switching regulators are GEFs and GAPs. While GEFs for different Ypt/Rabs are conserved between organisms, they do not share apparent sequence or structure similarity. In contrast, all known Ypt/Rab GAPs, which stimulate GTP hydrolysis, contain a conserved domain and their mechanism of action is thought to be similar. The two regulators of Ypt/Rab membrane recycling are GDI and GDF. While GDIs, which extract GDP-bound Ypt/Rabs from membranes, are not specific to individual Ypt/Rabs, GDFs, which recruit Ypt/Rabs to appropriate membranes, are expected to be Ypt/Rab specific. Currently, very little is known about GDFs.

The lifetime of a GTP-bound Ypt/Rab is thought to be dependent on the GAP concentration on the target membrane. When GTP is hydrolyzed and the Rab is converted to the inactive GDP-bound state, it is extracted from the membrane by GDI. The Rab-GDI complex in the cytoplasm allows the Ypt/Rab to recycle back to the donor membrane for a new round of activity.

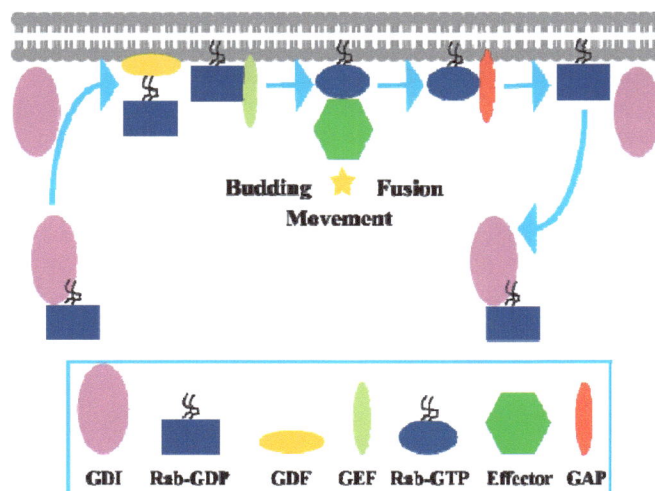

Figure 2: Regulation of Ypt/Rab nucleotide and membrane cycling. Ypt/Rab GTPases alternate between active GTP-bound and inactive GDP-bound conformations. The GDP-bound Ypt1/Rab in the cytosol is in complex with GDI. Each Ypt/Rab-GDP-GDI complex is targeted to a specific membrane that contains a cognate GDF. GDF replaces GDI by interacting with the prenyl groups and possibly a protein domain. The mechanism of Ypt/Rab targeting to a specific compartment remains unknown and represents a major challenge in the field. GDF transfers Ypt/Rab-GDP to the membrane where a GEF facilitates the nucleotide exchange reaction of GDP dissociation and GTP binding. The GTP-bound Ypt/Rab is biologically active and temporally interacts with multiple effectors to promote vesicle budding, movement along cytoskeleton, and fusion with target compartment. The GTP-bound state is transient and converts to inactive GDP-bound state by GTP hydrolysis, which is accelerated by a GAP. The GDP-bound Ypt/Rab can then be extracted from the membrane by GDI and the complex recycles to start a new cycle of activity.

2.1. Ypt/Rab GEFs

The identification of Ypt/Rab GEFs has been slow because GEFs for different Ypt/Rabs cannot be identified by sequence analysis. Ypt/Rab GEFs have been identified biochemically by determining the activities of proteins or complexes in stimulation of nucleotide exchange or interaction with the GDP-bound or nucleotie-free forms of Ypt/Rabs. It is now established that GEFs for the Rabs that have yeast Ypt homologs are conserved from yeast to humans. Currently, the mechanism by which Ypt/Rab GEFs stimulate nucleotide exchange is not completely understood [45, 46].

Ypt/Rab GEFs include diverse groups of proteins. Four families of Ypt/Rab GEFs have been well documented, including the TRAPP (transport protein particle) complexes, Sec2/Rabin proteins, Vps9 domain-containing proteins, and DENN (differentially expressed normal versus neoplastic) domain-containing proteins. Even though different families of GEFs share no sequence homology in the catalytic domain, structural studies of Rab GEFs in complex with nucleotide-free and GDP-bound Rabs suggest a common basic mechanism. The GEFs promote GDP dissociation from Rabs by opening switch I region and abrogating Mg^{2+} coordination [45]. The specific ways by which GEFs do that can be different. In the case of Vps9 domain-containing GEFs, a catalytic aspartic acid residue interacts with the β-phosphate of the GDP and a conserved lysine residue in the phosphate-binding loop of the Rab to release GDP and stabilize the nucleotide-free form [47]. In contrast, the TRAPP I complex employs catalytic patches present on multiple subunits to open the switch I region and block Mg^{2+} coordination [48].

Three TRAPP complexes were identified in yeast. All three complexes include the six TRAPP I subunits, Bet3, Bet5, Trs20, Trs23, Trs31, and Trs33. TRAPP II contains the subunits plus three additional subunits, Trs65, Trs120 and Trs130 [49]. TRAPP III contains one additional subunit, Trs85 [50]. The catalytic core of TRAPP I consists of four subunits: Bet3p, Bet5, Trs23 and Trs31. TRAPP I and TRAPP III act as GEFs for Ypt1 whereas TRAPP II acts as a Ypt31/32 GEF [50, 51]. TRAPP I plays a role in ER to Golgi transport, TRAPP III in autophagy, and TRAPP II in post-Golgi transport [49]. Only one mammalian

TRAPP complex has been identified and it acts as a GEF for Rab1 [52]. Chapter 2 provides more detailed discussion on the TRAPP complexes.

The Ypt/Rabs that regulate the last step of the exocytic pathway are Sec4 in yeast and Rab8 in higher eukaryotes. They control post-Golgi vesicle movement and fusion with the PM and regulate cell polarity. The GEFs that activate Sec4 and Rab8 are Sec2 and Rabin8, respectively. These GEFs are recruited to post-Golgi vesicles by interacting with the upstream Ypt/Rabs, Ypt31/Ypt32 and Rab11, respectively [30, 53]. These exocytic Rab-GEF activation cascades, together with endocytic Rab-GEF cascades, *e.g.*, the Rab22-Rabex-5-Rab5 [54] and Rab5-Mon1/Ccz1-Rab7 cascades [55, 56], exemplify an important concept of coordination of targeting and fusion by Rab GEFs,

Vps9 domain-containing GEFs activate members of the Rab5 GTPase subfamily, including Rab5, Rab21, Rab22, and Rab31, which regulate early endosome movement and fusion. The prototype of this family of GEFs is Vps9, which was originally identified in yeast [57] and was shown to act as a GEF for Vps21/Ypt51 (Rab5 homolog in yeast) [58]. In higher eukaryotes, Vps9 domain-containing proteins include RME-6 in *Caenorhabditis elegans*, Sprint in *Drosophila melanogaster*, and Rabex-5, RIN proteins, Alsin, and RME-6/GAPex-5 in mammals [59]. All the characterized Vps9 domain-containing proteins can act as Rab5 GEF. In addition, they exhibit distinct specificity towards other members of the subfamily. For example, Rabex-5 acts as a Rab21 GEF [60] while GAPex-5 acts as a Rab31 GEF [61]. In addition to the Vps9 domain, these GEFs also contain other functional domains including membrane-targeting domains for recruitment to endosomes. Different membrane targeting mechanisms and localization to distinct membrane domains can further contribute to the specificity of GEFs towards a given Rab. Chapter 7 provides more detailed discussion on the subject.

Another reported Rab GEF is the Ric1p/Rgp1p complex, which shows GEF activity towards Ypt6p in yeast [62]. Its homolog in higher eukaryotes is likely a GEF for Rab6.

DENN domain-containing GEFs target to various organelles and activate a diverse group of Rabs including Rab3, Rab9, Rab10, Rab12, Rab27, Rab28, Rab35, and Rab39 [63, 64]. In contrast to the other GEFs such as TRAPP complexes, Ric1/Rgp1 and Vps9, DENN domain-containing Rab GEFs are absent in the budding yeast *S. cerevisiae* [64]. However, they are conserved in evolution with one member in *Schizosaccharomyces pombe*, five in *Caenorhabditis elegans* and eighteen in humans [63]. Based on the Rab specificity of DENN domain-containing proteins, it seems that they act as GEFs for Rabs that do not have close yeast homologs.

2.2. Ypt/Rab GAPs

The first Ypt/Rab GAP, Gyp6, was identified in yeast in a genome-wide biochemical screen [65]. Since then, a large family of Ypt/Rab GAPs has been identified in both yeast and mammalian genomes based on sequence similarity. All known Ypt/Rab GAPs contain a GYP (yeast) or TBC (mammalian) domain. TBC is thought to stimulate GTP hydrolysis through a mechanism similar to that used by Ras and Rho GAPs, using "arginine and glutamine fingers" [45].

What hampered progress in elucidation of Ypt/Rab GAP function was the fact that in biochemical GTP hydrolysis assays, most Ypt GAPs seemed to be nonspecific for a certain Ypt [66]. Therefore, assignment of a GAP to a Ypt in yeast relies on genetic analysis, which assumes that GAPs are negative regulators of the Ypt. The problem is that GAPs for other GTPases, like Ras and Arfs, are thought to also act as downstream effectors [67, 68]. Because both GAPs and effectors interact with GTP-bound GTPases, it is sometimes hard to decipher the exact role of a certain GAP. For example, Gyp1 is considered a Ypt1 GAP based on genetic studies, but it also interacts with Ypt32 in its GTP-bound form. For this latter interaction, Gyp1 was suggested to act as a Ypt32 effector and to participate in a GAP cascade that define Ypt/Rab boundaries on compartments [31, 69].

A more recent *in vitro* analysis of the 38 human TBC-containing proteins with an extensive selection of Rabs suggests that at least some Rab GAPs exhibit Rab specificity [70-72]. Using altered experimental conditions for biochemical GAP assays with Gyps might allow defining Gyp specificity for Ypts as well.

There is at least one known non-TBC domain Rab GAP, Rab3GAP1/Rab3-GAP2 [45] that was identified based on purification of the active component [73]. This suggests that there might be more Ypt/Rab GAPs that are yet to be identified.

2.3. Membrane Cycling: GDI and GDF

Ypt/Rabs are attached to membranes through a C-terminal geranylgeranyl lipid group. In the cytoplasm Ypt/Rabs are found in a complex with special chaperones that protect the lipid in this hydrophilic environment. Newly synthesized Ypt/Rabs are first found in a complex with a Rab escort protein, REP, whereas recycling Ypt/Rabs are complexed with GDI, which can extract GDP-bound Ypt/Rabs from membranes. REP and GDI share sequence and structure similarities [74, 75]. In yeast there is one Ypt REP, Mrs6 [76], and one GDI, GDI1 (Sec19) [77]. In mammalian cells, there are at least three GDI isoforms (α, β and γ). GDIα is highly expressed in the brain while the latter two isoforms are ubiquitously expressed in all tissues [78]. GDI recognizes common structural features in the Rab proteins, *e.g.*, the conserved residues in the switch regions and the prenyl group [79], and binds to the GDP-bound forms of Rabs to form cytosolic Rab-GDI complexes, although the affinities may vary.

Each Ypt1/Rab-GDI complex is targeted to a specific membrane compartment such as ER, Golgi, or endosomes, but the mechanism is not well understood. It has been proposed that the hypervariable region upstream of the C-terminal cysteine motif of the Ypt/Rab contributes to specific membrane targeting [80]. However, potential binding sites on the membranes have not been identified. It is suggested that a GDF on the membrane can replace and release GDI to facilitate the insertion of the Rab prenyl groups into the membrane [81]. In agreement, the Yip3/PRA1 protein was shown to contain GDF activity towards endosomal Rabs such as Rab9 and Rab5 [82]. Yip3/PRA1 is a member of the Yip family of proteins and contains four membrane-associated domains with N- and C-termini facing the cytoplasm [83]. However, other members of the Yip family have not been reported to act as Ypt GDFs. In addition, a recent study postulates that GDF is not thermodynamically necessary for the membrane targeting of Rab-GDI complexes [84]. Once on the membrane, the Ypt/Rab is activated by its cognate GEF to release GDP and bind GTP. The membrane- and GTP-bound Ypt/Rab interacts with effectors and promotes downstream functions such as vesicle formation, movement and fusion.

3. DOWNSTREAM EFFECTORS

The way by which Ypt/Rabs can mediate multiple vesicular transport steps is through interaction with diverse effectors [20]. There are numerous examples of Ypt/Rab effectors that range from molecular motors to tethering factors and membrane fusion regulators (Fig. **3**). These effectors were mostly identified as specific Ypt/Rab interactors. When the identity of specific Ypt/Rab interaction domains on these effectors becomes clearer, it may be possible in the future to identify the whole Ypt/Rab "effectorome". Below are some examples of Ypt/Rab effectors that mediate the various vesicular transport steps.

3.1. Vesicle Formation: Cargo Sorting and Vesicle Budding

Formation of transport vesicles destined for different compartments requires sorting of cargo proteins into the vesicles and subsequent vesicle fission. Effectors of Rab GTPases play important roles in both processes; below are a number of examples.

Rab9 is localized on late endosomes (LEs) and regulates vesicular transport from LEs to the TGN (*trans* Golgi network) (Chapter 11). An important cargo protein transported in this pathway is the mannose 6-phosphate receptor (MPR). MPR is a *trans*-membrane protein that captures and delivers luminal lysosomal enzymes from the Golgi to endosomes and then returns to the Golgi for a new round of delivery. In LEs, Rab9 recruits the TIP47 effector and enhances its affinity to the cytoplasmic domain of MPR. Rab9 and TIP47 promote MPR sorting into the transport vesicles destined for the TGN [85]. Rab5 was implicated in sorting of transferrin into clathrin-coated endocytic vesicles using an *in vitro* budding assay [86] (Chapter 7). The Rab5 effector that mediates transferrin sorting is unknown. However, other cargo proteins are sorted into endocytic vesicles *via* direct interaction with Rab5. For example, Rab5 as well as the closely related Rab21 binds to a number of α subunits of the β1 integrins and promotes their endocytosis and PM recycling [87]. Integrin remodeling at the cell surface and their interaction with the extracellular matrix are important for cell adhesion and migration [87] (Chapter 7).

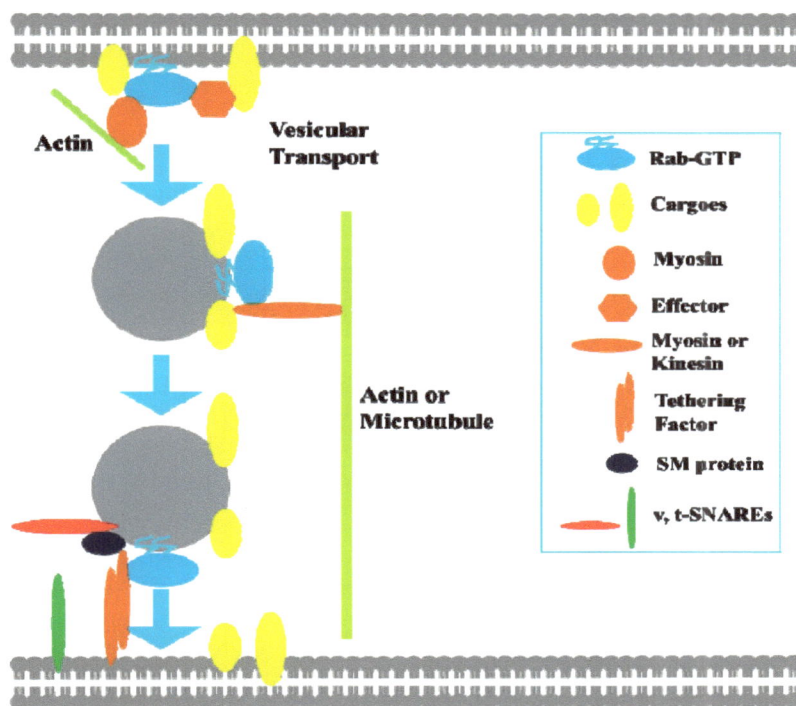

Figure 3: Ypt/Rab effectors and functions. Active Ypt/Rab-GTP molecules on the membrane temporally interact with a number of effectors to facilitate cargo selection/packaging, vesicle budding from a donor compartment, vesicle movement along actin or microtubule cables, and vesicle tethering to a target compartment. Some Rab effectors are cargo proteins that are packaged into transport vesicles *via* direct or indirect interaction with Rab-GTP. Myosins represent another group of Rab effectors that facilitate vesicle budding and movement along actin cytoskeleton. Along this line, some kinesins are also Rab effectors and recruited to the vesicle to facilitate the movement along microtubule cables. The best-characterized Rab effectors are arguably the tethering factors that dock transport vesicles to cognate acceptor compartments. This leads to formation of a *v*- and *t*-SNARE complex and ultimately to membrane fusion and delivery of cargo to the acceptor compartment.

Rab6 and its effector, the actin motor myosin II, were recently proposed to play a role in vesicle formation [88] (Chapter 3). Inhibition of Rab6 or myosin II function results in a fission block of the formation of Rab6-containing vesicles. Miserey-Lenkei *et al.* proposed that Rab6 recruits myosin II to the Golgi membrane and promotes anterograde and retrograde vesicle fission from this compartment [88]. Recruitment of molecular motors by Rab GTPases also plays a role in vesicle movement, as discussed below.

3.2. Vesicle Movement: Actin and Microtubule Motors

In addition to myosin II, other actin and microtubule motors were identified as Ypt/Rab effectors. These motors are recruited to various transport vesicles by Ypt/Rabs to facilitate vesicle movement along cytoskeleton cables. Myosin V motors are recruited by a number of Rabs to transport vesicles destined for the PM. In budding yeast, Myo2, a member of the myosin V family, is an effector of Ypt31/Ypt32 and is recruited by Ypt31/Ypt32 to post-Golgi vesicles [89] to facilitate movement to polarized budding sites on the plasma membrane [89-91]. In higher eukaryotes, myosin Va is recruited to specialized secretory vesicles by Rab27 in various exocytic cells of endocrine and exocrine origins, such as cytotoxic T lymphocytes (CTL) in the immune system and melanocytes in the skin [92, 93] (Chapters 5 and 6). Rab27 first binds to its effector melanophilin (Slac2) [94], which in turn recruits myosin Va to the secretory granules in CTL [95] or melanosomes in melanocytes [96, 97] to promote their movement towards the cell periphery. Inactivating mutations in Rab27 result in the autosomal recessive disorder Griscelli syndrome, which is characterized by defective melanosome transport and pigment dilution in skin and hair as well as defects in the immune system with uncontrolled CTL activation [98, 99]. Myosin Vb, on the other hand, is an effector for Rab11 [100] and Rab8 [101] and is recruited by these Rabs to recycling vesicles to facilitate movement towards the PM (Chapter 8).

Microtubule motors such as kinesins and dynein are also recruited to transport vesicles directly or indirectly by Rab GTPases to promote vesicle movement along minus- and plus-end microtubule cables. To maintain early endosomal distribution at the cell periphery, Rab5 activates hVPS34 (a PI 3-kinase) to produce PI3P in early endosomal membranes [102], which is required for recruitment of the plus-end kinesin KIF16B to early endosomes to maintain early endosomal distribution at the cell periphery [103] (Chapter 7). Rab7, on the other hand, recruits the minus-end motor dynein-dynactin complex to late endosomes *via* its effector RILP (Rab7-interacting lysosomal protein) and promotes endosomal movement towards the minus-end of microtubules and the perinuclear region of a cell [104, 105] (Chapter 10). Rab6 was shown to utilize a dynein motor for minus-end transport along microtubules. Rab6 recruits the dynein-dynactin complex to Golgi-derived vesicles directly or *via* its effectors BICD1 (Bicaudal-D1) and BICDR-1 (Bicaudal-D-related protein 1), to promote retrograde transport towards the centrosome [106, 107] (Chapter 3). In addition, Rab6 also directly binds Rabkinesin-6, a Golgi-associated kinesin-like protein, and this interaction is suggested to play a role in the trafficking of Golgi-derived vesicles and cytokinesis [108, 109].

3.3. Vesicle Attachment: Tethering Factors

Tethering factors are necessary for specific and efficient docking of transport vesicles to target membranes, prior to pairing of v- and t-SNAREs (SNAP receptors) and membrane fusion. Tethering factors were originally identified as Rab effectors and recruited by activated Rabs to the membrane to facilitate vesicle docking and fusion [110]. In general, there are two classes of tethering factors: long coiled-coil homodimers and large multi-subunit complexes.

EEA1 (early endosomal antigen 1), Rabenosyn-5, and p115/Uso1 are examples of coiled-coil homodimeric tethering factors that act as Ypt/Rab effectors. EEA1 and Rabenosyn-5 are both Rab5 effectors (Chapter 7). In addition to the Rab5-binding domain, they also contain a FYVE domain that binds to the membrane lipid PI3P (phosphotidylinositol 3-phosphate). In EEA1, the Rab5-binding domain is at the N-terminus and the FYVE domain is at the C-terminus [111], although there is another weak Rab5-binding domain upstream the FYVE domain. PI3P is enriched on early endosomal membrane and together with Rab5-GTP, recruits EEA1 onto early endosomes. The N-terminal Rab5-binding domain of EEA1 is thought to extend away from the membrane to capture and tether vesicles or other endosomes that contain Rab5-GTP to promote docking and fusion.

In Rabenosyn-5, the FYVE domain is at the N-terminus and the Rab5-binding domain is at the C-terminus [112, 113]. The extended C-terminal Rab5-binding domain may capture and tether endocytic vesicles or other endosomes containing Rab5-GTP to promote membrane fusion. Either EEA1 or Rabenosyn-5 is required for specific and efficient fusion between proteoliposomes reconstituted *in vitro* with early endosomal lipids, SNAREs, and Rab5 [114]. In addition, Rabenoysn-5 forms a complex with the SM (Sec1p-Munc18) protein, hVPS45, which is thought to control the SNARE function in membrane fusion [114].

Uso1 and p115 are orthologs in yeast and mammalian cells, respectively, and they act as tethering factors for ER-derived transport vesicles and the Golgi membrane. p115 binds to Rab1-GTP whereas Uso1 and Ypt1 interact genetically (Chapter 2). In addition, it has been reported that p115 also directly interacts with SNAREs, which together with Rab1 may recruit p115 to the ER vesicles [115]. However, its binding site on the Golgi membrane is not yet established.

The exocyst, and HOPS (homotypic fusion and vacuole protein sorting) complexes are examples of the multi-subunit tethering factors. The exocyst contains eight subunits, including Sec3, Sec5, Sec6, Sec8,

Sec10, Sec15, Exo70, and Exo84, and is involved in tethering of post-Golgi vesicles to specific regions on the PM (Chapter 4). Six of the eight subunits form a complex that is an effector for the Rab GTPase, Sec4, and is recruited to post-Golgi vesicles by the interaction of the Sec15 subunit with Sec4-GTP on the vesicle membrane. The remaining two subunits Sec3, and Exo70, are localized on the PM *via* interactions with the Rho/Cdc42 GTPases and PI [4,5] P_2 (phosphatidylinositol 4,5-bisphosphate). Assembly of exocyst subunits on the vesicle and those on the PM drives the vesicle tethering to the PM.

The HOPS complex consists of six subunits (Vps11, Vps16, Vps18, Vps33, Vps39 and Vps41) [116] (Chapter 10). The HOPS complex is an effector of Ypt7 and is recruited to the yeast vacuole *via* binding to Ypt7-GTP [117], as well as interaction with the SNAREs [118]. In this regard, the Vps33 subunit is actually a SM protein and may control the function of the SNAREs in the fusion reaction. These data suggest that the HOPS complex is both a tethering factor and a SNARE regulator [117].

3.4. Vesicle Fusion: SNARE and SM proteins

After vesicle tethering, the final execution of membrane fusion is mediated by complex formation between *v*-SNAREs on vesicles and *t*-SNAREs on target membrane compartments, which requires NSF (N-ethylmaleimide sensitive factor) and SNAP (soluble NSF attachment protein). NSF and SNAP were originally found as necessary soluble components for *in vitro* vesicle fusion [119, 120] whereas SNARE proteins were identified as membrane receptors for SNAP [121]. The *v*- and *t*-SNAREs form a stable four-helix bundle structure, the *trans*-SNAREpin, to bring apposing membranes close together to promote fusion [122]. After fusion, the remaining SNARE complex on the membrane is termed *cis*-SNARE complex, which is then disassembled by ATP hydrolysis catalyzed by the ATPase NSF and its adaptor SNAP to recycle the *v*- and *t*-SNAREs for another round of fusion [123].

SM (Sec1-Munc18) proteins facilitate SNAREpin formation by interacting with both *v*- and *t*-SNAREs and functioning as a clamp [123-126], which is suggested to stabilize and proofread the *trans*-SNAREpin and is necessary for fusion at physiological concentrations of SNAREs [123, 127]. The SM and SNARE proteins function downstream of the Rab-mediated membrane tethering and docking to promote vesicle fusion. At least some of the SM proteins are associated with the tethering factors that are Rab effectors, *e.g.*, Vps33 in the HOPS complex [117, 118] and hVPS45 associated with Rabenosyn-5 [114], suggesting that specific and efficient vesicle fusion in the cell requires Rab GTPases and their effectors.

4. FUNCTION

Ubiquitous Rabs that have close yeast homologs fall into two groups assigned to the exocytic and endocytic pathways (Fig. **1**). Both compartment specificity of and transport steps mediated by Ypts and Rabs are conserved from yeast to humans (Fig. **4**). The next two sections summarize our current knowledge of how Ypt/Rabs regulate transport in the exocytic and endocytic pathways with the help of their upstream regulators and downstream effectors. The last section is dedicated to Rabs with unknown function.

4.1. Exocytic Ypt/Rabs

Three Ypts are required for the early exocytic pathway in yeast cells: Ypt1 and the functional pair Ypt31/Ypt32. In addition to regulating Golgi entry and exit, these Ypts also regulate recycling processes: Ypt1 plays a role in autophagy, a pathway for shuttling cytosolic proteins and membranes for degradation in the lysosme, and Ypt31/32 in recycling of proteins between the PM and the Golgi. The closest homologs of Ypt1 and Ypt31/32, Rab1 and Rab11, respectively, were implicated in the corresponding transport processes in mammalian cells (Chapter 2). Ypt6 and its mammalian homolog Rab6 regulate transport at the Golgi complex (Chapter 3).

The last step of the exocytic pathway, exocytosis, is targeting of *trans*-Golgi derived vesicles to and fusion with the PM. Sec4 is required for this constitutive transport step in yeast cells (Chapter 4). In mammalian cells, exocytosis can be regulated by external signals, for example, secretion of hormones from endocrine cells and neurotransmitters from neurons. Rab3 and Rab27 play a role in this regulated exocytosis during granule secretion and synaptic transmission (Chapter 5). Rab27 also regulates cell-specific transport of melanosomes in melanocytes and of lytic granules in cytotoxic T lymphocytes (Chapter 6).

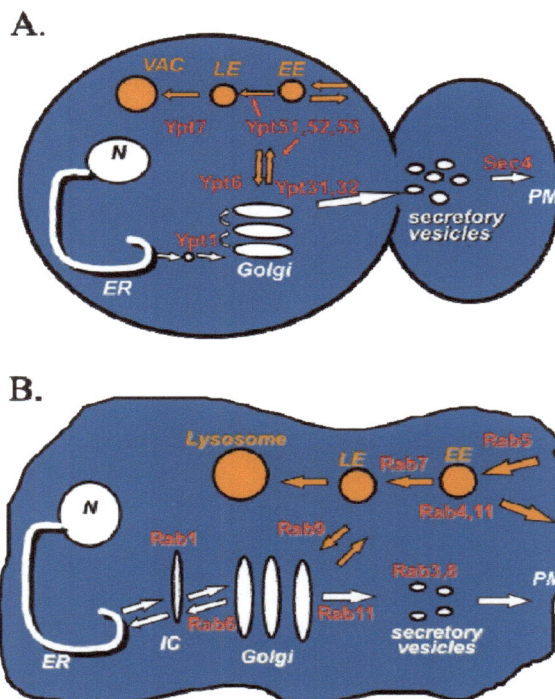

Figure 4: **Regulation of intracellular trafficking by Ypt/Rab GTPases.** A. In yeast, nine Ypts regulate the different steps of the exocytic (white) and endocytic (orange) pathways. The role of two other Ypts, Ypt10 and Ypt11, is not clear. B. In mammalian cells, a number of ubiquitous Rabs regulate the exocytic and endocytic pathways. ER-endoplasmic reticulum, PM-plasma membrane, EE-early endosomes, LE-late endosomes, vacuole is the yeast lysosome.

4.2. Endocytic Ypt/Rabs

Transport from the PM to early endosomes is regulated by the Rab5 subfamily, which includes Rab5, Rab21 and Rab22 in mammalian cells, and Ypt51/52/53 in yeast (Chapter 7). Transport from late endosomes to the lysosome is regulated by Rab7 in mammalian cells and Ypt7 in yeast (Chapter 10). In mammalian cells, recycling of proteins, like the MPR, from late lysosomes back to the Golgi is regulated by a homolog of Rab7, Rab9 (Chapter 11).

In mammalian cells, recycling among the PM, endosomes and the Golgi can be divided to slow and fast steps; Rab8 and Rab11 regulate the former and Rab4 the latter (Chapters 8 and 9). These Rabs fall into the exocytic Ypt/Rab group by sequence analysis (Fig. **1**). In addition to their role in PM recycling, there is also evidence for a role of Rab8, Rab11 and Rab4 in exocytosis [128-130]. Because Chapters 8 and 9 focus the role of these Rabs in recycling through endosomes, they are included in the second section of this eBook.

4.3. Rabs with Unknown Function

Until now most of the research done in this field concentrated on ubiquitous Rabs that have yeast homologs. While there is still much to learn about these Ypt/Rabs, more research will, no doubt, be devoted to the unknown Rabs. These Rabs are expected to play roles unique to mammalian cells (Chapter 12).

ACKNOWLEDGEMENT

We thank D. Taussig for text editing, and acknowledge support from the National Institutes of Health (GM45-444 for NS and GM074692 for GL).

CONFLICT OF INTEREST

There is no conflict of interest from any of the authors.

REFERENCES

[1]   Barbacid M. ras genes. Annu Rev Biochem 1987;56:779-827.
[2]   Gilman AG. G proteins: transducers of receptor-generated signals. Annu Rev Biochem 1987;56:615-49.
[3]   Segev N, Mulholland J, Botstein D. The yeast GTP-binding YPT1 protein and a mammalian counterpart are associated with the secretion machinery. Cell 1988 Mar 25;52(6):915-24.
[4]   Salminen A, Novick PJ. A ras-like protein is required for a post-Golgi event in yeast secretion. Cell 1987 May 22;49(4):527-38.
[5]   Segev N, Botstein D. The ras-like yeast YPT1 gene is itself essential for growth, sporulation, and starvation response. Mol Cell Biol 1987 Jul;7(7):2367-77.
[6]   Zahraoui A, Touchot N, Chardin P, Tavitian A. The human Rab genes encode a family of GTP-binding proteins related to yeast YPT1 and SEC4 products involved in secretion. J Biol Chem 1989 Jul 25;264(21):12394-401.
[7]   Bock JB, Matern HT, Peden AA, Scheller RH. A genomic perspective on membrane compartment organization. Nature 2001 Feb 15;409(6822):839-41.
[8]   Gotte M, Lazar T, Yoo JS, Scheglmann D, Gallwitz D. The full complement of yeast Ypt/Rab-GTPases and their involvement in exo- and endocytic trafficking. Subcell Biochem 2000;34:133-73.
[9]   Beranger F, Paterson H, Powers S, de Gunzburg J, Hancock JF. The effector domain of Rab6, plus a highly hydrophobic C terminus, is required for Golgi apparatus localization. Mol Cell Biol 1994 Jan;14(1):744-58.
[10] Haubruck H, Prange R, Vorgias C, Gallwitz D. The ras-related mouse ypt1 protein can functionally replace the YPT1 gene product in yeast. EMBO J 1989 May;8(5):1427-32.
[11] Bourne HR. Do GTPases direct membrane traffic in secretion? Cell 1988 Jun 3;53(5):669-71.
[12] Walworth NC, Brennwald P, Kabcenell AK, Garrett M, Novick P. Hydrolysis of GTP by Sec4 protein plays an important role in vesicular transport and is stimulated by a GTPase-activating protein in Saccharomyces cerevisiae. Mol Cell Biol 1992 May;12(5):2017-28.
[13] Jones S, Richardson CJ, Litt RJ, Segev N. Identification of regulators for Ypt1 GTPase nucleotide cycling. Mol Biol Cell 1998 Oct;9(10):2819-37.
[14] Richardson CJ, Jones S, Litt RJ, Segev N. GTP hydrolysis is not important for Ypt1 GTPase function in vesicular transport. Mol Cell Biol 1998 Feb;18(2):827-38.
[15] Jones S, Litt RJ, Richardson CJ, Segev N. Requirement of nucleotide exchange factor for Ypt1 GTPase mediated protein transport. J Cell Biol 1995 Sep;130(5):1051-61.
[16] Barbieri MA, Li G, Colombo MI, Stahl PD. Rab5, an early acting endosomal GTPase, supports *in vitro* endosome fusion without GTP hydrolysis. J Biol Chem 1994 Jul 22;269(29):18720-2.
[17] Li G, Stahl PD. Structure-function relationship of the small GTPase rab5. J Biol Chem 1993 Nov 15;268(32):24475-80.
[18] Stenmark H, Parton RG, Steele-Mortimer O, Lutcke A, Gruenberg J, Zerial M. Inhibition of rab5 GTPase activity stimulates membrane fusion in endocytosis. EMBO J 1994 Mar 15;13(6):1287-96.
[19] Segev N. Ypt/rab gtpases: regulators of protein trafficking. Sci STKE 2001 Sep 18;2001(100):re11.
[20] Segev N. Ypt and Rab GTPases: insight into functions through novel interactions. Curr Opin Cell Biol 2001 Aug;13(4):500-11.
[21] Stenmark H. Rab GTPases as coordinators of vesicle traffic. Nat Rev Mol Cell Biol 2009 Aug;10(8):513-25.
[22] Ali BR, Seabra MC. Targeting of Rab GTPases to cellular membranes. Biochem Soc Trans 2005 Aug;33(Pt 4):652-6.
[23] Pfeffer S, Aivazian D. Targeting Rab GTPases to distinct membrane compartments. Nat Rev Mol Cell Biol 2004 Nov;5(11):886-96.
[24] Goud B, Salminen A, Walworth NC, Novick PJ. A GTP-binding protein required for secretion rapidly associates with secretory vesicles and the plasma membrane in yeast. Cell 1988 Jun 3;53(5):753-68.
[25] Segev N. Mediation of the attachment or fusion step in vesicular transport by the GTP-binding Ypt1 protein. Science 1991 Jun 14;252(5012):1553-6.
[26] Guo W, Sacher M, Barrowman J, Ferro-Novick S, Novick P. Protein complexes in transport vesicle targeting. Trends Cell Biol 2000 Jun;10(6):251-5.

[27]   Pfeffer SR. GTP-binding proteins in intracellular transport. Trends Cell Biol 1992 Feb;2(2):41-6.

[28]   Jedd G, Mulholland J, Segev N. Two new Ypt GTPases are required for exit from the yeast trans-Golgi compartment. J Cell Biol 1997 May 5;137(3):563-80.

[29]   Jones S, Jedd G, Kahn RA, Franzusoff A, Bartolini F, Segev N. Genetic interactions in yeast between Ypt GTPases and Arf guanine nucleotide exchangers. Genetics 1999 Aug;152(4):1543-56.

[30]   Ortiz D, Medkova M, Walch-Solimena C, Novick P. Ypt32 recruits the Sec4p guanine nucleotide exchange factor, Sec2p, to secretory vesicles; evidence for a Rab cascade in yeast. J Cell Biol 2002 Jun 10;157(6):1005-15.

[31]   Rivera-Molina FE, Novick PJ. A Rab GAP cascade defines the boundary between two Rab GTPases on the secretory pathway. Proc Natl Acad Sci U S A 2009 Aug 25;106(34):14408-13.

[32]   Segev N. Coordination of intracellular transport steps by GTPases. Semin Cell Dev Biol 2011 Feb;22(1):33-8.

[33]   Chen SH, Chen S, Tokarev AA, Liu F, Jedd G, Segev N. Ypt31/32 GTPases and their novel F-box effector protein Rcy1 regulate protein recycling. Mol Biol Cell 2005 Jan;16(1):178-92.

[34]   Jedd G, Richardson C, Litt R, Segev N. The Ypt1 GTPase is essential for the first two steps of the yeast secretory pathway. J Cell Biol 1995 Nov;131(3):583-90.

[35]   Miaczynska M, Zerial M. Mosaic organization of the endocytic pathway. Exp Cell Res 2002 Jan 1;272(1):8-14.

[36]   Pfeffer S. A model for Rab GTPase localization. Biochem Soc Trans 2005 Aug;33(Pt 4):627-30.

[37]   Benado A, Nasagi-Atiya Y, Sagi-Eisenberg R. Protein trafficking in immune cells. Immunobiol 2009;214(6):403-21.

[38]   Corbeel L, Freson K. Rab proteins and Rab-associated proteins: major actors in the mechanism of protein-trafficking disorders. Eur J Pediatr 2008 Jul;167(7):723-9.

[39]   Baskys A, Bayazitov I, Zhu E, Fang L, Wang R. Rab-mediated endocytosis: linking neurodegeneration, neuroprotection, and synaptic plasticity? Ann N Y Acad Sci 2007 Dec;1122:313-29.

[40]   Subramani D, Alahari SK. Integrin-mediated function of Rab GTPases in cancer progression. Mol Cancer 2010;9:312.

[41]   Mitra S, Cheng KW, Mills GB. Rab GTPases implicated in inherited and acquired disorders. Semin Cell Dev Biol 2011 Feb;22(1):57-68.

[42]   Seabra MC, Mules EH, Hume AN. Rab GTPases, intracellular traffic and disease. Trends Mol Med 2002 Jan;8(1):23-30.

[43]   Agola JO, Jim PA, Ward HH, Basuray S, Wandinger-Ness A. Rab GTPases as regulators of endocytosis, targets of disease and therapeutic opportunities. Clin Genet 2011 Jun 8.

[44]   Li G. Rab GTPases, membrane trafficking and diseases. Curr Drug Targets 2011 Jul 1;12(8):1188-93.

[45]   Barr F, Lambright DG. Rab GEFs and GAPs. Curr Opin Cell Biol 2010 Aug;22(4):461-70.

[46]   Itzen A, Goody RS. GTPases involved in vesicular trafficking: structures and mechanisms. Semin Cell Dev Biol 2011 Feb;22(1):48-56.

[47]   Delprato A, Lambright DG. Structural basis for Rab GTPase activation by VPS9 domain exchange factors. Nat Struct Mol Biol 2007 May;14(5):406-12.

[48]   Cai Y, Chin HF, Lazarova D, et al. The structural basis for activation of the Rab Ypt1p by the TRAPP membrane-tethering complexes. Cell 2008 Jun 27;133(7):1202-13.

[49]   Sacher M, Kim YG, Lavie A, Oh BH, Segev N. The TRAPP complex: insights into its architecture and function. Traffic 2008 Dec;9(12):2032-42.

[50]   Lynch-Day MA, Bhandari D, Menon S, et al. Trs85 directs a Ypt1 GEF, TRAPPIII, to the phagophore to promote autophagy. Proc Natl Acad Sci U S A 2010 Apr 27;107(17):7811-6.

[51]   Morozova N, Liang Y, Tokarev AA, et al. TRAPPII subunits are required for the specificity switch of a Ypt-Rab GEF. Nat Cell Biol 2006 Nov;8(11):1263-9.

[52]   Yamasaki A, Menon S, Yu S, et al. mTrs130 is a component of a mammalian TRAPPII complex, a Rab1 GEF that binds to COPI-coated vesicles. Mol Biol Cell 2009 Oct;20(19):4205-15.

[53]   Knodler A, Feng S, Zhang J, et al. Coordination of Rab8 and Rab11 in primary ciliogenesis. Proc Natl Acad Sci U S A 2010 Apr 6;107(14):6346-51.

[54]   Zhu H, Liang Z, Li G. Rabex-5 is a Rab22 effector and mediates a Rab22-Rab5 signaling cascade in endocytosis. Mol Biol Cell 2009 Nov;20(22):4720-9.

[55]   Nordmann M, Cabrera M, Perz A, et al. The Mon1-Ccz1 complex is the GEF of the late endosomal Rab7 homolog Ypt7. Curr Biol 2010 Sep 28;20(18):1654-9.

[56]   Poteryaev D, Datta S, Ackema K, Zerial M, Spang A. Identification of the switch in early-to-late endosome transition. Cell 2010 Apr 30;141(3):497-508.

[57]   Burd CG, Mustol PA, Schu PV, Emr SD. A yeast protein related to a mammalian Ras-binding protein, Vps9p, is required for localization of vacuolar proteins. Mol Cell Biol 1996 May;16(5):2369-77.

[58]   Hama H, Tall GG, Horazdovsky BF. Vps9p is a guanine nucleotide exchange factor involved in vesicle-mediated vacuolar protein transport. J Biol Chem 1999 May 21;274(21):15284-91.

[59]   Carney DS, Davies BA, Horazdovsky BF. Vps9 domain-containing proteins: activators of Rab5 GTPases from yeast to neurons. Trends Cell Biol 2006 Jan;16(1):27-35.

[60]   Delprato A, Merithew E, Lambright DG. Structure, exchange determinants, and family-wide rab specificity of the tandem helical bundle and Vps9 domains of Rabex-5. Cell 2004 Sep 3;118(5):607-17.

[61]   Lodhi IJ, Chiang SH, Chang L, *et al.* Gapex-5, a Rab31 guanine nucleotide exchange factor that regulates Glut4 trafficking in adipocytes. Cell Metab 2007 Jan;5(1):59-72.

[62]   Siniossoglou S, Peak-Chew SY, Pelham HR. Ric1p and Rgp1p form a complex that catalyses nucleotide exchange on Ypt6p. EMBO J 2000 Sep 15;19(18):4885-94.

[63]   Marat AL, Dokainish H, McPherson PS. DENN domain proteins: regulators of Rab GTPases. J Biol Chem 2011 Apr 22;286(16):13791-800.

[64]   Yoshimura S, Gerondopoulos A, Linford A, Rigden DJ, Barr FA. Family-wide characterization of the DENN domain Rab GDP-GTP exchange factors. J Cell Biol 2010 Oct 18;191(2):367-81.

[65]   Strom M, Vollmer P, Tan TJ, Gallwitz D. A yeast GTPase-activating protein that interacts specifically with a member of the Ypt/Rab family. Nature 1993 Feb 25;361(6414):736-9.

[66]   Albert S, Gallwitz D. Two new members of a family of Ypt/Rab GTPase activating proteins. Promiscuity of substrate recognition. J Biol Chem 1999 Nov 19;274(47):33186-9.

[67]   McCormick F. ras GTPase activating protein: signal transmitter and signal terminator. Cell 1989 Jan 13;56(1):5-8.

[68]   Segev N. Focusing on Arf GAPs. Cell Logist 2011 Mar;1(2):47-8.

[69]   Nottingham RM, Pfeffer SR. Defining the boundaries: Rab GEFs and GAPs. Proc Natl Acad Sci U S A 2009 Aug 25;106(34):14185-6.

[70]   Fuchs E, Haas AK, Spooner RA, Yoshimura S, Lord JM, Barr FA. Specific Rab GTPase-activating proteins define the Shiga toxin and epidermal growth factor uptake pathways. J Cell Biol 2007 Jun 18;177(6):1133-43.

[71]   Haas AK, Yoshimura S, Stephens DJ, Preisinger C, Fuchs E, Barr FA. Analysis of GTPase-activating proteins: Rab1 and Rab43 are key Rabs required to maintain a functional Golgi complex in human cells. J Cell Sci 2007 Sep 1;120(Pt 17):2997-3010.

[72]   Yoshimura S, Egerer J, Fuchs E, Haas AK, Barr FA. Functional dissection of Rab GTPases involved in primary cilium formation. J Cell Biol 2007 Jul 30;178(3):363-9.

[73]   Nagano F, Sasaki T, Fukui K, Asakura T, Imazumi K, Takai Y. Molecular cloning and characterization of the noncatalytic subunit of the Rab3 subfamily-specific GTPase-activating protein. J Biol Chem 1998 Sep 18;273(38):24781-5.

[74]   Andres DA, Seabra MC, Brown MS, *et al.* cDNA cloning of component A of Rab geranylgeranyl transferase and demonstration of its role as a Rab escort protein. Cell 1993 Jun 18;73(6):1091-9.

[75]   Rak A, Pylypenko O, Niculae A, Pyatkov K, Goody RS, Alexandrov K. Structure of the Rab7:REP-1 complex: insights into the mechanism of Rab prenylation and choroideremia disease. Cell 2004 Jun 11;117(6):749-60.

[76]   Waldherr M, Ragnini A, Schweyer RJ, Boguski MS. MRS6--yeast homologue of the choroideraemia gene. Nat Genet 1993 Mar;3(3):193-4.

[77]   Garrett MD, Zahner JE, Cheney CM, Novick PJ. GDI1 encodes a GDP dissociation inhibitor that plays an essential role in the yeast secretory pathway. EMBO J 1994 Apr 1;13(7):1718-28.

[78]   Alory C, Balch WE. Organization of the Rab-GDI/CHM superfamily: the functional basis for choroideremia disease. Traffic 2001 Aug;2(8):532-43.

[79]   Rak A, Pylypenko O, Durek T, *et al.* Structure of Rab GDP-dissociation inhibitor in complex with prenylated YPT1 GTPase. Science 2003 Oct 24;302(5645):646-50.

[80]   Chavrier P, Gorvel JP, Stelzer E, Simons K, Gruenberg J, Zerial M. Hypervariable C-terminal domain of rab proteins acts as a targeting signal. Nature 1991 Oct 24;353(6346):769-72.

[81]   Dirac-Svejstrup AB, Sumizawa T, Pfeffer SR. Identification of a GDI displacement factor that releases endosomal Rab GTPases from Rab-GDI. EMBO J 1997 Feb 3;16(3):465-72.

[82]   Sivars U, Aivazian D, Pfeffer SR. Yip3 catalyses the dissociation of endosomal Rab-GDI complexes. Nature 2003 Oct 23;425(6960):856-9.

[83]   Lin J, Liang Z, Zhang Z, Li G. Membrane topography and topogenesis of prenylated Rab acceptor (PRA1). J Biol Chem 2001 Nov 9;276(45):41733-41.

[84]   Wu YW, Oesterlin LK, Tan KT, Waldmann H, Alexandrov K, Goody RS. Membrane targeting mechanism of Rab GTPases elucidated by semisynthetic protein probes. Nat Chem Biol 2010 Jul;6(7):534-40.

[85]   Carroll KS, Hanna J, Simon I, Krise J, Barbero P, Pfeffer SR. Role of Rab9 GTPase in facilitating receptor recruitment by TIP47. Science 2001 May 18;292(5520):1373-6.

[86]   McLauchlan H, Newell J, Morrice N, Osborne A, West M, Smythe E. A novel role for Rab5-GDI in ligand sequestration into clathrin-coated pits. Curr Biol 1998 Jan 1;8(1):34-45.

[87]   Pellinen T, Arjonen A, Vuoriluoto K, Kallio K, Fransen JA, Ivaska J. Small GTPase Rab21 regulates cell adhesion and controls endosomal traffic of beta1-integrins. J Cell Biol 2006 Jun 5;173(5):767-80.

[88]   Miserey-Lenkei S, Chalancon G, Bardin S, Formstecher E, Goud B, Echard A. Rab and actomyosin-dependent fission of transport vesicles at the Golgi complex. Nat Cell Biol Jul;12(7):645-54.

[89]   Casavola EC, Catucci A, Bielli P, *et al.* Ypt32p and Mlc1p bind within the vesicle binding region of the class V myosin Myo2p globular tail domain. Mol Microbiol 2008 Mar;67(5):1051-66.

[90]   Govindan B, Bowser R, Novick P. The role of Myo2, a yeast class V myosin, in vesicular transport. J Cell Biol 1995 Mar;128(6):1055-68.

[91]   Lipatova Z, Tokarev AA, Jin Y, Mulholland J, Weisman LS, Segev N. Direct interaction between a myosin V motor and the Rab GTPases Ypt31/32 is required for polarized secretion. Mol Biol Cell 2008 Oct;19(10):4177-87.

[92]   Fukuda M. Regulation of secretory vesicle traffic by Rab small GTPases. Cell Mol Life Sci 2008 Sep;65(18):2801-13.

[93]   Gomi H, Mori K, Itohara S, Izumi T. Rab27b is expressed in a wide range of exocytic cells and involved in the delivery of secretory granules near the plasma membrane. Mol Biol Cell 2007 Nov;18(11):4377-86.

[94]   Fukuda M, Kuroda TS, Mikoshiba K. Slac2-a/melanophilin, the missing link between Rab27 and myosin Va: implications of a tripartite protein complex for melanosome transport. J Biol Chem 2002 Apr 5;277(14):12432-6.

[95]   Stinchcombe JC, Barral DC, Mules EH, *et al.* Rab27a is required for regulated secretion in cytotoxic T lymphocytes. J Cell Biol 2001 Feb 19;152(4):825-34.

[96]   Hume AN, Collinson LM, Rapak A, Gomes AQ, Hopkins CR, Seabra MC. Rab27a regulates the peripheral distribution of melanosomes in melanocytes. J Cell Biol 2001 Feb 19;152(4):795-808.

[97]   Wu X, Rao K, Bowers MB, Copeland NG, Jenkins NA, Hammer JA, 3rd. Rab27a enables myosin Va-dependent melanosome capture by recruiting the myosin to the organelle. J Cell Sci 2001 Mar;114(Pt 6):1091-100.

[98]   Bahadoran P, Aberdam E, Mantoux F, *et al.* Rab27a: A key to melanosome transport in human melanocytes. J Cell Biol 2001 Feb 19;152(4):843-50.

[99]   Menasche G, Pastural E, Feldmann J, *et al.* Mutations in RAB27A cause Griscelli syndrome associated with haemophagocytic syndrome. Nat Genet 2000 Jun;25(2):173-6.

[100]   Lapierre LA, Kumar R, Hales CM, *et al.* Myosin vb is associated with plasma membrane recycling systems. Mol Biol Cell 2001 Jun;12(6):1843-57.

[101]   Roland JT, Kenworthy AK, Peranen J, Caplan S, Goldenring JR. Myosin Vb interacts with Rab8a on a tubular network containing EHD1 and EHD3. Mol Biol Cell 2007 Aug;18(8):2828-37.

[102]   Christoforidis S, Miaczynska M, Ashman K, *et al.* Phosphatidylinositol-3-OH kinases are Rab5 effectors. Nat Cell Biol 1999 Aug;1(4):249-52.

[103]   Hoepfner S, Severin F, Cabezas A, *et al.* Modulation of receptor recycling and degradation by the endosomal kinesin KIF16B. Cell 2005 May 6;121(3):437-50.

[104]   Johansson M, Rocha N, Zwart W, *et al.* Activation of endosomal dynein motors by stepwise assembly of Rab7-RILP-p150Glued, ORP1L, and the receptor betaIII spectrin. J Cell Biol 2007 Feb 12;176(4):459-71.

[105]   Jordens I, Fernandez-Borja M, Marsman M, *et al.* The Rab7 effector protein RILP controls lysosomal transport by inducing the recruitment of dynein-dynactin motors. Curr Biol 2001 Oct 30;11(21):1680-5.

[106]   Matanis T, Akhmanova A, Wulf P, *et al.* Bicaudal-D regulates COPI-independent Golgi-ER transport by recruiting the dynein-dynactin motor complex. Nat Cell Biol 2002 Dec;4(12):986-92.

[107]   Schlager MA, Kapitein LC, Grigoriev I, *et al.* Pericentrosomal targeting of Rab6 secretory vesicles by Bicaudal-D-related protein 1 (BICDR-1) regulates neuritogenesis. EMBO J 2010 May 19;29(10):1637-51.

[108]   Echard A, Jollivet F, Martinez O, *et al.* Interaction of a Golgi-associated kinesin-like protein with Rab6. Science 1998 Jan 23;279(5350):580-5.

[109]   Fontijn RD, Goud B, Echard A, *et al.* The human kinesin-like protein RB6K is under tight cell cycle control and is essential for cytokinesis. Mol Cell Biol 2001 Apr;21(8):2944-55.

[110]   TerBush DR, Maurice T, Roth D, Novick P. The Exocyst is a multiprotein complex required for exocytosis in Saccharomyces cerevisiae. EMBO J 1996 Dec 2;15(23):6483-94.

[111] Simonsen A, Lippe R, Christoforidis S, *et al.* EEA1 links PI(3)K function to Rab5 regulation of endosome fusion. Nature 1998 Jul 30;394(6692):494-8.

[112] Eathiraj S, Pan X, Ritacco C, Lambright DG. Structural basis of family-wide Rab GTPase recognition by rabenosyn-5. Nature 2005 Jul 21;436(7049):415-9.

[113] Nielsen E, Christoforidis S, Uttenweiler-Joseph S, *et al.* Rabenosyn-5, a novel Rab5 effector, is complexed with hVPS45 and recruited to endosomes through a FYVE finger domain. J Cell Biol 2000 Oct 30;151(3):601-12.

[114] Ohya T, Miaczynska M, Coskun U, *et al.* Reconstitution of Rab- and SNARE-dependent membrane fusion by synthetic endosomes. Nature 2009 Jun 25;459(7250):1091-7.

[115] Bentley M, Liang Y, Mullen K, Xu D, Sztul E, Hay JC. SNARE status regulates tether recruitment and function in homotypic COPII vesicle fusion. J Biol Chem 2006 Dec 15;281(50):38825-33.

[116] Price A, Seals D, Wickner W, Ungermann C. The docking stage of yeast vacuole fusion requires the transfer of proteins from a cis-SNARE complex to a Rab/Ypt protein. J Cell Biol 2000 Mar 20;148(6):1231-8.

[117] Hickey CM, Wickner W. HOPS initiates vacuole docking by tethering membranes before trans-SNARE complex assembly. Mol Biol Cell 2010 Jul;21(13):2297-305.

[118] Stroupe C, Collins KM, Fratti RA, Wickner W. Purification of active HOPS complex reveals its affinities for phosphoinositides and the SNARE Vam7p. EMBO J. 2006 Apr 19;25(8):1579-89.

[119] Clary DO, Griff IC, Rothman JE. SNAPs, a family of NSF attachment proteins involved in intracellular membrane fusion in animals and yeast. Cell 1990 May 18;61(4):709-21.

[120] Wilson DW, Wilcox CA, Flynn GC, *et al.* A fusion protein required for vesicle-mediated transport in both mammalian cells and yeast. Nature 1989 Jun 1;339(6223):355-9.

[121] Sollner T, Whiteheart SW, Brunner M, *et al.* SNAP receptors implicated in vesicle targeting and fusion. Nature 1993 Mar 25;362(6418):318-24.

[122] Sutton RB, Fasshauer D, Jahn R, Brunger AT. Crystal structure of a SNARE complex involved in synaptic exocytosis at 2.4 A resolution. Nature 1998 Sep 24;395(6700):347-53.

[123] Sudhof TC, Rothman JE. Membrane fusion: grappling with SNARE and SM proteins. Science 2009 Jan 23;323(5913):474-7.

[124] Dulubova I, Yamaguchi T, Gao Y, *et al.* How Tlg2p/syntaxin 16 'snares' Vps45. EMBO J 2002 Jul 15;21(14):3620-31.

[125] Misura KM, Scheller RH, Weis WI. Three-dimensional structure of the neuronal-Sec1-syntaxin 1a complex. Nature 2000 Mar 23;404(6776):355-62.

[126] Yamaguchi T, Dulubova I, Min SW, Chen X, Rizo J, Sudhof TC. Sly1 binds to Golgi and ER syntaxins *via* a conserved N-terminal peptide motif. Dev Cell 2002 Mar;2(3):295-305.

[127] Wickner W. Membrane fusion: five lipids, four SNAREs, three chaperones, two nucleotides, and a Rab, all dancing in a ring on yeast vacuoles. Annu Rev Cell Dev Biol 2010 Nov 10;26:115-36.

[128] Foster LJ, Klip A. Mechanism and regulation of GLUT-4 vesicle fusion in muscle and fat cells. Am J Physiol Cell Physiol 2000 Oct;279(4):C877-90.

[129] Nachury MV, Seeley ES, Jin H. Trafficking to the ciliary membrane: how to get across the periciliary diffusion barrier? Annu Rev Cell Dev Biol 2010 Nov 10;26:59-87.

[130] Ng EL, Tang BL. Rab GTPases and their roles in brain neurons and glia. Brain Res Rev 2008 Jun;58(1):236-46.

CHAPTER 2

The Golgi Gatekeepers: Ypt1-Rab1 and Ypt31/32-Rab11

David Taussig, Shu H. Chen and Nava Segev[*]

The University of Illinois at Chicago, USA

Abstract: The Golgi apparatus is the main sorting compartment of the secretory pathway. At the cis Golgi cisterna, proteins are transported from the endoplasmic reticulum (ER) and sorted for forward transport or for recycling back to the ER. At the trans Golgi cisterna, proteins arriving from the medial Golgi cisterna or from endosomes are sorted for transport to the plasma membrane (PM) or to endosomes. In yeast, the Ypt1 and functional pair Ypt31/32 GTPases are required for transport at the *cis* and *trans* Golgi, respectively. The mammalian homologues of Ypt1 and Ypt31/32, Rab1 and Rab11, respectively, play similar roles. Therefore, these GTPases are considered the Golgi gatekeepers. Here, we summarize our current knowledge of upstream regulation and downstream effectors of the Golgi Ypt/Rab GTPases. In addition, we discuss the role of the Golgi Ypt/Rab GTPases in human health and disease.

Keywords: GTPase, Ypt1, Ypt31/32, Rab1, Rab11, Golgi Ypt/Rabs.

1. INTRODUCTION

Transport of proteins and membranes between intracellular organelles is mediated by membrane-bound vesicles that form on one compartment and fuse with the next. Through the exocytic pathway, proteins and membrane are delivered to the plasma membrane (PM) and the cell's milieu. In addition, enzymes destined for endosomes and the lysosome also travel through this pathway. The Golgi apparatus is the sorting compartment of the exocytic pathway. On its cis side, cargo arriving from the endoplasmic reticulum (ER) is sorted between anterograde transport towards the PM and retrograde transport back to the ER. On its trans side, cargo arriving from the medial Golgi is sorted to the PM or to the endocytic pathway.

Two recycling processes interface with the exocytic pathway. On the ER side, mis-folded and aggregated proteins are sorted out of the exocytic pathway and shuttled to the lysosome for degradation *via* the autophagy pathway. On the PM side, transport machinery components and membrane proteins are recycled back to the Golgi *via* endosomes.

Ypt/Rab GTPases regulate the multiple steps of intracellular trafficking [1, 2]. These GTPases can attach to membranes and cycle between GTP and GDP-bound forms with the aid of their upstream regulators. Guanine-nucleotide exchange factors, GEFs, stimulate the GDP to GTP transition, whereas GTP-hydrolysis activating proteins, GAPs, stimulate the shift from the GTP to GDP bound form. When in the GTP-bound form, Ypt/Rabs interact with their downstream effectors and recruit them to membranes to mediate the multiple vesicular transport steps, from vesicle formation to their targeting and fusion [3].

Ypt/Rab GTPases of two groups, Ypt1-Rab1 and Ypt31/32-Rab11, regulate the complicated transport on the cis and trans sides of the Golgi, respectively. Ypt1 in yeast and Rab1 in mammalian cells regulate transport from the ER to the Golgi and through the autophagic pathway to the lysosome (vacuole in yeast). The functional pair Ypt31/32 in yeast and Rab11 in mammalian cells regulate transport from the Golgi to the PM and the recycling of PM proteins back to the Golgi (Fig. **1**).

The closest human homologs of Ypt1 belong to the Rab1 group, which includes Rab1A and Rab1B. Importantly, Rab1A was shown to be a functional homolog of Ypt1 [4]. The closest homologs of the

---
*Address correspondence to Nava Segev: Department of Biological Sciences, University of Illinois at Chicago, Chicago, IL 60607, USA; Tel: 312-355-0142; Fax: 312-413-2691; E-mail: nava@uic.edu

Guangpu Li and Nava Segev (Eds)

Ypt31/32 GTPases belong to the Rab11 group, which includes Rab11A, Rab11B and Rab25. The yeast Ypts and the human Rabs of the Ypt1-Rab1 and Ypt31/32-Rab11 groups form separate branches in the Ypt and Rab phylogenetic trees. In addition, in both the yeast and the human trees, the Ypt1-Rab1 and Ypt31/32-Rab11 groups fall into the sub-family of exocytic Ypt/Rabs (Figs. **2** and **3a**).

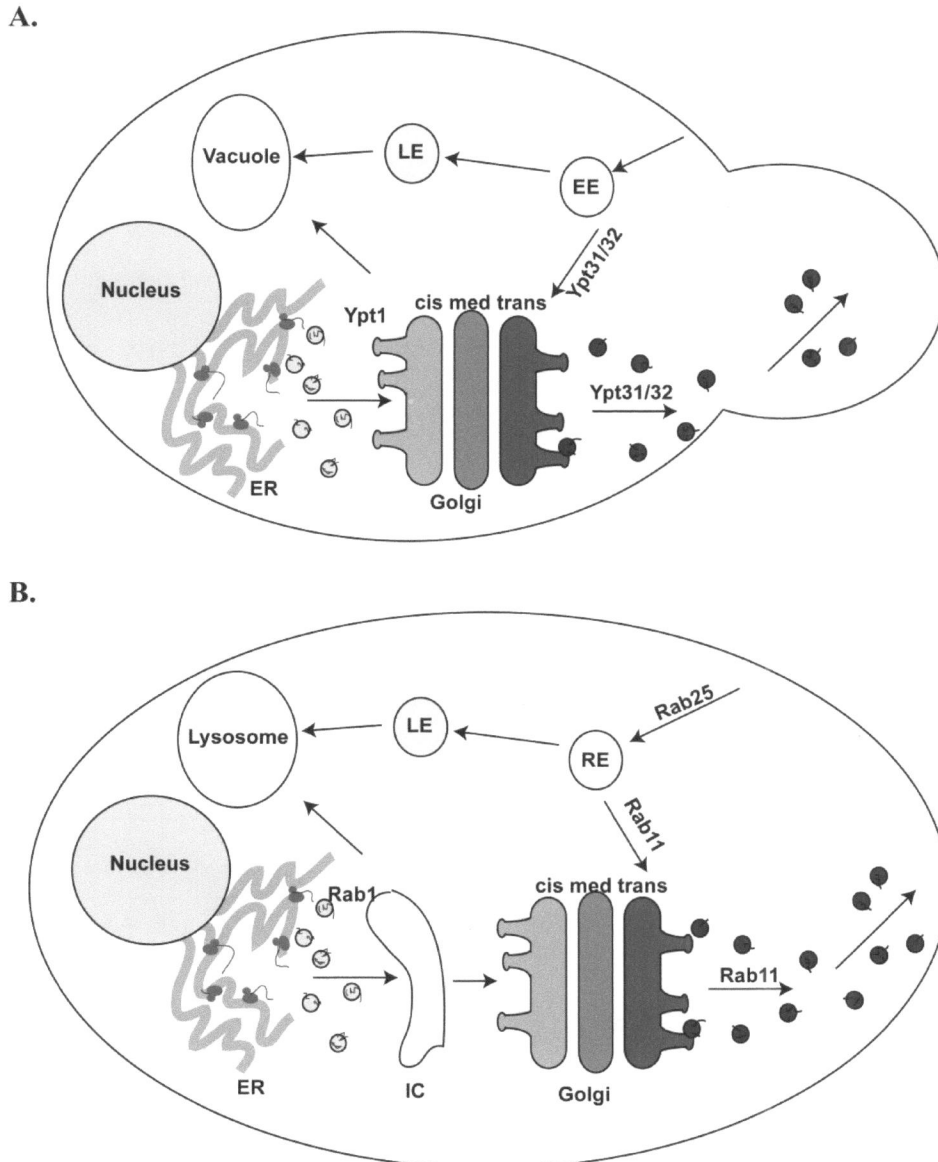

Figure 1: Regulation of intracellular transport steps by Ypt1-Rab1 and Ypt31/32-Rab11 GTPases. A. Yeast cells. **B.** Human cells. Abbreviations: endoplasmic reticulum, ER; intermediate compartment, IC; early endosomes, EE; late endosomes, LE. See text for discussion.

In this chapter we review our current knowledge regarding the transport steps that Ypt1-Rab1 and Ypt31/32-Rab11 GTPases regulate, and the upstream regulators and downstream effectors that help them accomplish these functions (Summarized in Table **1**). The role of Rab11A in slow recycling of PM receptors is further discussed in Chapter 9 (Goldenring *et al.*). In addition, we highlight the idea that the function of Ypt1-Rab1 and Ypt31/32-Rab11 is coordinated, and stress the importance of these GTPases in human health and disease.

A.

B.

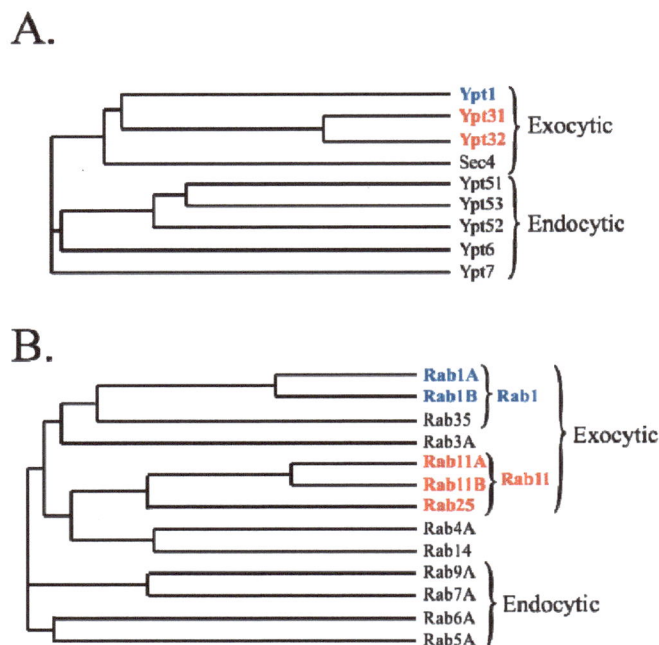

Figure 2: Multiple alignments of yeast Ypt and human Rab GTPases. A. Yeast Ypts. The phylogenetic tree of the nine *Saccharomyces cerevisiae* Ypts shown to function in intracellular trafficking. **B.** Human Rabs. Shown are members of the Rab1 and Rab11 groups. From the other Rab groups, only one member of each group was used for the multiple alignments. Sequences were retrieved from UniProt and aligned using ClustalW. Ypt1 and Rab1 are in Blue; Ypt31/32 and Rab11 in red.

Table 1: Regulators, effectors and functions of Ypt1-Rab1 and Ypt31/32-Rab11 GTPases.

Ypt/Rab	Transport step	GEF	GAP	Effectors
Ypt1/Rab1				
Ypt1	ER-to-Golgi Autophagy	TRAPP I TRAPP III	Gyp1,5,8	Cog2, Cog3
Rab1A/B	ER-to-Golgi Autophagy	mTRAPP	TBC1D20	Uso1, GM130, Golgin-84
Ypt31/32/Rab11				
Ypt31/32	Golgi-to-PM PM recycling	TRAPP II	Gyp2	Myo2,Sec2,Gyp1 Rcy1
Rab11A/B	Golgi-to-PM PM recycling	Huntingtin-complex	TBC1D15 EVI5	Rabin8 Rab11BP/ Rabphilin-11 FIPs 1-4
Rab25	PM recycling			Integrin α5β1

2. *CIS*-GOLGI: YPT1 AND RAB1

The first Ypt/Rab protein shown to be required for intracellular trafficking is the yeast Ypt1 GTPase. In the same paper that reported this it was also shown that Ypt1 functions at the Golgi and that its function is conserved from yeast to mammals [5]. Rab1 is the closest human homolog of Ypt1 and it can replace its essential function in yeast cells [4]. This functional conservation suggests that the essential interactions of

Ypt1 and Rab1 with their accessory proteins are also conserved. Indeed, the functions of Ypt1 and Rab1 seem to be conserved from yeast to human cells.

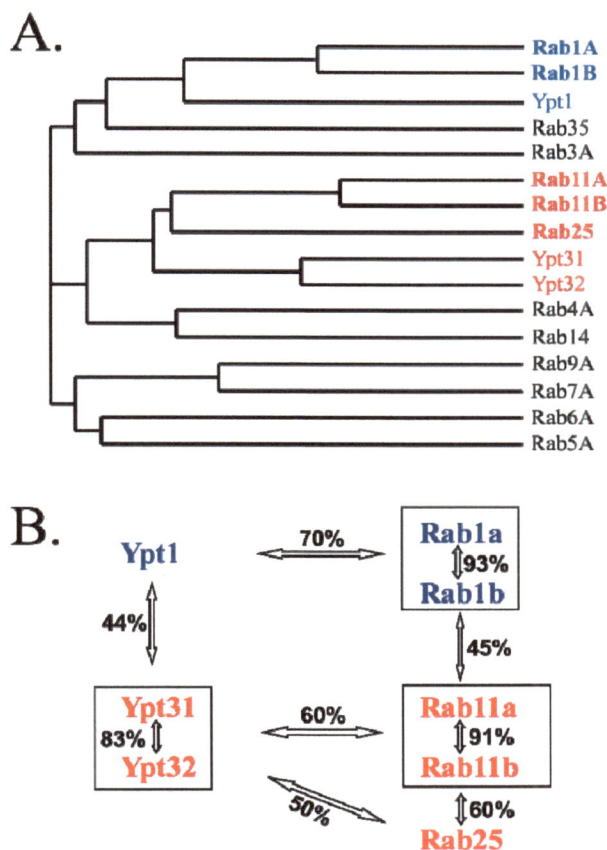

Figure 3: Multiple alignments of the Yp1/Rab1 and Ypt31/32/Rab11 groups. A. The yeast Ypt1 and Ypt31/32 protein sequences were aligned with the human Rab sequences used in Fig. **2**. Ypt1 is most similar to Rab1A and Rab1B, and Ypt31/32 fall in the same branch with Rab11A, Rab11B and Rab25. B. Percent of sequence identity shared between members of the Ypt-Rab1 and Ypt31/32-Rab11 groups using UniProt Alignment. Proteins considered functional homologs are boxed; Ypt1 and Rab1 are in Blue; Ypt31/32 and Rab11 in red.

2.1. Transport Steps: Golgi entry and Autophagy

Both Ypt1 and Rab1 play a role in two cellular trafficking pathways: ER-to-Golgi transport and autophagy. Rab1A and Rab1B share 93% identity at the amino acid level (Fig. **3B**), and currently it is not clear whether they have any non-overlapping functions. Therefore, we will consider them here as Rab1A/B. Rab35 is the closest human homolog of Rab1A/B [6, 7]. However, because Rab1A/B are closer to Ypt1 than to Rab35 (Fig. **3**), and because Rab35 seems to function away from the Golgi, in receptor recycling through endosomes [8, 9] and exosome secretion [10], it is not discussed in this chapter.

A role for Ypt1 in ER-to-Golgi transport was shown using localization, genetics, and biochemical analyses. First, Ypt1 was localized to the yeast Golgi using immuno-fluorescence and immuno-electron microscopy [5, 11]. Second, mutant cells carrying *ypt1* recessive mutations exhibit defects in trafficking in ER-to-Golgi transport [5, 12]. Finally, using *ypt1* mutations and anti-Ypt1 antibodies, it was shown that Ypt1 is required for a cell-free system that reconstitutes ER-to-Golgi transport [13, 14]. A role for Rab1 in ER-to-Golgi transport was suggested using anti-Rab1 antibodies in an *in vitro* reaction that reconstitutes ER-to-Golgi transport, and over-expression of dominant interfering Rab1 mutations *in vivo* [15, 16]. More recently, using siRNA experiments, depletion of both Rab1a and Rab1b was shown to be required for inhibition of

ER-to-Golgi delivery of herpes simplex virus envelope proteins [17]. In mammalian cells, a vesicular-tubular membrane cluster between the ER and the cis-Golgi is considered a pre-Golgi intermediate compartment, and Rab1 has been localized to these membranes [18].

A role for Ypt1 in autophagy was shown based on a defect of *ypt1* mutant cell survival in response to nitrogen starvation, a hallmark of autophagy in yeast [19, 20]. Using over-expression of dominant negative Rab1 mutant proteins and siRNA experiments, a role for Rab1 in autophagy was recently shown [21, 22]. In summary, the transport steps regulated by Ypt1 in yeast and Rab1 in mammalian cells are conserved.

2.2. Upstream Regulation

GEFs: GEFs are considered positive regulators of Ypt/Rabs because they stimulate the switch of these GTPases to the GTP-bound form, in which they can interact with their downstream effectors. Currently, identification of Ypt GEFs depends solely on biochemical studies because GEFs for different Ypts do not share sequence similarity. Even though the known yeast Ypt GEFs are conserved and have human homologs, there is not enough evidence to decipher whether the Ypt/Rab specificity of these GEFs is conserved as well. Recently, a family of Rab GEFs, which includes eighteen DENN-domain proteins, was identified [23]. DENN-domain proteins have not been identified in yeast and they act as GEFs for Rabs that do not have close yeast homologs [24].

TRAPP was identified as a GEF for Ypt1 through biochemical analyses. The biochemical characteristics of a purified Ypt1 GEF were reported to be similar to those of the TRAPP complex that resides on the Golgi [25, 26]. Subsequently, TRAPP purified from yeast cells was shown to act as a Ypt1 GEF [27, 28]. TRAPP is a modular complex and can be found in three different configurations in yeast cells; two of these complexes act as Ypt1 GEFs. TRAPP I, which contains five essential subunits, plays a role in ER-to-Golgi transport [29]. Four of the five essential TRAPP I subunits are required for its Ypt1 GEF activity [30]. The second complex that acts as a Ypt1 GEF is TRAPP III, which contains an additional subunit, Trs85. Trs85 plays a role in autophagy, and TRAPP III was suggested to act as a Ypt1 GEF in this pathway [19]. A third TRAPP complex, TRAPP II, contains three subunits in addition to the TRAPP I subunits and functions at the trans Golgi [29]. There is currently a debate whether TRAPP II acts as a Ypt1 or Ypt31/32 GEF [31] (see below).

All TRAPP subunits are conserved from yeast to human cells [31]. A role for mammalian TRAPP, mTRAPP, in ER-to-Golgi transport has been proposed [32, 33]. TRAPP complexes purified from mouse fibroblast cells using mTrs130 pull-down can act as a Rab1, but not Rab11 GEF. Since mTrs130 is a homolog of the yeast Trs130, a TRAPP II specific subunit, it was suggested that in mammalian cells TRAPP exists as a single complex and acts as a GEF for Rab1 [34]. Recently, two new mammalian-specific TRAPP subunits were identified and were shown to play a role in ER-to-Golgi transport [35]. Further studies of yeast and mammalian TRAPP subunits are required for determining whether yeast and mammalian TRAPP complexes have similar Ypt/Rab specificity and function.

GAPs: GAPs are considered negative regulators of Ypt/Rabs because they aid in switching these GTPases to their GDP-bound form, which is required for their subsequent extraction from membranes. All but one known Ypt/Rab GAPs contain a conserved domain termed GYP (GAP for Ypt proteins) in yeast and TBC (Tre-2/Bub2-Cdc16) in mammalian cells. There are eleven GYP-containing proteins in yeast and about forty TBC-containing proteins in mammalian cells [36]. Even though the identification of Ypt/Rab GAPs has been relatively straightforward because of the conserved GYP/TBC domain, assignment of GAPs to individual Ypts has lagged behind due to perceived promiscuity of GAP activity *in vitro* and the fact that Gyps are not essential for viability in yeast (Fig. **4**). Recently, similar analysis of Rab GAPs has shown higher GAP-Rab specificity [37].

Based on biochemical analysis, six of the ten yeast Gyps can stimulate GTP hydrolysis by Ypt1 (Fig. **4**). Ypt1 is the preferred substrate of Gyp5, and based on localization and genetic interactions at least three different Gyps were suggested to act on Ypt1 as GAPs: Gyp1, Gyp5 with Gyl1, and Gyp8 [38, 39]. Perhaps the most convincing evidence for the role for Gyp1, 5, and 8 as Ypt1 GAPs *in vivo* is their ability, when over-expressed, to suppress the cold-sensitive growth phenotype of mutant cells expressing Ypt1 restricted to the GTP bound form (Ypt1-Q67L) [38]. Because Ypt1 acts in ER-to-Golgi and autophagy, it is possible that it has more than one GAP.

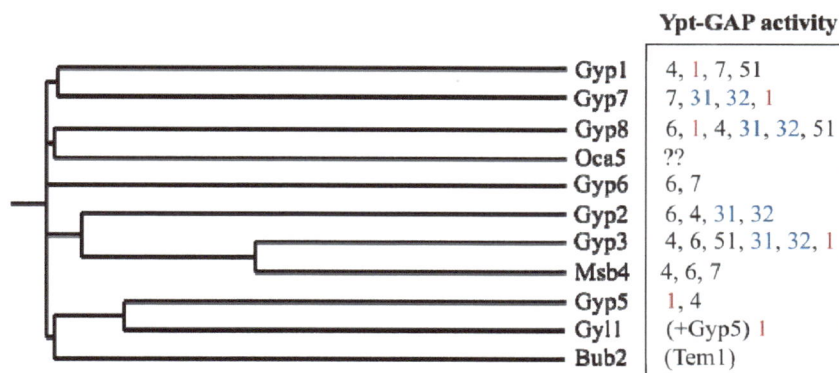

Ypt-GAP activity

Gyp1		4, 1, 7, 51
Gyp7		7, 31, 32, 1
Gyp8		6, 1, 4, 31, 32, 51
Oca5		??
Gyp6		6, 7
Gyp2		6, 4, 31, 32
Gyp3		4, 6, 51, 31, 32, 1
Msb4		4, 6, 7
Gyp5		1, 4
Gyl1		(+Gyp5) 1
Bub2		(Tem1)

Figure 4: Multiple alignment analysis and Ypt specificity of the yeast GYP-containing proteins. Sequences of yeast GYP-domain containing proteins were retrieved from UniProt and aligned using ClustalW (7/15/11). The ability of these proteins to stimulate GTP hydrolysis by the various Ypts is summarized on the right based on the following references: Gyp1 [49], Gyp7 [50], Gyp8 and Gyp5 [38], Gyp2 (Mdr1) and Gyp3 (Msb3) [51], Msb4 [52], Gyp6 [53], Gyl1 [54]. Bub2, together with Bfa1, acts as a Tim1 GAP, a GTPase that function in the spidle pole body [55]. Gyps that can activate Ypt1 are shown in red, Gyps that can activate Ypt31/32 are shown in blue.

As for Rab1, TBC1D20 was shown to act as a Rab1 GAP *in vitro*, and its localization to the ER as well as over-expression analysis add *in vivo* support to this suggestion [40]. The closest yeast homolog of TBC1D20 is Gyp8, one of the Gyps suggested to act on Ypt1. However, it is too early to decide whether Ypt/Rab-GAP specificity is conserved.

2.3. Downstream Effectors

A number of effectors are currently known for Ypt1 and Rab1; all of them are tethering factors. Tethering factors are proteins that play a role in the attachment of transport carriers to acceptor membranes. There are two major types of tethering factors: long coiled-coil proteins and large multi-subunit complexes [41]. The only verified effector of Ypt1 is the COG tethering complex, which is required for ER-to-Golgi transport. Two subunits of this complex, Cog3/Sec34 and Cog2/Sec35, were identified as effectors of Ypt1-GTP [42]. A number of coiled-coil tethering factors were identified as Rab1 effectors: P115, which is present on COPII vesicles [43], GM130 in complex with GRAP65 on the Golgi [44], and another Golgi protein Golgin-84 [45, 46]. Based on these Rab1 effectors it was suggested that Rab1 regulates the interaction of vesicle and compartment tethers [47].

Are the Ypt1-Rab1 effectors conserved? Genetic interaction between Uso1, the yeast homolog of p115, and Ypt1 was shown [48]. However, to our knowledge, there is no evidence that Uso1 is a Ypt1 effector. Likewise, the COG tethering complex is highly conserved, but currently there is no evidence that it acts as a Rab1 effector. At this time there are no known Ypt1 or Rab1 effectors that mediate autophagy. Thus, even though the function of Ypt1 and Rab1 is conserved, it is still unclear whether Ypt/Rab effector specificity is also conserved.

3. *TRANS*-GOLGI: YPT31/32 AND RAB11A/B

Yeast and human genomes contain two paralogs of the Ypt31/32-Rab11 cluster. These paralogs share 83% and 91% amino-acid sequence identity, respectively (Fig. **3B**). Even though a number of interactors were found specifically with one of the paralogs, currently there is no evidence that the two members of each pair interact with different proteins or have a separate function. Therefore, we will consider them here as Ypt31/32, and Rab11A/B.

3.1. Transport Steps: Golgi Exit and Recycling

Ypt31 and Ypt32 share 83% identity at the amino acid level (Fig. **3B**). Deletion of either does not affect cell growth, but deletion of both is lethal. Therefore, Ypt31 and Ypt32 are considered a functional pair [56]. To determine the role of Ypt31/32 in intracellular trafficking, we constructed temperature-sensitive mutant cells in which one gene is deleted and the other is mutated [12, 57]. Analysis of *ypt31Δ32ts* mutant cells showed that transport of cargo is blocked at the trans-Golgi and these cells accumulate large Golgi. Based on this evidence it was concluded that Ypt31/32 play a role in exit from the Golgi [56]. More recently, it has been shown that Ypt31/32 play another role in recycling of proteins back to the Golgi. Using two cargo proteins, Kex2, a furin homolog, and Snc1, a vSNARE, it was demonstrated that the endosome-to-Golgi transport step is defective in *ypt31Δ/32ts* mutant cells [58].

Rab11A and Rab11B share 91% sequence identity, which is even higher than the similarity between Ypt31 and Ypt32 (Fig. **3B**). A role for Rab11 in Golgi-to-PM transport was suggested based on localization and Rab11 mutant analyses [59, 60]. Over-expression and localization analyses further revealed Rab11-regulated exocytosis in neurons [61]. More recently, siRNA was used to show a role for Rab11 in directional transport of proteins to the surface of cilia. In this case, both Rab11A and Rab11B were depleted to show the effect [62]. In *Drosophila*, Rab11 was shown to mediate post-Golgi trafficking of rhodopsin to the apical membrane of photoreceptors [63]. Rab25 was shown to play a role in transport of integrin α5β1 to the PM at the pseudopodial tips in a human ovarian carcinoma cell line [64]. Integrins are trans-membrane extra-cellular matrix receptors and are important for cell shape and motility.

More evidence currently points to an additional role for Rab11 in PM recycling [63, 65, 66]. Rab11 is localized to recycling endosomes together with Rab4, but in distinct membrane domains [67]. Most of the identified Rab11 effectors play a role in PM recycling. Likewise, Rab25 also localizes to and plays a role in recycling endosomes [68].

In summary, like the functional conservation of the Ypt1-Rab1 group, the two separate transport steps regulated by Ypt31/32 and Rab11A/B are conserved from yeast to humans.

3.2. Upstream Regulation

GEFs: It was shown that TRAPP can act as a GEF for Ypt31/32 *in vitro* [27]. Further analysis of TRAPP complexes suggested that whereas TRAPP I acts as a Ypt1 GEF, TRAPP II acts as a Ypt31/32 GEF. The biochemical evidence was supported by genetic interactions between Ypts and TRAPP subunits and by the effect of TRAPP II-specific mutations on Ypt31/32 intracellular localization [69]. In agreement, a role for the TRAPP II-specific subunit Trs130 was shown in late Golgi transport [29]. Since another group published negative biochemical data regarding a Ypt31/32 GEF activity using TRAPP complexes purified from yeast lysates, this idea is currently considered controversial [70]. Huntingtin, a protein associated with Huntington's disease, was suggested to play a role in nucleotide exchange by Rab11. While huntingtin does not seem to interact directly with Rab11, it was suggested that it is important for Rab11 GEF activity on recycling endosomes [71]. The identity of a GEF for Rab25 is currently unknown.

GAPs: Using biochemical GTP hydrolysis assays with different Gyps, four of the ten known yeast Gyps stimulated GTP hydrolysis by Ypt31/32: Gyp7, Gyp8, Gyp2, and Gyp3 (Fig. **4**). However, none of these Gyps use Ypt31/32 as a preferred substrate. Gyp2 has a relatively low GAP activity on Ypt31/32, about five-fold compared to ~300-fold activity on Ypt6 and Sec4 [51]. Interestingly, deletion of *GYP2*, but not *GYP1*, can rescue the growth and secretory defects of *ypt32Δ32ts* mutant cells. Moreover, like over-expression of Ypt31/32 (but not of Ypt6 and Sec4), deletion of *GYP2*, but not *GYP1*, can rescue the growth and trafficking phenotypes of *trs130ts* mutant cells [57]. Therefore, Gyp2 depletion has a positive effect on the functions of Ypt31/32 and its TRAPP II GEF *in vivo*, and is an excellent candidate for a Ypt31/32 GAP.

Currently, it is unknown which TBC-containing protein acts as a Rab11 GAP. Two such proteins were reported to act as GAPs for Rab11 *in vitro*. First, TBC1D15, like its closest yeast homolog Gyp7, acts as a GAP for Rab7 and, to a lesser extent, for Rab11 [72]. In agreement, the rice homolog of TBC1D15,

OsGAP1, acts as a GAP on OsRab11, a homolog of the human Rab11 [73]. However, *in vivo* studies suggest that TBC1D15 is a Rab7 GAP [74]. The second TBC-containing protein reported to have a Rab11 GAP activity is EVI5 [75]. Interestingly, the closest yeast homolog of EVI5 is Gyp5 (our BLAST analysis), which does not act as a Ypt31/32 GAP (Fig. **4**). Therefore, it is still unclear whether Ypt/Rab-GAP specificity is conserved.

3.3. Downstream Effectors

There are four known effectors of Ypt31/32; three of these effectors act in the Golgi-to-PM transport. First, it was shown that Ypt31/32 interact directly with the myosin V actin motor, Myo2, to allow polarized delivery of vesicles to the growing bud [76]. Second, Ypt32 recruits Sec2, the GEF for Sec4, to these vesicles [77]. Since Sec4 functions in the fusion of trans-Golgi vesicles with the PM, this finding suggests a coordination of Ypt31/32-mediated vesicle formation and targeting with Sec4-mediated vesicle fusion. Third, Gyp1 was identified as a Ypt32 effector. Because Gyp1 is considered a Ypt1 GAP, its interaction with Ypt32 was suggested to play a role in the inactivation of Ypt1 in late Golgi [78]. Finally, the F-box protein Rcy1 was identified as a Ypt31/32 effector, and together these proteins play a role in recycling of proteins from the PM through endosomes to the Golgi [58]. Since F-box proteins act as adaptors for ubiquitin ligases, we looked at the effect of Ypt31/32 and Rcy1 on the ubiquitination of the recycling vSNARE Snc1. We have recently shown that ubiquitination of Snc1 is regulated by Ypt31/32 and Rcy1, and that this ubiquitination is important for the recycling of Snc1 through the Golgi [79]. This defines a novel role for Ypt/Rab GTPases in marking cargo for sorting.

One effector of Rab11, which acts in polarized transport in primary cilia, is Rabin8, a GEF for Rab8 [62]. Interestingly, the yeast homologs of these proteins are Ypt31/32, Sec2, and Sec4, respectively, which were also shown to interact (see above). Another effector of Rab11 is Myosin Vb [80]. Even though this interaction is analogous to the Ypt31/32-Myo2 interaction in yeast, its suggested role is in recycling endosomes and not in Golgi-to-PM transport as was shown in yeast cells. Rab11BP/Rabphilin-11 was identified as a Rab11 effector that acts in PM recycling [81, 82]. Another set of proteins, FIP1-4 and RCP, was identified originally based on the ability of these proteins to interact with Rab11. It was later determined that these proteins interact with multiple Rab and Arf GTPases and they seem to play roles in multiple membrane trafficking events [83]. Integrin α5β1 was identified as a Rab25 effector and this interaction was suggested to play a role in delivery of this integrin to the PM at pseudopodial tips [64].

In summary, it seems that the function of Ypt31/32 and Rab11A/B in Golgi-to-PM membrane and PM recycling is conserved, and there are at least two homologous effectors of Ypt31/32 and Rab11: Sec2/Rabin8 and Myosin V.

4. COORDINATION

A significant amount of data currently points to a role for GTPases in the coordination of intracellular trafficking pathways [84]. A role for Ypt1-Rab1 and Ypt31/32-Rab11 and their upstream regulators in trafficking coordination has been suggested at two levels: Coordination of Ypt function within the Golgi and integration of Ypt/Rab mediated vesicular-transport steps. In other words, the Golgi Ypts may coordinate the nucleotide-bound state of each other, and thereby coordinate the trafficking steps that each regulate.

4.1. Coordination of Golgi Ypt Function

The upstream regulation of Ypt1 and Ypt31/32 was suggested to be coordinated both at the activation and inhibition levels. First, the finding that the modular complex TRAPP acts as a GEF for both Ypt1 and Ypt31/32 suggests that this complex could potentially coordinate sequential activation of the Golgi gatekeepers [27, 69]. Thus, conversion of TRAPP I, the GEF for Ypt1, to TRAPP II, the GEF for Ypt31/32, can coordinate Ypt1-mediated Golgi entry with Ypt31/32-mediated Golgi exit. Second, a role for a GAP cascade was recently proposed to ensure that there is only one Ypt active at each Golgi compartment. Specifically, Gyp1, a Ypt1 GAP, is recruited as an effector of Ypt32 to the late Golgi [78]. Whether mammalian cells exhibit similar coordination between Golgi Rabs is still unknown.

4.2. Integration of Vesicular Transport Steps

The Golgi Ypt/Rabs have been implicated in transport-step coordination together with their own GEFs and with Arf GEFs.

In one type of cascade, one Ypt/Rab recruits the GEF of the Ypt/Rab that functions after it. A Ypt32-Sec2-Sec4 cascade was suggested to integrate Ypt32-mediated trans-Golgi vesicle targeting and their subsequent Sec4-mediated fusion with the PM. Here, Sec2, the GEF for Sec4, is recruited to the vesicles by Ypt32 [77]. As mentioned above, a similar cascade of interactions was proposed for the mammalian homologs of these proteins: Rab11-Rabin8-Rab8 in primary ciliogenesis [62].

The second type of a Ypt/Rab cascade recruits GEFs for Arfs. Arf GTPases play a role in assembly of vesicle coats. Arf GEFs contain a conserved Sec7 domain, which is required for their GEF activity [85]. In yeast, a cascade of alternating Golgi Ypts – Ypt1 and Ypt31/32 – and Golgi Arf-GEFs –Gea2, Sec7 and Syt1 – was proposed to coordinate transport through the Golgi based on over-expression analysis [86]. An insight into the mechanism for this cascade came from mammalian cells. The mammalian homolog of Ypt1, Rab1, in its GTP-bound form was shown to interact with the mammalian homolog of Gea2, GBF1 [87]. Both Rab1 and GBF1 localize to the ER-Golgi intermediate compartment and the proposed role of the Rab1-GBF1-Arf1 cascade is to coordinate exchange of COPII-to-COPI coat proteins on the membranes [41].

5. INVOLVEMENT OF RAB1/11/25 IN DISEASE

Rab GTPases regulate intracellular trafficking in human cells, and as such are associated with human diseases that involve this process. Rabs 1, 11, and 25 are particularly important due to their central roles in anterograde transport and recycling of PM-associated proteins. The diseases most strongly associated with these Rabs and their accessory factors are discussed below.

5.1. Acquired Disorders

Diabetes: Diabetes is caused by improper regulation of sugar uptake from the bloodstream to the cell as a response to insulin. Much of the research regarding treatment for diabetes has been focused on the glucose transporter GLUT4. This transporter functions at the PM, but has well-characterized endocytosis and recycling steps in its trafficking repertoire. As a result, Rab11 has been the subject of much research as a potential modifier of GLUT4 localization.

First, over-expression of the Rab11-binding domain of the Rab11 effector RabBP was used as a dominant inhibitor of Rab11. Using this approach, Rab11 was shown to play a role in the localization of GLUT4 to GLUT4-specific-vesicles (GSV) in adipocytes [88]. In addition, over-expression of Rab11 along with GLUT4 in cardiomyocytes reduces the level of PM-associated GLUT4 by 50%. Similarly, over-expression of dominant negative Rab11N124I lowers the insulin-stimulated GLUT4 transport by the same amount [89]. More recently it has been shown that Rab11A is activated as a response to insulin [90], and knockdown of Rab11A or its effector FIP2 caused an increase of GLUT4 localized to the PM [91]. Together, these results indicate that GLUT4 cycling between the PM and intracellular compartments is tightly controlled by Rab11, which therefore might be a target for the treatment of diabetes.

Neurodegenerative Disorders: Parkinson's disease is associated with the accumulation of misfolded α-synuclein in Lewy bodies. How Lewy bodies cause neurodegeneration is still debated, but one popular view is that misfolded α-synuclein builds up in the ER, causing ER stress, and possibly overloading the general or specific autophagy pathways [92, 93]. Accordingly, Rab1 has been found to be closely involved in the pathology of Parkinson's. Studies using a yeast model for Parkinson's disease showed that over-expression of α-synuclein is lethal in yeast, and this lethality can be rescued by over-expression of Ypt1 [94]. In addition, accumulation of α-synuclein blocks ER-to-Golgi tethering, the step mediated by Ypt1 [95]. Furthermore, over-expression of Rab1 was shown to rescue neuron loss in *Drosophila, C. elegans*, and rat models of the disease [94].

Huntington's disease is a genetic disorder caused by an increase in the number of CAG repeats in the huntingtin gene, thereby lengthening a stretch of glutamines within the protein, which is proportionally toxic in neurons. Since it was discovered almost two decades ago, much research has been done to elucidate the function of the huntingtin protein, a task that is still incomplete. Defective Rab11-mediated transport has recently been implicated in neuronal loss in Huntington's models. In 2008, the huntingtin protein was reported to be associated with a GEF for Rab11 [71]. More recently, it was shown that Rab11 activity is impaired in a mouse model of the disease, and expression of a constitutively active form of Rab11 can improve the survival of neurons. This rescue was suggested to be a result of overcoming defective recycling of the glutamate transporter EAAC1 [71]. In accordance with this theory, endocytic recycling was shown to be impaired in cultured neurons expressing a mutant huntingtin construct, and Rab11 over-expression rescues neuronal cell death in a *Drosophila* model of the disease [96].

Alzheimer's disease, the most common form of dementia, is characterized by a number of cellular anomalies, and which of these is the cause of neurodegeneration is still not known. One of the most defining symptoms of Alzheimer's is the accumulation of aβ, a peptide derived from cleavage of the amyloid precursor protein (APP), into distinctive extracellular plaques. Understanding the processing of this protein may lead to more effective treatment or prevention of Alzheimer's. In 2003 it was shown that both APP and PAK3, a protein kinase that interacts with APP and mediates apoptosis, co-localize mainly with Rab11-containing vesicles [97]. Furthermore, Rab11 is known to interact directly with presenilin 1, mutations of which cause most of the early-onset genetic cases of Alzheimer's [98]. Therefore, it is likely that factors involved in Alzheimer's are under Rab11 control.

Multiple sclerosis: Multiple sclerosis (MS) is an autoimmune disease caused by inflammation of neurons in the brain and spinal cord. While the root causes of MS are still not fully understood, most researchers agree that there are both genetic and environmental factors contributing to the disease. One of the genetic mutations tightly associated with MS is in EVI5 [99]. While its contribution to MS is not understood, EVI5 binds to Rab11 and acts as its GAP [75, 100].

Cancer: Cancer, the term used for all diseases marked by unregulated cell growth, is usually associated with changes in the expression of a plethora of genes. Not surprisingly, many of these genes are involved in protein transport. Cancer cells are usually desensitized to signaling pathways and have down-regulated degradation pathways, both of which require endocytosis. Furthermore, the rapid cell division observed in tumor cells requires anterograde transport at a higher rate than normal. As a rule, albeit not without exception, most cancer cells exhibit repression of Rabs involved in endocytosis, and hyper expression or activation of Rabs involved in anterograde transport and exocytosis [101]. There have been many examples of changes in Rab expression in a variety of cancer cell types, and a sampling of those pertinent to this chapter are briefly detailed below.

Rab1 isoforms have been shown to be over-expressed in multiple cancer types, *e.g.*, Rab1A in tongue cancer [102] and Rab1B along with Rab25 in liver cancer [103]. Other examples of Rab25 over-expression are found in prostate [104], ovarian, and breast cancers [104, 105]. Rab25 over-expression is correlated with increased proliferation of cultured tumor cells, while its RNAi knockdown results in the opposite effect [105]. The interaction of Rab25 with its integrin effector was suggested to play a role in cell migration, which is crucial for tumor invasiveness [64].

Rab11, while not having been shown to exhibit altered expression itself, has also been associated with cancer. EVI5, a suggested Rab11 GAP, is an oncogene [100]. In addition, an effector of Rab11, Rab-coupling protein, RCP, is frequently amplified in breast cancer. Over-expression of this protein in normal human cells results in several tumorigenic properties, and RNAi knockdown of RCP inhibits tumor formation in a mouse xenograft model [106].

While many observations have been made of Rab association with cancer, cause-and-effect relationships are still poorly understood. A more complete grasp of how Rabs, their regulators, and effectors are involved in cancer development may yield new therapeutic targets.

5.2. Inherited Diseases

Cystic fibrosis is a common life-shortening recessive inherited disease. It is caused by mutations in the CFTR gene, which encodes an anion channel functioning at the PM. The most common mutation, ΔF508, results in decreased stability of the protein and, therefore, impaired anion and water transport through the PM. Research has been done to study the trafficking of the CFTR protein, as modifiers of this transport might be used to increase the amount of CFTR at the PM, and, thereby, alleviate symptoms of cystic fibrosis. It has been shown that Rab11 plays a crucial role in CFTR transport. It was reported that over-expression of Rab11 increased the amount of PM-associated wild-type or ΔF508 CFTR [107]. Furthermore, Rab11A was among the proteins found to co-precipitate with CFTR purified from human epithelial cell lysates [108]. Another study showed that CFTR is localized to Rab11-containing vesicles, and that inhibition of Rab11 by RNAi or expression of dominant negative Rab11bS25N reduced CFTR activity at the PM, while expression of constitutively active Rab11bQ70L had the opposite effect [109].

Subunits of TRAPP, the GEF for the Golgi Ypt/Rabs, are conserved from yeast to humans. The only subunit currently implicated in human disease is Trs20/SEDL. X-linked spondyloepiphyseal dysplasia tarda (SEDT) is a rare genetic disorder that causes defective development of bone and cartridge and results in dwarfism. It is caused by mutations in SEDL, a gene that encodes the human functional homolog of the yeast Trs20, a TRAPP subunit [110]. It is not clear how the SEDL mutations affect cellular trafficking and lead to SEDT symptoms.

5.3. Infectious Diseases

Rabs are involved in multiple diseases caused by infectious bacteria and viruses that use the intracellular trafficking machinery for their own insidious ends. Most pathogens use the endocytic machinery as their mode of entry into a host cell. As such, the Rabs regulating endocytosis are those primarily associated with infectious diseases. However, there are many examples of pathogens, already inside the cell, using anterograde trafficking steps for their assembly and propagation. Therefore, elements controlling these steps could be viable drug targets for the treatment of such diseases.

Influenza is one of the most common and deadly pathogens worldwide. Recently it was shown that this virus uses Rab11A-regulated trafficking to transport its proteins to the PM for assembly. First, viral proteins co-localize with Rab11A. Second, perturbing Rab11A function, using siRNA or expression of a dominant negative Rab11A mutant protein, disrupts accumulation of viral particles in the cell [111]. Rab11-mediated traffic is also utilized by mouse polyomavirus. This was shown by co-localization of virions with transferrin as well as Rab11 itself by fluorescence resonance energy transfer (FRET) [112]. Rather than using it for transportation, the pathogen *Trypanosoma cruzi* was shown to decrease expression of Rab11 in infected cardiomyocytes. This effect was proposed to weaken the innate immune response of the host cell, which is the greatest threat to this pathogen [113].

Rab1 has also been associated with pathogens. For example, a screen for host proteins that interact with Hepatitis C protein NS5A yielded TBC1D20, a Rab1 GAP. RNAi knockdown of TBC1D20 inhibited viral accumulation [114], and knockdown of Rab1 itself was shown to reduce Hepatitis C RNA levels [115]. However, the exact nature of the Rab1 role in Hepatitus C development is still poorly understood. The bacterial pathogen Legionella pneumophila provides an example of how a pathogen manipulates Rab1 function to allow bacterial propagation. This intracellular parasite secretes two proteins that modify Rab1 function: SidM, a Rab1 GEF that recruits Rab1 to Legionella-containing vacuole (LCV), and LepB, a Rab1 GAP that inactivates it [116]. In early stages of infection, SidM-activated Rab1 is exploited to fetch ER vesicles as a LCV membrane source. In a later stage, LepB-inactivated Rab1 is removed from the LCV membrane to allow its maturation [117].

6. CONCLUSIONS AND PERSPECTIVES

Currently, it is clear that the Golgi Ypt/Rabs and their functions are conserved. In addition, the GEFs, GAPs, and some effectors are also conserved from yeast to mammals. However, it is still unclear whether the specificity of Ypt/Rab interactions with their regulators and effectors are conserved as well.

Ypt1 and Rab1 were the first set of conserved GTPases shown to regulate intracellular trafficking, and Ypt31/32 along with Rab11 followed soon after. However, there are still multiple open questions regarding the mechanisms by which these Ypt/Rabs act. The identity of the GEFs and GAPs of the Golgi Ypt/Rabs is still uncertain. While a number of effectors have been identified for the Golgi Ypt/Rabs, we expect that more effectors will be found. For example, there are no known effectors for Ypt1 or Rab1 in autophagy or for Ypt31/32 in vesicle formation. As for coordination, in spite of multiple hints based on protein interactions, there is still no evidence for a physiological role of such coordination.

The Golgi Rabs as well as their regulators and effectors have been implicated in many infectious, inherited and acquired human diseases. Therefore, we expect that these proteins will be targets of diagnostic and therapeutic studies on human diseases in the future.

CONFLICT OF INTEREST

There is no conflict of interest from any of the authors.

REFERENCES

[1] Segev N. Ypt/rab GTPases: regulators of protein trafficking. Sci STKE 2001 Sep 18;2001(100):re11.
[2] Stenmark H. Rab GTPases as coordinators of vesicle traffic. Nat Rev Mol Cell Biol 2009 Aug;10(8):513-25.
[3] Segev N. Ypt and Rab GTPases: insight into functions through novel interactions. Curr Opin Cell Biol 2001 Aug;13(4):500-11.
[4] Haubruck H, Prange R, Vorgias C, Gallwitz D. The ras-related mouse ypt1 protein can functionally replace the YPT1 gene product in yeast. EMBO J 1989 May;8(5):1427-32.
[5] Segev N, Mulholland J, Botstein D. The yeast GTP-binding YPT1 protein and a mammalian counterpart are associated with the secretion machinery. Cell 1988 Mar 25;52(6):915-24.
[6] Bock JB, Matern HT, Peden AA, Scheller RH. A genomic perspective on membrane compartment organization. Nature 2001 Feb 15;409(6822):839-41.
[7] Pereira-Leal JB, Seabra MC. Evolution of the Rab family of small GTP-binding proteins. J Mol Biol 2001 Nov 2;313(4):889-901.
[8] Allaire PD, Marat AL, Dall'Armi C, Di Paolo G, McPherson PS, Ritter B. The Connecdenn DENN domain: a GEF for Rab35 mediating cargo-specific exit from early endosomes. Mol Cell 2010 Feb 12;37(3):370-82.
[9] Patino-Lopez G, Dong X, Ben-Aissa K, *et al.* Rab35 and its GAP EPI64C in T cells regulate receptor recycling and immunological synapse formation. J Biol Chem 2008 Jun 27;283(26):18323-30.
[10] Hsu C, Morohashi Y, Yoshimura S, *et al.* Regulation of exosome secretion by Rab35 and its GTPase-activating proteins TBC1D10A-C. J Cell Biol 2010 Apr 19;189(2):223-32.
[11] Preuss D, Mulholland J, Franzusoff A, Segev N, Botstein D. Characterization of the Saccharomyces Golgi complex through the cell cycle by immunoelectron microscopy. Mol Biol Cell 1992 Jul;3(7):789-803.
[12] Jedd G, Richardson C, Litt R, Segev N. The Ypt1 GTPase is essential for the first two steps of the yeast secretory pathway. J Cell Biol 1995 Nov;131(3):583-90.
[13] Baker D, Wuestehube L, Schekman R, Botstein D, Segev N. GTP-binding Ypt1 protein and Ca2+ function independently in a cell-free protein transport reaction. Proc Natl Acad Sci U S A 1990 Jan;87(1):355-9.
[14] Segev N. Mediation of the attachment or fusion step in vesicular transport by the GTP-binding Ypt1 protein. Science 1991 Jun 14;252(5012):1553-6.
[15] Plutner H, Cox AD, Pind S, *et al.* Rab1b regulates vesicular transport between the endoplasmic reticulum and successive Golgi compartments. J Cell Biol 1991 Oct;115(1):31-43.
[16] Tisdale EJ, Bourne JR, Khosravi-Far R, Der CJ, Balch WE. GTP-binding mutants of rab1 and rab2 are potent inhibitors of vesicular transport from the endoplasmic reticulum to the Golgi complex. J Cell Biol 1992 Nov;119(4):749-61.
[17] Zenner HL, Yoshimura SI, Barr FA, Crump CM. Analysis of Rab GTPase-activating proteins indicates that Rab1a/b and Rab43 are important for HSV-1 secondary envelopment. J Virol 2011 Jun 15.
[18] Marie M, Dale HA, Sannerud R, Saraste J. The function of the intermediate compartment in pre-Golgi trafficking involves its stable connection with the centrosome. Mol Biol Cell 2009 Oct;20(20):4458-70.

[19]   Lynch-Day MA, Bhandari D, Menon S, *et al.* Trs85 directs a Ypt1 GEF, TRAPPIII, to the phagophore to promote autophagy. Proc Natl Acad Sci U S A 2010 Apr 27;107(17):7811-6.

[20]   Segev N, Botstein D. The ras-like yeast YPT1 gene is itself essential for growth, sporulation, and starvation response. Mol Cell Biol 1987 Jul;7(7):2367-77.

[21]   Huang J, Birmingham CL, Shahnazari S, *et al.* Antibacterial autophagy occurs at PI(3)P-enriched domains of the endoplasmic reticulum and requires Rab1 GTPase. Autophagy 2011 Jan;7(1):17-26.

[22]   Zoppino FC, Militello RD, Slavin I, Alvarez C, Colombo MI. Autophagosome formation depends on the small GTPase Rab1 and functional ER exit sites. Traffic 2010 Sep;11(9):1246-61.

[23]   Marat AL, Dokainish H, McPherson PS. DENN domain proteins: regulators of Rab GTPases. J Biol Chem 2011 Apr 22;286(16):13791-800.

[24]   Yoshimura S, Gerondopoulos A, Linford A, Rigden DJ, Barr FA. Family-wide characterization of the DENN domain Rab GDP-GTP exchange factors. J Cell Biol 2010 Oct 18;191(2):367-81.

[25]   Jones S, Richardson CJ, Litt RJ, Segev N. Identification of regulators for Ypt1 GTPase nucleotide cycling. Mol Biol Cell 1998 Oct;9(10):2819-37.

[26]   Sacher M, Barrowman J, Schieltz D, Yates JR, 3rd, Ferro-Novick S. Identification and characterization of five new subunits of TRAPP. Eur J Cell Biol 2000 Feb;79(2):71-80.

[27]   Jones S, Newman C, Liu F, Segev N. The TRAPP complex is a nucleotide exchanger for Ypt1 and Ypt31/32. Mol Biol Cell 2000 Dec;11(12):4403-11.

[28]   Wang W, Sacher M, Ferro-Novick S. TRAPP stimulates guanine nucleotide exchange on Ypt1p. J Cell Biol 2000 Oct 16;151(2):289-96.

[29]   Sacher M, Barrowman J, Wang W, *et al.* TRAPP I implicated in the specificity of tethering in ER-to-Golgi transport. Mol Cell 2001 Feb;7(2):433-42.

[30]   Kim YG, Raunser S, Munger C, *et al.* The architecture of the multisubunit TRAPP I complex suggests a model for vesicle tethering. Cell 2006 Nov 17;127(4):817-30.

[31]   Sacher M, Kim YG, Lavie A, Oh BH, Segev N. The TRAPP complex: insights into its architecture and function. Traffic 2008 Dec;9(12):2032-42.

[32]   Loh E, Peter F, Subramaniam VN, Hong W. Mammalian Bet3 functions as a cytosolic factor participating in transport from the ER to the Golgi apparatus. J Cell Sci 2005 Mar 15;118(Pt 6):1209-22.

[33]   Yu S, Satoh A, Pypaert M, Mullen K, Hay JC, Ferro-Novick S. mBet3p is required for homotypic COPII vesicle tethering in mammalian cells. J Cell Biol 2006 Jul 31;174(3):359-68.

[34]   Yamasaki A, Menon S, Yu S, *et al.* mTrs130 is a component of a mammalian TRAPPII complex, a Rab1 GEF that binds to COPI-coated vesicles. Mol Biol Cell 2009 Oct;20(19):4205-15.

[35]   Scrivens PJ, Noueihed B, Shahrzad N, Hul S, Brunet S, Sacher M. C4orf41 and TTC-15 are mammalian TRAPP components with a role at an early stage in ER-to-Golgi trafficking. Mol Biol Cell 2011 Jun;22(12):2083-93.

[36]   Gao X, Jin C, Xue Y, Yao X. Computational analyses of TBC protein family in eukaryotes. Protein Pept Lett 2008;15(5):505-9.

[37]   Barr F, Lambright DG. Rab GEFs and GAPs. Curr Opin Cell Biol 2010 Aug;22(4):461-70.

[38]   De Antoni A, Schmitzova J, Trepte HH, Gallwitz D, Albert S. Significance of GTP hydrolysis in Ypt1p-regulated endoplasmic reticulum to Golgi transport revealed by the analysis of two novel Ypt1-GAPs. J Biol Chem 2002 Oct 25;277(43):41023-31.

[39]   Lafourcade C, Galan JM, Gloor Y, Haguenauer-Tsapis R, Peter M. The GTPase-activating enzyme Gyp1p is required for recycling of internalized membrane material by inactivation of the Rab/Ypt GTPase Ypt1p. Mol Cell Biol 2004 May;24(9):3815-26.

[40]   Haas AK, Yoshimura S, Stephens DJ, Preisinger C, Fuchs E, Barr FA. Analysis of GTPase-activating proteins: Rab1 and Rab43 are key Rabs required to maintain a functional Golgi complex in human cells. J Cell Sci 2007 Sep 1;120(Pt 17):2997-3010.

[41]   Sztul E, Lupashin V. Role of vesicle tethering factors in the ER-Golgi membrane traffic. FEBS Lett 2009 Dec 3;583(23):3770-83.

[42]   Suvorova ES, Duden R, Lupashin VV. The Sec34/Sec35p complex, a Ypt1p effector required for retrograde intra-Golgi trafficking, interacts with Golgi SNAREs and COPI vesicle coat proteins. J Cell Biol 2002 May 13;157(4):631-43.

[43]   Allan BB, Moyer BD, Balch WE. Rab1 recruitment of p115 into a cis-SNARE complex: programming budding COPII vesicles for fusion. Science 2000 Jul 21;289(5478):444-8.

[44]   Moyer BD, Allan BB, Balch WE. Rab1 interaction with a GM130 effector complex regulates COPII vesicle cis-- Golgi tethering. Traffic 2001 Apr;2(4):268-76.

[45]   Diao A, Rahman D, Pappin DJ, Lucocq J, Lowe M. The coiled-coil membrane protein golgin-84 is a novel rab effector required for Golgi ribbon formation. J Cell Biol 2003 Jan 20;160(2):201-12.

[46]   Satoh A, Wang Y, Malsam J, Beard MB, Warren G. Golgin-84 is a rab1 binding partner involved in Golgi structure. Traffic 2003 Mar;4(3):153-61.

[47]   An Y, Chen CY, Moyer B, Rotkiewicz P, Elsliger MA, Godzik A, *et al*. Structural and functional analysis of the globular head domain of p115 provides insight into membrane tethering. J Mol Biol 2009 Aug 7;391(1):26-41.

[48]   Sapperstein SK, Lupashin VV, Schmitt HD, Waters MG. Assembly of the ER to Golgi SNARE complex requires Uso1p. J Cell Biol 1996 Mar;132(5):755-67.

[49]   Du LL, Novick P. Yeast rab GTPase-activating protein Gyp1p localizes to the Golgi apparatus and is a negative regulator of Ypt1p. Mol Biol Cell 2001 May;12(5):1215-26.

[50]   Vollmer P, Will E, Scheglmann D, Strom M, Gallwitz D. Primary structure and biochemical characterization of yeast GTPase-activating proteins with substrate preference for the transport GTPase Ypt7p. Eur J Biochem 1999 Feb;260(1):284-90.

[51]   Albert S, Gallwitz D. Two new members of a family of Ypt/Rab GTPase activating proteins. Promiscuity of substrate recognition. J Biol Chem 1999 Nov 19;274(47):33186-9.

[52]   Albert S, Gallwitz D. Msb4p, a protein involved in Cdc42p-dependent organization of the actin cytoskeleton, is a Ypt/Rab-specific GAP. Biol Chem 2000 May-Jun;381(5-6):453-6.

[53]   Strom M, Vollmer P, Tan TJ, Gallwitz D. A yeast GTPase-activating protein that interacts specifically with a member of the Ypt/Rab family. Nature 1993 Feb 25;361(6414):736-9.

[54]   Friesen H, Colwill K, Robertson K, Schub O, Andrews B. Interaction of the Saccharomyces cerevisiae cortical actin patch protein Rvs167p with proteins involved in ER to Golgi vesicle trafficking. Genetics 2005 Jun;170(2):555-68.

[55]   Geymonat M, Spanos A, Smith SJ, *et al*. Control of mitotic exit in budding yeast. *In vitro* regulation of Tem1 GTPase by Bub2 and Bfa1. J Biol Chem 2002 Aug 9;277(32):28439-45.

[56]   Jedd G, Mulholland J, Segev N. Two new Ypt GTPases are required for exit from the yeast trans-Golgi compartment. J Cell Biol 1997 May 5;137(3):563-80.

[57]   Sciorra VA, Audhya A, Parsons AB, Segev N, Boone C, Emr SD. Synthetic genetic array analysis of the PtdIns 4-kinase Pik1p identifies components in a Golgi-specific Ypt31/rab-GTPase signaling pathway. Mol Biol Cell 2005 Feb;16(2):776-93.

[58]   Chen SH, Chen S, Tokarev AA, Liu F, Jedd G, Segev N. Ypt31/32 GTPases and their novel F-box effector protein Rcy1 regulate protein recycling. Mol Biol Cell 2005 Jan;16(1):178-92.

[59]   Chen W, Feng Y, Chen D, Wandinger-Ness A. Rab11 is required for trans-golgi network-to-plasma membrane transport and a preferential target for GDP dissociation inhibitor. Mol Biol Cell 1998 Nov;9(11):3241-57.

[60]   Urbe S, Huber LA, Zerial M, Tooze SA, Parton RG. Rab11, a small GTPase associated with both constitutive and regulated secretory pathways in PC12 cells. FEBS Lett 1993 Nov 15;334(2):175-82.

[61]   Khvotchev MV, Ren M, Takamori S, Jahn R, Sudhof TC. Divergent functions of neuronal Rab11b in Ca2+-regulated versus constitutive exocytosis. J Neurosci 2003 Nov 19;23(33):10531-9.

[62]   Knodler A, Feng S, Zhang J, *et al*. Coordination of Rab8 and Rab11 in primary ciliogenesis. Proc Natl Acad Sci U S A 2010 Apr 6;107(14):6346-51.

[63]   Satoh AK, O'Tousa JE, Ozaki K, Ready DF. Rab11 mediates post-Golgi trafficking of rhodopsin to the photosensitive apical membrane of Drosophila photoreceptors. Development 2005 Apr;132(7):1487-97.

[64]   Caswell PT, Spence HJ, Parsons M, *et al*. Rab25 associates with alpha5beta1 integrin to promote invasive migration in 3D microenvironments. Dev Cell 2007 Oct;13(4):496-510.

[65]   Hsu VW, Prekeris R. Transport at the recycling endosome. Curr Opin Cell Biol 2010 Aug;22(4):528-34.

[66]   Ng EL, Tang BL. Rab GTPases and their roles in brain neurons and glia. Brain Res Rev 2008 Jun;58(1):236-46.

[67]   Sonnichsen B, De Renzis S, Nielsen E, Rietdorf J, Zerial M. Distinct membrane domains on endosomes in the recycling pathway visualized by multicolor imaging of Rab4, Rab5, and Rab11. J Cell Biol 2000 May 15;149(4):901-14.

[68]   Wang X, Kumar R, Navarre J, Casanova JE, Goldenring JR. Regulation of vesicle trafficking in madin-darby canine kidney cells by Rab11a and Rab25. J Biol Chem 2000 Sep 15;275(37):29138-46.

[69]   Morozova N, Liang Y, Tokarev AA, *et al*. TRAPPII subunits are required for the specificity switch of a Ypt-Rab GEF. Nat Cell Biol 2006 Nov;8(11):1263-9.

[70] Cai Y, Chin HF, Lazarova D, *et al.* The structural basis for activation of the Rab Ypt1p by the TRAPP membrane-tethering complexes. Cell 2008 Jun 27;133(7):1202-13.

[71] Li X, Sapp E, Valencia A, *et al.* A function of huntingtin in guanine nucleotide exchange on Rab11. Neuroreport 2008 Oct 29;19(16):1643-7.

[72] Zhang XM, Walsh B, Mitchell CA, Rowe T. TBC domain family, member 15 is a novel mammalian Rab GTPase-activating protein with substrate preference for Rab7. Biochem Biophys Res Commun 2005 Sep 16;335(1):154-61.

[73] Heo JB, Rho HS, Kim SW, *et al.* OsGAP1 functions as a positive regulator of OsRab11-mediated TGN to PM or vacuole trafficking. Plant Cell Physiol 2005 Dec;46(12):2005-18.

[74] Peralta ER, Martin BC, Edinger AL. Differential effects of TBC1D15 and mammalian Vps39 on Rab7 activation state, lysosomal morphology, and growth factor dependence. J Biol Chem 2010 May 28;285(22):16814-21.

[75] Dabbeekeh JT, Faitar SL, Dufresne CP, Cowell JK. The EVI5 TBC domain provides the GTPase-activating protein motif for RAB11. Oncogene 2007 Apr 26;26(19):2804-8.

[76] Lipatova Z, Tokarev AA, Jin Y, Mulholland J, Weisman LS, Segev N. Direct interaction between a myosin V motor and the Rab GTPases Ypt31/32 is required for polarized secretion. Mol Biol Cell 2008 Oct;19(10):4177-87.

[77] Ortiz D, Medkova M, Walch-Solimena C, Novick P. Ypt32 recruits the Sec4p guanine nucleotide exchange factor, Sec2p, to secretory vesicles; evidence for a Rab cascade in yeast. J Cell Biol 2002 Jun 10;157(6):1005-15.

[78] Rivera-Molina FE, Novick PJ. A Rab GAP cascade defines the boundary between two Rab GTPases on the secretory pathway. Proc Natl Acad Sci U S A 2009 Aug 25;106(34):14408-13.

[79] Chen SH, Shah AH, Segev N. Ypt31/32 GTPases and their F-Box effector Rcy1 regulate ubiquitination of recycling proteins. Cell Logist 2011 Jan;1(1):21-31.

[80] Hales CM, Vaerman JP, Goldenring JR. Rab11 family interacting protein 2 associates with Myosin Vb and regulates plasma membrane recycling. J Biol Chem 2002 Dec 27;277(52):50415-21.

[81] Mammoto A, Ohtsuka T, Hotta I, Sasaki T, Takai Y. Rab11BP/Rabphilin-11, a downstream target of rab11 small G protein implicated in vesicle recycling. J Biol Chem 1999 Sep 3;274(36):25517-24.

[82] Zeng J, Ren M, Gravotta D, *et al.* Identification of a putative effector protein for rab11 that participates in transferrin recycling. Proc Natl Acad Sci U S A 1999 Mar 16;96(6):2840-5.

[83] Horgan CP, McCaffrey MW. The dynamic Rab11-FIPs. Biochem Soc Trans 2009 Oct;37(Pt 5):1032-6.

[84] Segev N. Coordination of intracellular transport steps by GTPases. Semin Cell Dev Biol 2011 Feb;22(1):33-8.

[85] Cox R, Mason-Gamer RJ, Jackson CL, Segev N. Phylogenetic analysis of Sec7-domain-containing Arf nucleotide exchangers. Mol Biol Cell 2004 Apr;15(4):1487-505.

[86] Jones S, Jedd G, Kahn RA, Franzusoff A, Bartolini F, Segev N. Genetic interactions in yeast between Ypt GTPases and Arf guanine nucleotide exchangers. Genetics 1999 Aug;152(4):1543-56.

[87] Monetta P, Slavin I, Romero N, Alvarez C. Rab1b interacts with GBF1 and modulates both ARF1 dynamics and COPI association. Mol Biol Cell 2007 Jul;18(7):2400-10.

[88] Zeigerer A, Lampson MA, Karylowski O, *et al.* GLUT4 retention in adipocytes requires two intracellular insulin-regulated transport steps. Mol Biol Cell 2002 Jul;13(7):2421-35.

[89] Uhlig M, Passlack W, Eckel J. Functional role of Rab11 in GLUT4 trafficking in cardiomyocytes. Mol Cell Endocrinol 2005 May 12;235(1-2):1-9.

[90] Schwenk RW, Eckel J. A novel method to monitor insulin-stimulated GTP-loading of Rab11a in cardiomyocytes. Cell Signal 2007 Apr;19(4):825-30.

[91] Schwenk RW, Luiken JJ, Eckel J. FIP2 and Rip11 specify Rab11a-mediated cellular distribution of GLUT4 and FAT/CD36 in H9c2-hIR cells. Biochem Biophys Res Commun 2007 Nov 9;363(1):119-25.

[92] Chu CT. Diversity in the regulation of autophagy and mitophagy: lessons from Parkinson's disease. Parkinsons Dis 2011;2011:789431.

[93] Vives-Bauza C, Przedborski S. Mitophagy: the latest problem for Parkinson's disease. Trends Mol Med 2011 Mar;17(3):158-65.

[94] Cooper AA, Gitler AD, Cashikar A, *et al.* Alpha-synuclein blocks ER-Golgi traffic and Rab1 rescues neuron loss in Parkinson's models. Science 2006 Jul 21;313(5785):324-8.

[95] Gitler AD, Bevis BJ, Shorter J, *et al.* The Parkinson's disease protein alpha-synuclein disrupts cellular Rab homeostasis. Proc Natl Acad Sci U S A 2008 Jan 8;105(1):145-50.

[96] Richards P, Didszun C, Campesan S, *et al.* Dendritic spine loss and neurodegeneration is rescued by Rab11 in models of Huntington's disease. Cell Death Differ 2011 Feb;18(2):191-200.

[97] McPhie DL, Coopersmith R, Hines-Peralta A, *et al.* DNA synthesis and neuronal apoptosis caused by familial Alzheimer disease mutants of the amyloid precursor protein are mediated by the p21 activated kinase PAK3. J Neurosci 2003 Jul 30;23(17):6914-27.

[98]   Dumanchin C, Czech C, Campion D, *et al.* Presenilins interact with Rab11, a small GTPase involved in the regulation of vesicular transport. Hum Mol Genet 1999 Jul;8(7):1263-9.

[99]   Hoppenbrouwers IA, Aulchenko YS, Ebers GC, *et al.* EVI5 is a risk gene for multiple sclerosis. Genes Immun 2008 Jun;9(4):334-7.

[100] Westlake CJ, Junutula JR, Simon GC, *et al.* Identification of Rab11 as a small GTPase binding protein for the Evi5 oncogene. Proc Natl Acad Sci U S A 2007 Jan 23;104(4):1236-41.

[101] Mosesson Y, Mills GB, Yarden Y. Derailed endocytosis: an emerging feature of cancer. Nat Rev Cancer 2008 Nov;8(11):835-50.

[102] Shimada K, Uzawa K, Kato M, *et al.* Aberrant expression of RAB1A in human tongue cancer. Br J Cancer 2005 May 23;92(10):1915-21.

[103] He H, Dai F, Yu L, *et al.* Identification and characterization of nine novel human small GTPases showing variable expressions in liver cancer tissues. Gene Expr 2002;10(5-6):231-42.

[104] Calvo A, Xiao N, Kang J, *et al.* Alterations in gene expression profiles during prostate cancer progression: functional correlations to tumorigenicity and down-regulation of selenoprotein-P in mouse and human tumors. Cancer Res 2002 Sep 15;62(18):5325-35.

[105] Cheng KW, Lahad JP, Kuo WL, *et al.* The RAB25 small GTPase determines aggressiveness of ovarian and breast cancers. Nat Med 2004 Nov;10(11):1251-6.

[106] Zhang J, Liu X, Datta A, *et al.* RCP is a human breast cancer-promoting gene with Ras-activating function. J Clin Invest 2009 Aug;119(8):2171-83.

[107] Gentzsch M, Chang XB, Cui L, *et al.* Endocytic trafficking routes of wild type and DeltaF508 cystic fibrosis transmembrane conductance regulator. Mol Biol Cell 2004 Jun;15(6):2684-96.

[108] Swiatecka-Urban A, Talebian L, Kanno E, *et al.* Myosin Vb is required for trafficking of the cystic fibrosis transmembrane conductance regulator in Rab11a-specific apical recycling endosomes in polarized human airway epithelial cells. J Biol Chem 2007 Aug 10;282(32):23725-36.

[109] Silvis MR, Bertrand CA, Ameen N, *et al.* Rab11b regulates the apical recycling of the cystic fibrosis transmembrane conductance regulator in polarized intestinal epithelial cells. Mol Biol Cell 2009 Apr;20(8):2337-50.

[110] Gecz J, Shaw MA, Bellon JR, de Barros Lopes M. Human wild-type SEDL protein functionally complements yeast Trs20p but some naturally occurring SEDL mutants do not. Gene 2003 Nov 27;320:137-44.

[111] Eisfeld AJ, Kawakami E, Watanabe T, Neumann G, Kawaoka Y. RAB11A is essential for transport of the influenza virus genome to the plasma membrane. J Virol 2011 Jul;85(13):6117-26.

[112] Liebl D, Difato F, Hornikova L, Mannova P, Stokrova J, Forstova J. Mouse polyomavirus enters early endosomes, requires their acidic pH for productive infection, and meets transferrin cargo in Rab11-positive endosomes. J Virol 2006 May;80(9):4610-22.

[113] Batista DG, Silva CF, Mota RA, *et al.* Trypanosoma cruzi modulates the expression of Rabs and alters the endocytosis in mouse cardiomyocytes *in vitro*. J Histochem Cytochem 2006 Jun;54(6):605-14.

[114] Sklan EH, Staschke K, Oakes TM, *et al.* A Rab-GAP TBC domain protein binds hepatitis C virus NS5A and mediates viral replication. J Virol 2007 Oct;81(20):11096-105.

[115] Sklan EH, Serrano RL, Einav S, Pfeffer SR, Lambright DG, Glenn JS. TBC1D20 is a Rab1 GTPase-activating protein that mediates hepatitis C virus replication. J Biol Chem 2007 Dec 14;282(50):36354-61.

[116] Neunuebel MR, Chen Y, Gaspar AH, Backlund PS, Jr., Yergey A, Machner MP. De-AMPylation of the small GTPase Rab1 by the pathogen Legionella pneumophila. Science 2011 Jul 22;333(6041):453-6.

[117] Joshi AD, Swanson MS. Secrets of a successful pathogen: legionella resistance to progression along the autophagic pathway. Front Microbiol 2011;2:138.

CHAPTER 3

Rab6 GTPase

Bruno Goud[1,*] and Anna Akhmanova[2]

[1]*Institut Curie, France and* [2]*Utrecht University, The Netherlands*

Abstract: Rab6 belongs to the most conserved Rab GTPases. It localizes to the Golgi apparatus and cytoplasmic vesicles and interacts with numerous effectors, including golgins and vesicle tethering complexes, adaptor proteins, as well as microtubule-based and actin-based motors. Rab6 regulates trafficking from endosomes to the Golgi apparatus, from the Golgi to ER and to the plasma membrane, as well as intra-Golgi transport and Golgi homeostasis. Recent studies demonstrate that Rab6 can coordinate all the sequential events underlying transport between two compartments, including budding/fission, transport and docking/tethering of membrane carriers.

Keywords: Golgi Complex, Vesicular Transport, Endocytosis, Exocytosis, Molecular Motors.

1. INTRODUCTION: EVOLUTIONARY CONSERVATION OF RAB6 AND RAB6 ISOFORMS

Ypt6/RAB6 is one of five ancestral *RAB* genes conserved from yeast to humans. One Rab6 homologue is present in yeasts (Ypt6p in *S. cerevisiae* and Ryh1 in *S. pombe*) and flies [1]. In several eukaryotic lineages, *RAB6* family was independently expanded by gene duplication, resulting in the presence of two *Rab6* genes in *C. elegans*, five in *A. thaliana* (AtRabH1a-H1e) and two genes (*RAB6A* and *RAB6B*) in vertebrates. The human *RAB6A* and *RAB6B* genes are encoded on chromosome 11 and chromosome 3, respectively. Of note, in mouse and human, *RAB6A* generates two alternatively spliced variants called *RAB6A* and *RAB6A'* [2]. This mechanism represents, to our knowledge, the only example of a splicing-based functional diversification within the *RAB* gene family. Isoform Rab6A' appears to be the closest functional equivalent to the yeast homolog Ypt6.

All these Rab6 isoforms are canonical members of Rab GTPase family with a C-terminal CXC prenylation motif. In addition, a divergent Rab6 isoform named *RAB6C* or *WTH3* has been identified in two independent studies as a novel cDNA from a human fetal kidney library and as a differentially methylated DNA fragment [3, 4]. Rab6C shows highest identity to Rab6A' (97%), but possesses a carboxyl-terminal extension of 46 amino acids (Fig. **1**). We recently reported that *RAB6C* is in fact a primate-specific intronless gene likely derived by retroposition from a fully processed *RAB6A'* transcript [5]. *RAB6C* is transcribed in human brain, testis, prostate and breast; it is considerably less abundant and less stable than Rab6A'. The GTP-binding motif of Rab6C displays significant amino acid differences compared to Rab6A', resulting in a greatly reduced GTP-binding affinity. Rab6C displays no membrane localization; instead, the noncanonical GTP-binding domain of Rab6C mediates its targeting to the centrosome. The overexpression of Rab6C results in G1 arrest, while its specific depletion generates tetraploid cells with supernumerary centrosomes, suggesting a role for Rab6C in cell cycle progression [5]. Thus, *RAB6C* might be considered as a rare example of a recently emerged retrogene, encoding a functional protein with altered characteristics compared to other Rab6 isoforms.

2. RAB6 EXPRESSION AND LOCALIZATION

The alternatively spliced mammalian variants Rab6A and Rab6A' are ubiquitously expressed. They differ by only three amino acids due to substitutions in regions flanking the third conserved GDP/GTP-binding domain (G3): Val62>Ile and Thr87/Val88>Ala/Ala. Rab6B is preferentially expressed in neuronal cells [6]. It shows 91% identity to Rab6A, the main amino acid differences being located in the C-terminal hypervariable region of the protein (Fig. **1**).

*****Address correspondence to Bruno Goud:** Institut Curie, CNRS UMR144, 26 rue d'Ulm, 75248 Paris cedex 05, France; Tel: 33-1-56-24-63-98; Fax: 33-1-56-24-64-21; E-mail: bruno.goud@curie.fr

Guangpu Li and Nava Segev (Eds)

```
                              G1                        G2
                            GXXXXKT                     T

Hum_Rab6A    MSTGGDFGNPLRKFKLVFL GEQSVGKT SLITRFMYDSFDNTYQA I IGIDFLSKTMYLEDR 60
Hum_Rab6A'   MSTGGDFGNPLRKFKLVFL GEQSVGKT SLITRFMYDSFDNTYQA I IGIDFLSKTMYLEDR 60
Hum_Rab6B    MSAGGDFGNPLRKFKLVFL GEQSVGKT SLITRFMYDSFDNTYQA I IGIDFLSKTMYLEDR 60

Hum_Rab6C    MSAGGDFGNPLRKFKLVFL GEQSVAKT SLITRFRYDSFDNTYQA I IGIDFLSKTMYLEDG 60

                      G3
                    DXXGQ

Hum_Rab6A    TVRLQLW DTAGQ ERFRSLIPSYIRDS TV AVVVYDITNVNSFQQTTKWIDDVRTERGSDVI 120
Hum_Rab6A'   TIRLQLW DTAGQ ERFRSLIPSYIRDS AA AVVVYDITNVNSFQQTTKWIDDVRTERGSDVI 120
Hum_Rab6B    TVRLQLW DTAGQ ERFRSLIPSYIRDSTVAVVVYDITNLNSFQQTSKWIDDVRTERGSDVI 120

Hum_Rab6C    TIGLRLW DTAGQ ERLRSLIPRYIRDSAAAVVVYDITNVNSFQQTTKWIDDVRTERGSDVI 120

             G4                           G5
           NKXD                          ETSAK

Hum_Rab6A    IMLVG NKTDL ADKRQVSIEEGERKAKELNVMFI ETSAK AGYNVKQLFRRVAAALPGMEST 180
Hum_Rab6A'   IMLVG NKTDL ADKRQVSIEEGERKAKELNVMFI ETSAK AGYNVKQLFRRVAAALPGMEST 180
Hum_Rab6B    IMLVG NKTDL ADKRQITIEEGEQRAKELSVMFI ETSAK TGYNVKQLFRRVASALPGMENV 180

Hum_Rab6C    ITLVG NRTDL ADKRQVSVEEGERKAKGLNVTFI ETRAK AGYNVKQLFRRVAAALPGMEST 180

                                Prenylation

Hum_Rab6A    QDRSREDMIDIKLEKPQEQPVSEGG CSC  208
Hum_Rab6A'   QDRSREDMIDIKLEKPQEQPVSEGG CSC  208
Hum_Rab6B    QEKSKEGMIDIKLDKPQEPPASEGG CSC  208

Hum_Rab6C    QDGSREDMSDIKLEKPQEQTVSEGGCSCYSPMSSSTLPQKPPYSFIDCSVNIGLNLFPSLIT
             FCNSSLLPVSWR 254
```

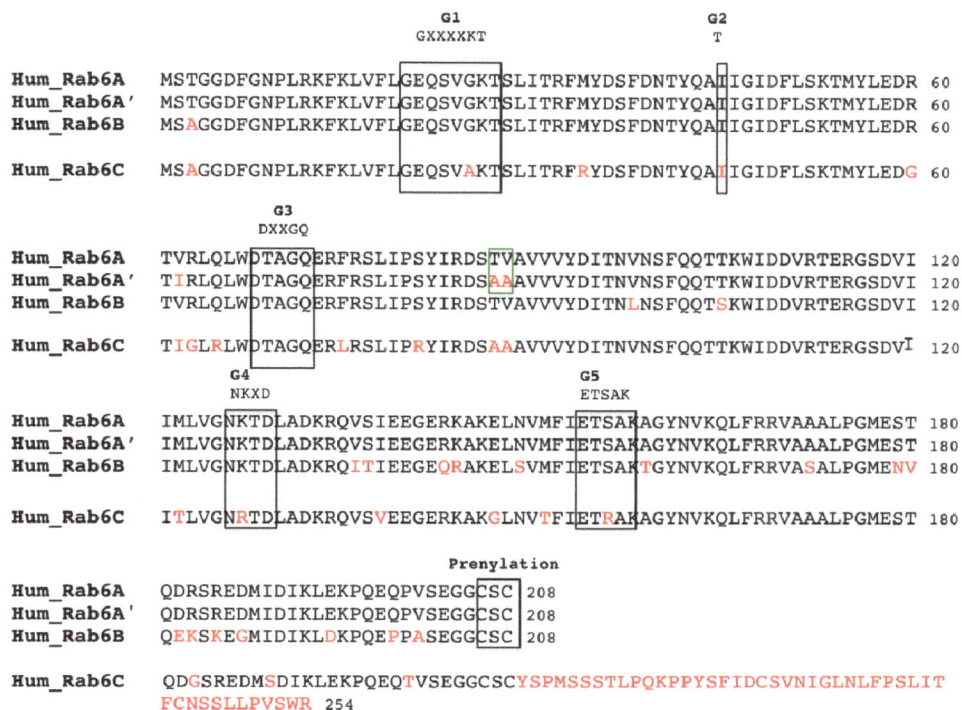

Figure 1: Rab6 isoforms. Shown are amino acid sequence alignments of human Rab6A, Rab6A', Rab6B, and Rab6C. The boxed regions indicate the conserved GTP/GDP-binding motifs, the Thr residue in G2 loop, and the C-terminal Cys motif. Residues in red indicate divergence from the consensus sequence of Rab6A.

Rab6A/A'as well as Rab6B localize to the *medial* and *trans* Golgi cisternae and *trans*-Golgi network (TGN), as well as itinerant tubulovesicular carriers that move along microtubules, mainly in the plus-end direction [7-10]. Rab6A and Rab6A' appear to localize to exactly the same structures [11].

The overall localization pattern of Rab6 is highly conserved in evolution: for example, Rab6 is bound to *trans*-Golgi and transport vesicles also in *Toxoplasma gondii* [12]. In addition, Rab6 has also been found associated with specialized organelles in various cell types, such as α–granules in platelets [13], synaptic vesicles in hypothalamic neurons [14], immature rod outer segment (ROS) membranes in photoreceptor cells [15] and dense granule membranes in *T. gondii* [12].

3. RAB6 INTERACTING PROTEINS AND EFFECTORS

Very little is known about the proteins that regulate Rab6 GTP/GDP cycle. A complex between two proteins, Ric1p and Rgp1p, has been shown to stimulate GTP/GDP exchange on the yeast Ypt6p [16]. So far, no mammalian Rab6 guanine nucleotide exchange factor (GEF) has been identified. No obvious orthologs of Ric1p are present in mammals. In addition, we were unable to detect any interaction between the putative human ortholog of Rgp1p, KIAA0258 [17] and human Rab6 (unpublished results).

Rab6A has the lowest GTPase activity among Rab GTPases, suggesting that GAP (GTPase activating) protein(s) is(are) required to stimulate GTP hydrolysis by Rab6 [18]. The GAPCenA protein (also known as TBCD11) was originally identified as a GAP for Rab6, and to a lesser extent for Rab4 [19]. GAPCenA consists of an N-terminal phosphotyrosine binding (PTB) domain, a central TBC domain conserved in all Rab GAP proteins [20], and a C-terminal coiled-coiled region containing the domain which binds to Rab6 and possibly some other Rabs [21]. GAPCenA was recently shown to also interact with Rab36, a Golgi-associated Rab of unknown function [22]. In the same study, the authors reported that GAPCenA displays only a weak GAP activity towards Rab36, but also towards Rab6A and Rab4A [22]. Another report has also challenged

GAPCenA activity on Rab6, proposing instead that GAPCenA is the Rab4A GAP [21]. Thus, further work is needed to clarify these contradictory findings and to identify the exact function of GAPCenA.

A number of potential Rab6 effectors have been described (listed in Table **1**). Similar to effectors of other Rabs, they constitute a group of structurally and functionally diverse proteins. The affinities of Rab6A/GTP for three effector domains (Bicaudal D2, p150[Glued] and PIST, see Table **1**) have been estimated: they are in the low micromolar range and the interactions are characterized by relatively fast off-rates [23].

Table 1: Rab6-binding proteins.

Interacting and effectors proteins (in human)	Rab interactions	Proposed function(s)	References
RABAC1/PRA1	Rab6; several other Rabs	GDI displacement factor	[81-83]
GAPCenA/RABGAP1/TBC1D11	Rab6A/A'/B; Rab36	Rab6 GTPase-activating protein	[19, 22]
Rab6IP1/DENND5A	Rab6A/A'/B; Rab11A	Tethering of endosome-derived vesicles to TGN membranes	[43, 44]
ELKS/ERC1	Rab6A/A'/B	Tethering/docking of Rab6 exocytotic vesicles to plasma membrane	[7, 71]
Bicaudal D1/D2	Rab6A/A'/B	Recruitment of the dynein/dynactin complex to Golgi membranes; retrograde transport in neuronal cells	[67, 68, 74]
Bicaudal D-related protein 1 (BICDR1)	Rab6A/B	Pericentrosomal targeting of Rab6 secretory vesicles during neuritogenesis	[69]
p150[Glued]	Rab6A/A'	Recruitment of dynactin to membranes; Regulation of dynein activity in mitosis	[68, 79]
DYNLRB1	Rab6A/A'/B	Dynein light chain; regulation of dynein activity?	[84, 85]
KIF20A/MKlp2/Rabkinesin-6	Rab6A	Movement of Rab6 vesicles between Golgi and ER? Cytokinesis	[86-89]
Myosin II	Rab6A/A'/B	Fission of Rab6 transport vesicles from Golgi membranes	[11]
OCRL	Rab6A/A'/B; several other Rabs	5'-phosphatase mutated in Oculo-Cerebro-Renal Syndrome of Lowe; function in Rab6 pathways unknown	[90]
INPP5B	Rab6A/B; several other Rabs	5'-phosphatase; function in Rab6 pathways unknown	[54]
Giantin	Rab6A; Rab1A	Golgin; function in Rab6 pathways unknown	[50]
COG6	Rab6A/B, Rab1A/B, Rab41	Golgin, subunit of the octomeric COG (Conserved Oligomeric Golgi) complex; intra-Golgi retrograde transport?	[54]
SCYL1BP1	Rab6A	Golgin mutated in gerodermia osteodysplastica; function in Rab6 pathways unknown	[91]
Golgin-97/GOLGA1	Rab6A	TGN Golgin; function in Rab6 pathways unknown	[49]
GCC185/GCC2	Rab6B	TGN Golgin; function in Rab6 pathways unknown	[25, 27, 51]
Vps52	Rab6A/A'	Subunit of the GARP complex; retrograde transport between endosomes and Golgi	[37]

Table 1: cont.....

TMF/ARA160	Rab6A/A'/B	TATA element modulatory factor/ Androgen receptor-coactivator of 160 kDa; retrograde transport between endosomes and Golgi and between Golgi and ER.	[42]
Mint1,2,3/X11a,b,g	Rab6A/A'/B	Proteins with PTB (phosphotyrosine binding) and PDZ domains; intracellular processing of APP (amyloid precursor protein)	[92]
PIST	Rab6A	Protein with a PDZ domain; function in Rab6 pathways unknown	[23]
PMM1	Rab6A/A'/B	Phosphomannomutase 1; function in Rab6 pathways unknown	[93]

The minimal Rab6 binding domains (BD) of Rab6 effectors consist of parallel α-helices and short coiled-coil segments. This was recently confirmed by solving the 3D structure of Rab6A (Q72L mutant) in complex with the Rab6 binding domain of Rab6IP1 [24]. The mode of binding of Rab6A to Rab6IP1 BD resembles that described for Rab6A bound to GCC185 Rab6 BD, although both Rab6 BDs bear no compositional or structural resemblance to each other [25, 26]. It should be pointed out that the functional relevance of Rab6A/GCC185 interaction is a matter of debate [25, 27].

4. RAB6-MEDIATED PATHWAYS

Numerous studies have implicated Rab6 in the regulation of several transport pathways at the level of the Golgi complex. In addition, Rab6 appears to play an important role in mitosis and cytokinesis.

4.1. Endosome to Golgi Transport

The first evidence for a role of Rab6 in endosome to Golgi pathway comes from the inhibitory effect of anti-Rab6 antibody on the transport of internalized B-fragment of Shiga toxin (STxB) between endosomes and the TGN in a permeabilized cell system [28, 29]. In the same study, the overexpression of dominant negative (T27N mutants) forms of Rab6A' and to a lesser extent of Rab6A was also shown to inhibit the delivery of STxB to TGN membranes and blocks STxB in endosomal compartments [29]. The role of Rab6 in this pathway was later confirmed by siRNA-mediated Rab6 depletion [21, 30]. In addition, Del Nery *et al.* showed that the specific depletion of Rab6A', but not that of Rab6A, inhibited transport of STxB, suggesting that Rab6A' plays a predominant role in endosome to Golgi transport [30]. However, both Rab6A and A' isoforms appear to be required for retrograde transport of another glycolipid-bound toxin, ricin [31]. Other toxins, such as subtilase cytotoxin, also utilize a Rab6-dependent retrograde trafficking route [32].

In addition to transport of internalized STxB to TGN, Rab6 has also been implicated in recycling of endogenous markers, such as TGN38/46, Mannose-6-phosphate receptor (MPR46), γ-adaptin and transferrin receptor (TfR) [33, 34]. Such a role of Rab6 in endosome to TGN retrieval pathway(s) is similar to the one attributed to yeast Rab GTPAses Ypt6p and Ryh1p [35, 36], indicating that this Rab6 function has been conserved throughout evolution; this notion is further supported by Rab6 involvement in retrograde dense granule to Golgi pathway in *T. gondii* [12].

As Ypt6p, Rab6 is likely involved in targeting/tethering of endosome-derived vesicles to the TGN membranes. The proposed Rab6 effector in this pathway is the human GARP (Golgi-associated retrograde protein also known as VFT) complex composed of the Vps52, Vps53 and Vps54 proteins. Rab6 directly binds to the Vps52 subunit [37]. It was recently shown that GARP specifically and directly interacts with SNAREs that participate in the endosome-to-TGN retrograde route regulated by Rab6 (*i.e.* Syntaxin6, Syntaxin16 and Vamp4) [29, 38]. Of note, the yeast GARP complex binds to the Syntaxin6 ortholog Tlg1p, *via* another subunit, Vps51p [39]. The human orthologue of Vps51, also named Ang2, has been recently identified [40].

The two other Rab6 effectors that are likely involved in this pathway are TMF/ARA160 and Rab6IP1A/B. TMF/ARA160 is the human ortholog of yeast Smg1p, a coiled-coil protein recruited by Ypt6p on Golgi membranes [41]. TMF/ARA160 binds to the three Rab6 isoforms and its depletion blocks retrograde transport of STxB [42]. Similar to Smg1p, TMF/ARA160 could function as a tether on TGN membranes for endosome-derived vesicles. Rab6IP1A/B are large cytosolic proteins that can be recruited on Golgi membranes by Rab6 [43]. Rab6IP1A/B also bind to Rab11 present on the membranes of the endosomal recycling compartment (ERC), suggesting that Rab6IP1A/B couple the function of these two GTPases. Rab6IP1 was also recently shown to associate with Sorting Nexin 1 (SNX1), a subunit of the retromer complex that functions to sort various cargoes from early endosomes back to the TGN [44]. In particular, the retromer participates in the transport of STxB from early endosomes to the TGN [31, 45]. The interaction between Rab6IP1 and the retromer could be involved in the spatial organization of the retromer-mediated pathways [44]. However, multiple retrieval pathways have been now described between endosomes and TGN [46]. More work is needed to better characterize the compartments involved, in particular the role of the ERC, and to better understand the function of Rab6 and its effectors in these pathways.

4.2. Intra-Golgi Transport and Golgi Homeostasis

The major pool of Rab6 is located on the Golgi membranes, where Rab6 associates with several golgins, large coiled coil proteins that are predicted to adopt elongated, rod-like conformations. Golgins were proposed to maintain the structure of the Golgi stacks and organize them into ribbons, participate in intra-Golgi vesicle tethering and serve as scaffolds for other Golgi-associated proteins [47]. *Drosophila* Rab6 was shown to interact with two golgins, dGCC88 and dGolgin-97 [48]. Mammalian Rab6 also binds to golgin-97 [49], as well as giantin [50] and GCC185 [25, 51], although it should be noted that the latter interaction is disputed [27]. An interaction with the golgin GC5 has also been reported for the *Arabidopsis* Rab6 homologs AtRabH1b/c [52], suggesting that this feature is highly conserved in evolution. The functional significance of Rab6-golgin interaction could be two fold: Rab6 could be needed to recruit certain golgins to the membranes, or the golgins could use their interaction with Rab6 for vesicle tethering. Recently, it was shown that golgins typically contain multiple Rab-binding sites [48, 51], and therefore, the role of Rab6 might be redundant with that of other Golgi Rabs. This might explain why Rab6 is not essential for maintaining Golgi structure.

Another binding partner of Rab6 at the Golgi is the Conserved Oligomeric Golgi (COG) complex: purified yeast COG binds to Ypt6p [53], and mammalian Rab6 interacts with the COG6 in a yeast two-hybrid assay [54]. COG complex is preferentially associated with cis- and medial Golgi cisternae; it binds to COPI coats and Golgi SNAREs, and participates in the retrograde intra-Golgi trafficking. The depletion of COG subunits causes Golgi fragmentation, which can be suppressed by simultaneous Rab6 depletion [55]. Similarly, knockdown of Rab6 also suppresses Golgi fragmentation caused by the depletion of another Golgi tethering complex, ZW10/RINT-1 [55]. COG and ZW10/RINT-1 likely act in two independent intra-Golgi trafficking pathways, because microinjection of Rab6-binding BICD2 C-terminus inhibited Golgi fragmentation induced by ZW10 depletion, but did not rescue the effect of COG inactivation [55]. This might be due to the fact that both BICD and ZW10 act through pathways that involve cytoplasmic dynein, while COG activity is mechanistically different. Golgi disruption induced by the depletion of Golgin-84 or by *Chlamydia* infection could be also counteracted by Rab6 knockdown [56].

Golgi homeostasis can be profoundly perturbed by different stress conditions, such as the exposure of cells to hypotonic media, low temperature (15° C for mammalian cells) or the drug brefeldin A, which all cause formation of Golgi-derived tubules [57, 58]. Interestingly, cold-induced tubules are strongly Rab6-positive [58]. Further, brefeldin A-induced tubule formation is sensitive to overexpression of GDP-restricted Rab6A or A', and similarly, formation of tubules initiated by hypotonic shock is suppressed by GDP-Rab6A [57]. Golgi tubulation is microtubule-dependent, and likely requires at least some of the Rab6 effectors that are also involved in the transport of normal Rab6-decorated membrane carriers. Interestingly, microtubule-independent Golgi scattering induced by microtubule-depolymerizing drug nocodazole is also inhibited by GDP-Rab6A, suggesting a potential role of some Golgi-stack associated microtubule-independent Rab6 effectors [57]. The picture that emerges from all these studies suggests that although the presence of Rab6 is not required for maintaining Golgi integrity, it makes the Golgi more dynamic and thus more susceptible to fragmentation or disruption due to different treatments or stress conditions.

Currently, very little is known about the factors that control Rab6 association with the Golgi. Interestingly, overexpression of a GTP-restricted Rab33B, which is localized to the medial Golgi, reduced Rab6 accumulation at the Golgi by ~50%, suggesting that it might act together with a Rab6 GAP in a retrograde intra-Golgi route [59]. Another factor that was shown to partially affect Rab6 association with the Golgi is Yip1A (a mammalian homologue of the yeast Yip1p), a transmembrane protein localized to the ER-Golgi intermediate compartment (ERGIC) [60]. This effect is, however likely to be indirect, because Yip1A and Rab6 bind to different Golgi subcompartments and do not interact directly [60]. The knowledge of Rab6 GEFs and GAPs is urgently needed to understand the mechanistic basis of Rab6 recruitment and function on the Golgi membranes.

4.3. Golgi to Endoplasmic Reticulum Transport

The hypothesis that Rab6 controls a Golgi to ER retrograde transport pathway was originally formulated to explain the redistribution of Golgi resident enzymes such as the β-1,4-galacosyltransferase into the ER following acute (vaccinia system) overexpression of the GTPase-deficient Rab6A mutant (Rab6A Q72L) [61]. This brefeldin A-like effect was not observed with the dominant negative mutant Rab6A T27N and required the integrity of microtubules [61]. A role for Rab6A is this pathway was further supported by the identification in live cells of GFP-Rab6A-positive dynamic transport carriers containing STxB, which moved along microtubules from the Golgi to the cell margin. At the cell periphery, these structures were shown to merge with the peripheral ER, suggesting that they corresponded to ER entry sites for Shiga toxin [8]. The finding that microinjection of anti-COPI antibodies does not block the delivery of STxB to ER whereas Rab6A T27N did lead to the definition of a COPI-independent route between Golgi and ER used by Shiga toxin and at least some Golgi resident enzymes to recycle back to the ER [62].

The existence of a Rab6-mediated Golgi to ER route, at least in its original formulation, has been challenged later on. First, Young *et al.* provided evidence that the bulk of Golgi-to-ER recycling process regulated by Rab6 takes place at the proximity of the Golgi *via* a direct route connecting *trans*-Golgi/TGN with the ER [34]. Such a connection has been documented a long time ago by Novikoff and co-workers (the "GERL hypothesis", [63, 64]). The close apposition of ER membranes to *trans*-cisternae has been recently confirmed by 3D electron tomography [65]. The proposed Rab6 effectors in this pathway are Bicaudal D and the dynein/dynactin complex [34]. Second, re-investigation of the dynamics of GFP-Rab6A positive vesicles by live cell imaging showed that in cells incubated with STxB, only a minority (less than 10%) of Rab6 positive vesicles contain STxB [7]. In contrast, the majority could be stained with exogenous secretory markers, such as VSV-G or neuropeptide Y (NPY), indicating that most, if not all Rab6 positive vesicles moving along microtubules from the Golgi to the cell periphery are in fact post-Golgi secretory vesicles (see paragraph on Golgi to plasma membrane transport).

Thus, although the evidence in favor of a role for Rab6 in a Golgi to ER recycling route appears solid and has been obtained in several laboratories [8, 34, 57, 61, 62], its exact nature remains to be established. One way to go would be to identify Rab6 effectors involved in this pathway. Another issue that remains to be clarified concerns the respective role of Rab6A and Rab6A'. Based on overexpression experiments of Rab6A and Rab6A' constructs, it has been proposed that only the Rab6A isoform plays a role in Golgi to ER pathway [2]. However, several observations have challenged this hypothesis: specific silencing of Rab6A does not significantly affect transport of STxB between Golgi and ER [30] and overexpression of GTP-locked mutants of both isoforms can redistribute Golgi resident enzymes into the ER [34]. Furthermore, specific silencing of Rab6A' also interferes with the recycling process of ERGIC-53, a marker of pre-Golgi/intermediate compartments [34]. The exact function of different Rab6 isoforms in Golgi to ER transport is thus still unclear.

4.4. Golgi to Plasma Membrane Transport

Because of the involvement of Rab6 is Golgi homeostasis and Golgi-related retrieval pathways, it is not surprising that Rab6 function is important for normal secretion [66]. However, a more direct role of Rab6 in plasma membrane-directed trafficking was uncovered by imaging of carriers of constitutive secretion [7]. It transpired that in different cultured cell lines the majority of such carriers are decorated by Rab6A/A', and in turn, the majority of Rab6A/A' vesicles that emerge from the Golgi and move to the cell periphery contain different secretion markers and can be directly observed to fuse with the plasma membrane. Importantly, depletion of Rab6A/A' delays, but does not block constitutive secretion. However, the carriers of constitutive exocytosis formed in the absence of Rab6A/A' show a strongly altered behavior, as their microtubule-based movement and fusion with the plasma membrane are affected.

Rab6 is required for highly processive motility of exocytotic vesicles to microtubule plus ends. This process likely involves Rab6A/A'/B effectors Bicaudal D1 and 2 (BICD1/2), coiled coil proteins, which bind to Rab6 through their C-terminal segments [67. 68] and weakly associate with kinesin-1 (KIF5B), one of the major motors responsible for Rab6 vesicle transport, through their central domain [7]. Another motor complex associated with Rab6 vesicles is dynein-dynactin. It binds to Rab6 directly through the dynactin large subunit p150^{Glued}, and indirectly through the N-terminal part of BICD1/2 [67, 68]. Importantly, inhibition of both kinesin-1 and dynein affects the speed and processivity of vesicle movements but does not abolish their motility, indicating that additional motors must be involved [7].

Interestingly, both vertebrates and flies encode a homologue of BICD1/2, BICDR-1, another Rab6 effector which participates in motor recruitment [69]. Similar to BICD1/2, BICDR-1 also interacts with dynein-dynactin complex; however, instead of kinesin-1 it associates with the kinesin-3 KIF1C. The balance of motor activities is shifted on BICDR-1- compared to BICD1/2- bound Rab6 vesicles: their microtubule minus-end directed motility predominates, and instead of the cell periphery they accumulate around the centrosome. BICDR-1 is expressed in developing neurons, and plays a role in regulation of neuritogenesis [69].

Fusion of Rab6 vesicles with the plasma membrane is controlled by yet another Rab6 effector, ELKS (also known as CAST2, ERC1 and Rab6IP2) [7, 70, 71]. This protein, which is best known for its role in the organization of neuronal secretory sites (the cytomatrix at the active zone) [72] is predominantly localized to cortical patches in cultured fibroblasts. These patches are important for microtubule stabilization and attachment to the plasma membrane at the leading edge of migrating cells [73]. ELKS is not essential for Rab6 vesicle exocytosis, but promotes vesicle fusion with the plasma membrane and can thus affect spatial distribution of exocytotic events, coordinating microtubule organization with vesicle fusion [7].

Importantly, detailed analysis showed that GFP-Rab6 does not exchange on the exocytotic carriers. This result indicates that Rab6 GEFs and GAPs are unlikely to come into play from the time the Rab6-decorated carrier has formed and budded from the Golgi until it fuses with the plasma membrane [7].

A recent study has extended the function of Rab6A/A' to the formation of the Rab6 vesicles [11]. This function involves a novel Rab6 effector, Myosin IIA, that can be recruited by Rab6 on Golgi/TGN membranes. The inhibition of Myosin II function impairs the fission of Rab6 vesicles from Golgi membranes and trafficking of both anterograde and retrograde cargoes [11].

All the above studies mostly focused on mammalian Rab6A/A' isoforms. Rab6B has been studied much less extensively as it is predominantly neuronal; however, it is likely to have a similar function in secretion as it interacts with many of the same effectors as Rab6A/A', localizes to the Golgi and vesicular carriers, colocalizes with exocytotic cargo and plays a role in neurite outgrowth [6, 69, 74].

Direct participation in exocytosis is an evolutionary conserved Rab6 feature: genetic studies of the *S. pombe* Ryh1 showed that it is required for the Golgi to plasma membrane transport [36]. Furthermore, detailed studies in flies showed that Rab6 is required for exocytosis during fly oogenesis: the correct delivery of secreted proteins such as Gurken (a TGFα homologue) and nurse cell membrane generation are affected in Rab6-null egg chambers [75, 76]. The interaction between Rab6 and BicD is also conserved in flies, and BicD participates in Rab6-mediated membrane trafficking and polarization pathways. The potentially direct involvement of *Drosophila* Rab6 in polarized exocytosis is supported by the similarities observed between the defects of egg chambers lacking Rab6 and Sec5, a subunit of the multiprotein complex exocyst, which is important for vesicle fusion with the plasma membrane [76]. Participation of Rab6 in rhodopsin transport [77] and modulation of the Notch signaling pathway [78] might also be related to the direct function of Rab6 in secretion, although in both cases the involvement in endosome-to-Golgi route might be equally important.

4.5. Rab6 in Mitosis and Cytokinesis

Several studies have highlighted a specific role for Rab6 and some of Rab6-interacting proteins Rab6 during mitosis and cytokinesis. Rab6A' as well as GAPCenA depletion arrest cells in metaphase and activate the Mad2-spindle checkpoint [79]. In arrested cells, p150^{Glued} remains attached to kinetochores, suggesting that Rab6A' and GAPCenA are involved in the dynamics of the dynein/dynactin complex at kinetochores. Surprisingly, Rab6A':GTP was found to be cytosolic in metaphase-arrested cells. In addition, Golgi fragmentation, which is critical for mitotic progression, appeared normal in Rab6A'-depleted cells [79]. This suggests that the Rab6A' regulated pathway acting at the metaphase–anaphase transition may not be directly linked to membrane trafficking events.

The depletion of another Rab6 effector, Rab6IP1, also leads to a block in metaphase and activation of the Mad2-spindle checkpoint [43]. In addition, Rab6IP1-depleted cells (likely those that passed the metaphase block) display cytokinesis defects, which is not observed after Rab6 depletion. As mentioned above, Rab6IP1 also interacts with Rab11, and good evidence exists that Rab11-mediated pathways participate in the completion of cytokinesis, likely by delivering proteins and/or lipids required for the abscission of daughter cells [80]. This raises the interesting hypothesis of a coupling between Rab6 and Rab11 functions, not only in metaphase (see above) but also during cytokinesis. However, the exact role of Rab6-mediated trafficking events, for instance, between Golgi and the midbody, remains to be established.

5. CONCLUSIONS: RAB6 FUNCTION

Despite the apparent complexity of Rab6 function, several key features have now emerged. First, as other Rab GTPases, Rab is involved in docking/tethering events of transport vesicles on acceptor membranes, either at the plasma membrane (for Golgi-derived secretory vesicles) or at TGN membranes (for endosomes-derived vesicles) (Fig. **2**). Rab6 might also be involved in docking/tethering of Golgi-derived vesicles to ER membranes, but this remains to be established. Second, an important Rab6 function is to recruit, either directly or through adaptor proteins, microtubule and actin-based motors. Association with motor proteins plays a role in budding, fission and transport of Rab6-decorated membrane carriers. It appears, therefore, that Rab6 is able to coordinate all the sequential events underlying transport between two compartments, *e.g.*, budding/fission, movement along cytoskeletal tracks, and docking/tethering processes. This explains the facilitating role of Rab6 in both retrograde (intra-Golgi and Golgi to ER) and anterograde (Golgi to PM) transport pathways.

After 20 years of studies carried out in several laboratories, Rab6 now appears to be a general organizer of Golgi/TGN associated transport pathways. However, several important questions remain unanswered. One of them concerns the identification of GEFs and GAPs that can activate/inactivate Rab6. Their identification will allow for better understanding of Rab6 function at various steps of intracellular transport. Another issue is the respective role of the two Rab6 isoforms, Rab6A and Rab6A'. Although their localization appears to be identical, several studies have suggested that they play non-overlapping roles. For instance, Rab6A and Rab6A' are individually required for fission of Rab6 transport vesicles from the Golgi complex [11]. A tentative hypothesis is that at least some effectors need to bind to both Rab6 isoforms in order to fulfill their functions. Alternatively, some effectors may have better affinities *in vivo* to one or the other isoform. Finally, virtually nothing is known about the role of Rab6A isoforms as well as Rab6B in mammalian development and mammalian tissue function. Whole animal and tissue-specific mouse knockout approaches will be needed to address this question.

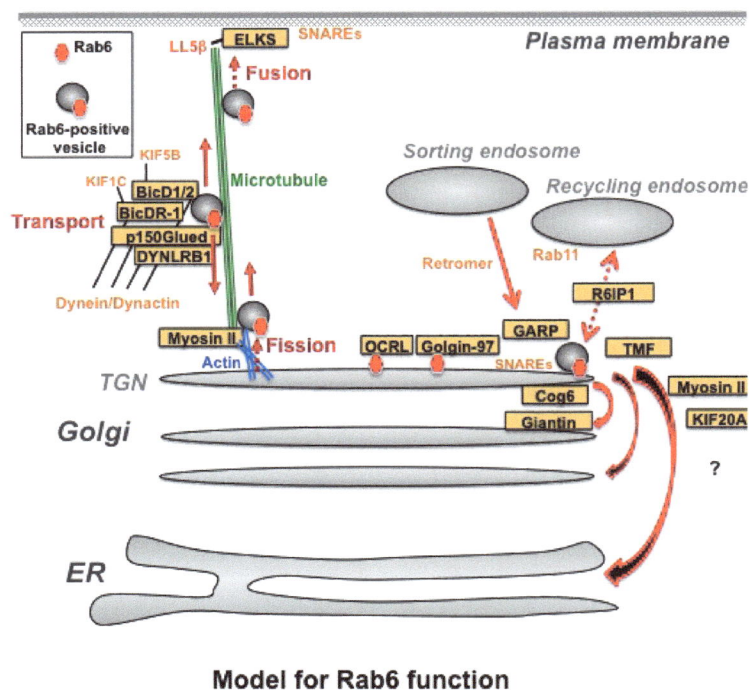

Model for Rab6 function

Figure 2: Model for Rab6 function. Rab6 regulates several transport pathways at the level of the Golgi complex (red arrows): endosome to Golgi transport; intra-Golgi transport and Golgi homeostasis; Golgi to endoplasmic reticulum transport; and Golgi to plasma membrane transport. Known Rab6 effectors functioning in these pathways are indicated in the yellow boxes.

CONFLICT OF INTEREST

There is no conflict of interest from any of the authors.

REFERENCES

[1]   Pereira-Leal JB, Seabra MC. Evolution of the Rab family of small GTP-binding proteins. J Mol Biol 2001 Nov 2;313(4):889-901.

[2]   Echard A, Opdam FJ, de Leeuw HJ, *et al.* Alternative splicing of the human Rab6A gene generates two close but functionally different isoforms. Mol Biol Cell 2000 Nov;11(11):3819-33.

[3]   Simpson JC, Wellenreuther R, Poustka A, Pepperkok R, Wiemann S. Systematic subcellular localization of novel proteins identified by large-scale cDNA sequencing. EMBO Rep 2000 Sep;1(3):287-92.

[4]   Shan J, Yuan L, Budman DR, Xu HP. WTH3, a new member of the Rab6 gene family, and multidrug resistance. Biochim Biophys Acta 2002 Apr 3;1589(2):112-23.

[5]   Young J, Menetrey J, Goud B. RAB6C is a retrogene that encodes a centrosomal protein involved in cell cycle progression. J Mol Biol 2010 Mar 19;397(1):69-88.

[6]   Opdam FJ, Echard A, Croes HJ, *et al.* The small GTPase Rab6B, a novel Rab6 subfamily member, is cell-type specifically expressed and localised to the Golgi apparatus. J Cell Sci 2000 Aug;113 (Pt 15):2725-35.

[7]   Grigoriev I, Splinter D, Keijzer N, *et al.* Rab6 regulates transport and targeting of exocytotic carriers. Dev Cell 2007 Aug;13(2):305-14.

[8]   White J, Johannes L, Mallard F, *et al.* Rab6 coordinates a novel Golgi to ER retrograde transport pathway in live cells. J Cell Biol 1999 Nov 15;147(4):743-60.

[9]   Nizak C, Monier S, del Nery E, Moutel S, Goud B, Perez F. Recombinant antibodies to the small GTPase Rab6 as conformation sensors. Science 2003 May 9;300(5621):984-7.

[10] Racine V, Sachse M, Salamero J, Fraisier V, Trubuil A, Sibarita JB. Visualization and quantification of vesicle trafficking on a three-dimensional cytoskeleton network in living cells. J Microsc 2007 Mar;225(Pt 3):214-28.

[11]   Miserey-Lenkei S, Chalancon G, Bardin S, Formstecher E, Goud B, Echard A. Rab and actomyosin dependent fission of transport vesicles at the Golgi complex. Nat Cell Biol 2010 Jul;12(7):645-54.

[12]   Stedman TT, Sussmann AR, Joiner KA. Toxoplasma gondii Rab6 mediates a retrograde pathway for sorting of constitutively secreted proteins to the Golgi complex. J Biol Chem 2003 Feb 14;278(7):5433-43.

[13]   Karniguian A, Zahraoui A, Tavitian A. Identification of small GTP-binding rab proteins in human platelets: thrombin-induced phosphorylation of rab3B, rab6, and rab8 proteins. Proc Natl Acad Sci U S A 1993 Aug 15;90(16):7647-51.

[14]   Tixier-Vidal A, Barret A, Picart R, *et al.* The small GTP-binding protein, Rab6p, is associated with both Golgi and post-Golgi synaptophysin-containing membranes during synaptogenesis of hypothalamic neurons in culture. J Cell Sci 1993 Aug;105 (Pt 4):935-47.

[15]   Deretic D, Papermaster DS. Rab6 is associated with a compartment that transports rhodopsin from the trans-Golgi to the site of rod outer segment disk formation in frog retinal photoreceptors. J Cell Sci 1993 Nov;106 (Pt 3):803-13.

[16]   Siniossoglou S, Peak-Chew SY, Pelham HR. Ric1p and Rgp1p form a complex that catalyses nucleotide exchange on Ypt6p. EMBO J 2000 Sep 15;19(18):4885-94.

[17]   Brass AL, Dykxhoorn DM, Benita Y, *et al.* Identification of host proteins required for HIV infection through a functional genomic screen. Science 2008 Feb 15;319(5865):921-6.

[18]   Bergbrede T, Pylypenko O, Rak A, Alexandrov K. Structure of the extremely slow GTPase Rab6A in the GTP bound form at 1.8A resolution. J Struct Biol 2005 Dec;152(3):235-8.

[19]   Cuif MH, Possmayer F, Zander H, *et al.* Characterization of GAPCenA, a GTPase activating protein for Rab6, part of which associates with the centrosome. EMBO J 1999 Apr 1;18(7):1772-82.

[20]   Neuwald AF. A shared domain between a spindle assembly checkpoint protein and Ypt/Rab-specific GTPase-activators. Trends Biochem Sci 1997 Jul;22(7):243-4.

[21]   Fuchs E, Haas AK, Spooner RA, Yoshimura S, Lord JM, Barr FA. Specific Rab GTPase-activating proteins define the Shiga toxin and epidermal growth factor uptake pathways. J Cell Biol 2007 Jun 18;177(6):1133-43.

[22]   Kanno E, Ishibashi K, Kobayashi H, Matsui T, Ohbayashi N, Fukuda M. Comprehensive screening for novel rab-binding proteins by GST pull-down assay using 60 different mammalian Rabs. Traffic 2010 Apr;11(4):491-507.

[23]   Bergbrede T, Chuky N, Schoebel S, *et al.* Biophysical analysis of the interaction of Rab6a GTPase with its effector domains. J Biol Chem 2009 Jan 30;284(5):2628-35.

[24]   Recacha R, Boulet A, Jollivet F, *et al.* Structural basis for recruitment of Rab6-interacting protein 1 to Golgi *via* a RUN domain. Structure 2009 Jan 14;17(1):21-30.

[25]   Burguete AS, Fenn TD, Brunger AT, Pfeffer SR. Rab and Arl GTPase family members cooperate in the localization of the golgin GCC185. Cell 2008 Jan 25;132(2):286-98.

[26]   Fernandes H, Franklin E, Recacha R, Houdusse A, Goud B, Khan AR. Structural aspects of Rab6-effector complexes. Biochem Soc Trans 2009 Oct;37(Pt 5):1037-41.

[27]   Houghton FJ, Chew PL, Lodeho S, Goud B, Gleeson PA. The localization of the Golgin GCC185 is independent of Rab6A/A' and Arl1. Cell 2009 Aug 21;138(4):787-94.

[28]   Mallard F, Antony C, Tenza D, Salamero J, Goud B, Johannes L. Direct pathway from early/recycling endosomes to the Golgi apparatus revealed through the study of shiga toxin B-fragment transport. J Cell Biol 1998 Nov 16;143(4):973-90.

[29]   Mallard F, Tang BL, Galli T, *et al.* Early/recycling endosomes-to-TGN transport involves two SNARE complexes and a Rab6 isoform. J Cell Biol 2002 Feb 18;156(4):653-64.

[30]   Del Nery E, Miserey-Lenkei S, Falguieres T, *et al.* Rab6A and Rab6A' GTPases play non-overlapping roles in membrane trafficking. Traffic 2006 Apr;7(4):394-407.

[31]   Utskarpen A, Slagsvold HH, Iversen TG, Walchli S, Sandvig K. Transport of ricin from endosomes to the Golgi apparatus is regulated by Rab6A and Rab6A'. Traffic 2006 Jun;7(6):663-72.

[32]   Smith RD, Willett R, Kudlyk T, *et al.* The COG complex, Rab6 and COPI define a novel Golgi retrograde trafficking pathway that is exploited by SubAB toxin. Traffic 2009 Oct;10(10):1502-17.

[33]   Medigeshi GR, Schu P. Characterization of the *in vitro* retrograde transport of MPR46. Traffic 2003 Nov;4(11):802-11.

[34]   Young J, Stauber T, del Nery E, Vernos I, Pepperkok R, Nilsson T. Regulation of microtubule-dependent recycling at the trans-Golgi network by Rab6A and Rab6A'. Mol Biol Cell 2005 Jan;16(1):162-77.

[35]   Luo Z, Gallwitz D. Biochemical and genetic evidence for the involvement of yeast Ypt6-GTPase in protein retrieval to different Golgi compartments. J Biol Chem 2003 Jan 10;278(2):791-9.

[36]   He Y, Sugiura R, Ma Y, *et al.* Genetic and functional interaction between Ryh1 and Ypt3: two Rab GTPases that function in S. pombe secretory pathway. Genes Cells 2006 Mar;11(3):207-21.

[37]   Liewen H, Meinhold-Heerlein I, Oliveira V, *et al.* Characterization of the human GARP (Golgi associated retrograde protein) complex. Exp Cell Res 2005 May 15;306(1):24-34.

[38]   Perez-Victoria FJ, Bonifacino JS. Dual roles of the mammalian GARP complex in tethering and SNARE complex assembly at the trans-golgi network. Mol Cell Biol 2009 Oct;29(19):5251-63.

[39]   Conibear E, Cleck JN, Stevens TH. Vps51p mediates the association of the GARP (Vps52/53/54) complex with the late Golgi t-SNARE Tlg1p. Mol Biol Cell 2003 Apr;14(4):1610-23.

[40]   Perez-Victoria FJ, Schindler C, Magadan JG, *et al.* Ang2/fat-free is a conserved subunit of the Golgi-associated retrograde protein complex. Mol Biol Cell 2010 Oct 1;21(19):3386-95.

[41]   Siniossoglou S, Pelham HR. An effector of Ypt6p binds the SNARE Tlg1p and mediates selective fusion of vesicles with late Golgi membranes. EMBO J 2001 Nov 1;20(21):5991-8.

[42]   Yamane J, Kubo A, Nakayama K, Yuba-Kubo A, Katsuno T, Tsukita S. Functional involvement of TMF/ARA160 in Rab6-dependent retrograde membrane traffic. Exp Cell Res 2007 Oct 1;313(16):3472-85.

[43]   Miserey-Lenkei S, Waharte F, Boulet A, *et al.* Rab6-interacting protein 1 links Rab6 and Rab11 function. Traffic 2007 Oct;8(10):1385-403.

[44]   Wassmer T, Attar N, Harterink M, *et al.* The retromer coat complex coordinates endosomal sorting and dynein-mediated transport, with carrier recognition by the trans-Golgi network. Dev Cell 2009 Jul;17(1):110-22.

[45]   Bujny MV, Popoff V, Johannes L, Cullen PJ. The retromer component sorting nexin-1 is required for efficient retrograde transport of Shiga toxin from early endosome to the trans Golgi network. J Cell Sci 2007 Jun 15;120(Pt 12):2010-21.

[46]   Bonifacino JS, Rojas R. Retrograde transport from endosomes to the trans-Golgi network. Nat Rev Mol Cell Biol 2006 Aug;7(8):568-79.

[47]   Goud B, Gleeson PA. TGN golgins, Rabs and cytoskeleton: regulating the Golgi trafficking highways. Trends Cell Biol 2010 Mar 12.

[48]   Sinka R, Gillingham AK, Kondylis V, Munro S. Golgi coiled-coil proteins contain multiple binding sites for Rab family G proteins. J Cell Biol 2008 Nov 17;183(4):607-15.

[49]   Barr FA. A novel Rab6-interacting domain defines a family of Golgi-targeted coiled-coil proteins. Curr Biol 1999 Apr 8;9(7):381-4.

[50]   Rosing M, Ossendorf E, Rak A, Barnekow A. Giantin interacts with both the small GTPase Rab6 and Rab1. Exp Cell Res 2007 Jul 1;313(11):2318-25.

[51]   Hayes GL, Brown FC, Haas AK, Nottingham RM, Barr FA, Pfeffer SR. Multiple Rab GTPase binding sites in GCC185 suggest a model for vesicle tethering at the trans-Golgi. Mol Biol Cell 2009 Jan;20(1):209-17.

[52]   Latijnhouwers M, Gillespie T, Boevink P, Kriechbaumer V, Hawes C, Carvalho CM. Localization and domain characterization of Arabidopsis golgin candidates. J Exp Bot 2007;58(15-16):4373-86.

[53]   Suvorova ES, Duden R, Lupashin VV. The Sec34/Sec35p complex, a Ypt1p effector required for retrograde intra-Golgi trafficking, interacts with Golgi SNAREs and COPI vesicle coat proteins. J Cell Biol 2002 May 13;157(4):631-43.

[54]   Fukuda M, Kanno E, Ishibashi K, Itoh T. Large scale screening for novel rab effectors reveals unexpected broad Rab binding specificity. Mol Cell Proteomics 2008 Jun;7(6):1031-42.

[55]   Sun Y, Shestakova A, Hunt L, Sehgal S, Lupashin V, Storrie B. Rab6 regulates both ZW10/RINT-1 and conserved oligomeric Golgi complex-dependent Golgi trafficking and homeostasis. Mol Biol Cell 2007 Oct;18(10):4129-42.

[56]   Rejman Lipinski A, Heymann J, Meissner C, *et al.* Rab6 and Rab11 regulate Chlamydia trachomatis development and golgin-84-dependent Golgi fragmentation. PLoS Pathog 2009 Oct;5(10):e1000615.

[57]   Jiang S, Storrie B. Cisternal rab proteins regulate Golgi apparatus redistribution in response to hypotonic stress. Mol Biol Cell 2005 May;16(5):2586-96.

[58]   Martinez-Alonso E, Ballesta J, Martinez-Menarguez JA. Low-temperature-induced Golgi tubules are transient membranes enriched in molecules regulating intra-Golgi transport. Traffic 2007 Apr;8(4):359-68.

[59]   Starr T, Sun Y, Wilkins N, Storrie B. Rab33b and Rab6 are functionally overlapping regulators of Golgi homeostasis and trafficking. Traffic 2010 May;11(5):626-36.

[60]   Kano F, Yamauchi S, Yoshida Y, *et al.* Yip1A regulates the COPI-independent retrograde transport from the Golgi complex to the ER. J Cell Sci 2009 Jul 1;122(Pt 13):2218-27.

[61]   Martinez O, Antony C, Pehau-Arnaudet G, Berger EG, Salamero J, Goud B. GTP-bound forms of rab6 induce the redistribution of Golgi proteins into the endoplasmic reticulum. Proc Natl Acad Sci U S A 1997 Mar 4;94(5):1828-33.

[62]   Girod A, Storrie B, Simpson JC, *et al*. Evidence for a COP-I-independent transport route from the Golgi complex to the endoplasmic reticulum. Nat Cell Biol 1999 Nov;1(7):423-30.

[63]   Novikoff AB, Essner E, Quintana N. Golgi Apparatus and Lysosomes. Fed Proc 1964 Sep-Oct;23:1010-22.

[64]   Novikoff PM, Novikoff AB, Quintana N, Hauw JJ. Golgi apparatus, GERL, and lysosomes of neurons in rat dorsal root ganglia, studied by thick section and thin section cytochemistry. J Cell Biol 1971 Sep;50(3):859-86.

[65]   Mogelsvang S, Marsh BJ, Ladinsky MS, Howell KE. Predicting function from structure: 3D structure studies of the mammalian Golgi complex. Traffic 2004 May;5(5):338-45.

[66]   Martinez O, Schmidt A, Salamero J, Hoflack B, Roa M, Goud B. The small GTP-binding protein rab6 functions in intra-Golgi transport. J Cell Biol 1994 Dec;127(6 Pt 1):1575-88.

[67]   Matanis T, Akhmanova A, Wulf P, *et al*. Bicaudal-D regulates COPI-independent Golgi-ER transport by recruiting the dynein-dynactin motor complex. Nat Cell Biol 2002;4(12):986-92.

[68]   Short B, Preisinger C, Schaletzky J, Kopajtich R, Barr FA. The Rab6 GTPase regulates recruitment of the dynactin complex to Golgi membranes. Curr Biol 2002;12(20):1792-5.

[69]   Schlager MA, Kapitein LC, Grigoriev I, *et al*. Pericentrosomal targeting of Rab6 secretory vesicles by Bicaudal-D-related protein 1 (BICDR-1) regulates neuritogenesis. EMBO J 2010 May 19;29(10):1637-51.

[70]   Nakata T, Kitamura Y, Shimizu K, *et al*. Fusion of a novel gene, ELKS, to RET due to translocation t(10;12)(q11;p13) in a papillary thyroid carcinoma. Genes Chromosomes Cancer 1999 Jun;25(2):97-103.

[71]   Monier S, Jollivet F, Janoueix-Lerosey I, Johannes L, Goud B. Characterization of novel Rab6-interacting proteins involved in endosome-to-TGN transport. Traffic 2002 Apr;3(4):289-97.

[72]   Ohtsuka T, Takao-Rikitsu E, Inoue E, *et al*. Cast: a novel protein of the cytomatrix at the active zone of synapses that forms a ternary complex with RIM1 and munc13-1. J Cell Biol 2002 Aug 5;158(3):577-90.

[73]   Lansbergen G, Grigoriev I, Mimori-Kiyosue Y, *et al*. CLASPs attach microtubule plus ends to the cell cortex through a complex with LL5beta. Dev Cell 2006 Jul;11(1):21-32.

[74]   Wanschers BF, van de Vorstenbosch R, Schlager MA, *et al*. A role for the Rab6B Bicaudal-D1 interaction in retrograde transport in neuronal cells. Exp Cell Res 2007 Oct 1;313(16):3408-20.

[75]   Januschke J, Nicolas E, Compagnon J, Formstecher E, Goud B, Guichet A. Rab6 and the secretory pathway affect oocyte polarity in Drosophila. Development 2007 Oct;134(19):3419-25.

[76]   Coutelis JB, Ephrussi A. Rab6 mediates membrane organization and determinant localization during Drosophila oogenesis. Development 2007 Apr;134(7):1419-30.

[77]   Shetty KM, Kurada P, O'Tousa JE. Rab6 regulation of rhodopsin transport in Drosophila. J Biol Chem 1998 Aug 7;273(32):20425-30.

[78]   Purcell K, Artavanis-Tsakonas S. The developmental role of warthog, the notch modifier encoding Drab6. J Cell Biol 1999 Aug 23;146(4):731-40.

[79]   Miserey-Lenkei S, Couedel-Courteille A, Del Nery E, *et al*. A role for the Rab6A' GTPase in the inactivation of the Mad2-spindle checkpoint. EMBO J 2006 Jan 25;25(2):278-89.

[80]   Montagnac G, Echard A, Chavrier P. Endocytic traffic in animal cell cytokinesis. Curr Opin Cell Biol 2008 Aug;20(4):454-61.

[81]   Janoueix-Lerosey I, Jollivet F, Camonis J, Marche PN, Goud B. Two-hybrid system screen with the small GTP-binding protein Rab6. Identification of a novel mouse GDP dissociation inhibitor isoform and two other potential partners of Rab6. J Biol Chem 1995 Jun 16;270(24):14801-8.

[82]   Abdul-Ghani M, Gougeon PY, Prosser DC, Da-Silva LF, Ngsee JK. PRA isoforms are targeted to distinct membrane compartments. J Biol Chem 2001 Mar 2;276(9):6225-33.

[83]   Sivars U, Aivazian D, Pfeffer SR. Yip3 catalyses the dissociation of endosomal Rab-GDI complexes. Nature 2003 Oct 23;425(6960):856-9.

[84]   Kail M, Barnekow A. Identification and characterization of interacting partners of Rab GTPases by yeast two-hybrid analyses. Methods Mol Biol 2008;440:111-25.

[85]   Wanschers B, van de Vorstenbosch R, Wijers M, Wieringa B, King SM, Fransen J. Rab6 family proteins interact with the dynein light chain protein DYNLRB1. Cell Motil Cytoskeleton 2008 Mar;65(3):183-96.

[86]   Echard A, Jollivet F, Martinez O, *et al*. Interaction of a Golgi-associated kinesin-like protein with Rab6. Science 1998 Jan 23;279(5350):580-5.

[87]   Hill E, Clarke M, Barr FA. The Rab6-binding kinesin, Rab6-KIFL, is required for cytokinesis. EMBO J 2000 Nov 1;19(21):5711-9.

[88]   Fontijn RD, Goud B, Echard A, *et al*. The human kinesin-like protein RB6K is under tight cell cycle control and is essential for cytokinesis. Mol Cell Biol 2001 Apr;21(8):2944-55.

[89]   Neef R, Preisinger C, Sutcliffe J, *et al.* Phosphorylation of mitotic kinesin-like protein 2 by polo-like kinase 1 is required for cytokinesis. J Cell Biol 2003 Sep 1;162(5):863-75.

[90]   Hyvola N, Diao A, McKenzie E, Skippen A, Cockcroft S, Lowe M. Membrane targeting and activation of the Lowe syndrome protein OCRL1 by rab GTPases. EMBO J 2006 Aug 23;25(16):3750-61.

[91]   Hennies HC, Kornak U, Zhang H, *et al.* Gerodermia osteodysplastica is caused by mutations in SCYL1BP1, a Rab-6 interacting golgin. Nat Genet 2008 Dec;40(12):1410-2.

[92]   Teber I, Nagano F, Kremerskothen J, Bilbilis K, Goud B, Barnekow A. Rab6 interacts with the mint3 adaptor protein. Biol Chem 2005 Jul;386(7):671-7.

[93]   Barnekow A, Thyrock A, Kessler D. Chapter 5: rab proteins and their interaction partners. Int Rev Cell Mol Biol 2009;274:235-74.

CHAPTER 4

Polarized Exocytosis in Yeast: Sec4p

Andreas Knödler, Vishnu Ganesan and Wei Guo[*]

University of Pennsylvania, USA

Abstract: Sec4p is one of the founding members of the Rab family of small GTPases, which controls post-Golgi exocytosis in the budding yeast Saccharomyces cerevisiae. Studying Sec4p in the facile yeast genetic system has provided important insight into the molecular mechanisms of Rab proteins function and regulation. Sec4p is activated by its guanine nucleotide exchange factor Sec2p, which is in turn recruited by Ypt31p and Ypt32p, the upstream Rab proteins that regulate vesicle budding from the tans-Golgi network. The downstream effectors of Sec4p include the exocyst, which mediates vesicle tethering, and the lgl proteins Sro7/77p that regulate SNARE assembly and membrane fusion. In addition, activation of Sec4p is required for the engagement of secretory vesicles to actin cables for directional transport of cargos to the daughter cell. Thus, through the control of exocytosis machinery and interaction with actin cytoskeleton, Sec4p plays an important role in asymmetric cell growth in budding yeast.

Keywords: Sec4, Exocyst, Sec2/Rabin8, Exocytosis, Cell Polarity.

1. INTRODUCTION

SEC4 was one of the 23 genes discovered in the classic yeast screen for mutants that showed a block in secretion [1]. Since that seminal paper 30 years ago, the '*sec*' mutants have been studied in-depth and proven vital to our understanding of vesicle trafficking and membrane fusion. Analysis of Sec4p showed that it shares sequence homology with GTP-binding proteins of the Ras superfamily, including a 30% amino acid sequence identity with Ras [2, 3]. Most of the similarity is confined to four domains important for the binding and hydrolysis of GTP. Further studies confirmed Sec4p's GTP binding capability [4], and placed Sec4p as one of the founding members of the Rab family of small GTP-binding proteins in eukaryotic cells. Mutations in the *SEC4* gene led to intracellular accumulation of 80-100 nm post-Golgi secretory vesicles and it was thus speculated to control the final stage of the secretory pathway in budding yeast [1].

2. THE LOCALIZATION OF SEC4P

Sec4p is distributed to three distinct pools in yeast cells: secretory vesicles, plasma membrane, and cytosol. Newly synthesized Sec4p rapidly associates with secretory vesicles, which are transported to and eventually fused with the plasma membrane. The plasma membrane-bound pool of Sec4p can then be extracted by Rab GDP-dissociation inhibitor (GDI) into the cytosol for later recycle of Sec4p to new secretory vesicles for another round of vesicle fusion [5-7]. Attachment of Sec4p to the membrane is facilitated by geranylgeranylation, the addition of isoprenyl lipid anchor to two cysteines at the very C-terminus of the protein. This modification is catalyzed by geranylgeranyl transferase II [8, 9]. It has been shown that this double prenylation motif not only ensures proper anchoring of Rab proteins to the membrane but is also important for the correct localization of Rab proteins to their subcellular compartment [10]. The membrane protein *YIP1* can specifically recognize di- versus mono-geranylgeranylated Rab proteins [10]. It is believed that *YIP1* family members can actively compete with Rab-GDI for Rab protein interaction, possibly helping recruit Rab proteins to their correct target membrane [11-13]. Upon attachment of Sec4p to the membrane, Rab-GDI is displaced by a Rab-GDI-displacement factor (GDF). Then, a guanine nucleotide exchange factor (GEF) activates Sec4p by exchanging GDP with GTP (Fig. **1**) and renders it resistant to GDI. Sec4p-GTP then mediates the docking of the transport vesicle to the plasma

*Address corresponding to Wei Guo: Department of Biology, University of Pennsylvania, Philadelphia, PA 19104-6018, USA; Tel: 215-898-9384; Fax: 215-898-8780; E-mail: guowei@sas.upenn.edu

Guangpu Li and Nava Segev (Eds)

membrane in concert with its downstream effectors. A specific GTPase-activating protein (GAP) then catalyzes the hydrolysis of GTP to GDP (Fig. **1**), which allows Rab-GDI to bind to Sec4p followed by extraction of Sec4p from the plasma membrane, thus completing the Rab cycle.

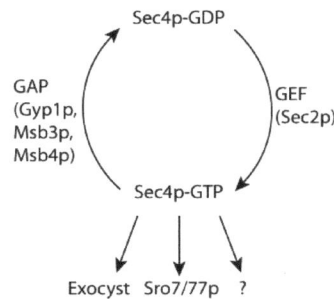

Figure 1: Sec4p cycles between the GTP-bound active form and the GDP-bound inactive form. Upon attachment of Sec4p to the secretory vesicle, its guanine nucleotide exchange factor (GEF) Sec2p exchanges GDP with GTP and converts Sec4p into its active form. Sec4p-GTP then regulates the tethering and docking of the vesicle to the plasma membrane through its downstream effectors, including the exocyst and Sro7/77p. Sec4p may have other unidentified downstream effectors (indicated by the question mark). Switching Sec4p to its inactive form is achieved by hydrolyzing GTP to GDP. This is facilitated by the GTPase activating proteins (GAPs). Msb3p, Msb4p and Gyp1 have been shown to have GAP activity towards Sec4p.

3. SEC2P IS THE GUANINE NUCLEOTIDE EXCHANGE FACTOR (GEF) FOR SEC4P

Sec2p was found in the same genetic screen that identified Sec4p [1]. Sec2p catalyzes the dissociation of GDP as well as the reloading of GTP (Fig. **1**) and is highly potent, 100-1000 times more effective than other GEFs such as Mss4p and Rabex-5 [14, 15]. The Sec2p GEF domain lies within its N-terminus (aa 1-160). Structural studies show that this region forms a homodimeric parallel coiled-coil structure [15]. Within this GEF domain, interaction with Sec4p occurs at a small stretch between residues 94 and 119. It was proposed that Sec2 structurally reorganizes switch I, switch II, and the P loop of Sec4p and leads to the decreased nucleotide affinity for GDP [15]. *sec2* mutants display an accumulation of vesicles randomly distributed throughout the cell, implying that the activation of Sec4p by Sec2p is needed for the polarized delivery of secretory vesicles [16, 17].

Proper localization of Sec2p to sites of polarized growth is critical for its function. An early study showed that a C-terminal truncated mutant version of Sec2p displays temperature sensitivity and a secretion defect [18]. Further studies revealed a stretch of 59 amino acids within the C-terminus that is necessary for accurate Sec2p localization to the plasma membrane. Deletion of these 59 amino acids (aa 450–508) showed a weaker association with microsomal membranes in biochemical fractionation studies. Additionally these truncated mutants show dramatic loss of polarized localization of Sec2p to the bud tip [14]. Recent studies from the Novick lab demonstrated the interaction of Sec2p with PI(4)P enriched on the Golgi membrane [19].

Ypt31p and Ypt32p are functionally redundant Rab GTPases that function in vesicle budding from the Golgi [20]. Novick and colleagues showed that over-expression of Ypt31 or Ypt32p suppressed secretory defects caused by mutations in the C-terminal domain of Sec2p [17]. This led to the discovery of an interaction between GTP-bound Ypt31/32p and the C-terminus of Sec2p. Further experiments demonstrated that this interaction brings Sec2p to the secretory vesicle in proximity to Sec4p [17]. The interaction of Sec2p with both Ypt32p and PI(4)P represents a coincidence detection mechanism that is commonly found in the small GTPase field. Once recruited to the vesicles, Sec2p can then activate Sec4p, and the GTP-bound Sec4p in turn interacts with its effectors.

Different organelles are defined by their own specific sets of Rab proteins, which in return specifically regulate different branches of intracellular membrane transport (see recent review by Stenmark [21]). Novick and colleagues proposed a "Rab cascade" model linking different membrane transport steps together [17]. This model postulates that the GEF of a downstream Rab GTPase also serves as an effector of an upstream Rab protein. As discussed above, the Rab GTPase Ypt32p directly interacts with the Sec4p GEF Sec2p and may promote Sec2p function indirectly by bringing it into close proximity with its substrate, Sec4p [17]. This suggests that two Rab family proteins, Sec4p and Ypt32p, might be functionally linked in a regulatory cascade through the GEF protein Sec2p. Recently, it was shown that such a cascade exists in mammalian cells. The mammalian homolog of Sec4p is Rab8, and its GEF is Rabin8. Similar to the findings in yeast, Rabin8 not only acts as the GEF for Rab8, but also interacts with the GTP-bound form of Rab11, the mammalian counterpart of Ypt32p. However, rather than simply recruiting Rabin8 to sites of catalytic requirement, Rab11 also stimulates the GEF activity of Rabin8 towards Rab8 [22]. These molecular interactions provide a mechanism for the coupling of cargo exit from the TGN/recycling endosomes (involving Rab11) to later vesicle docking and fusion at the plasma membrane (involving Rab8) in mammalian cells.

The Rab cascade is not only conserved in exocytic trafficking, but also exists in other stages of membrane trafficking [23-26]. For example, using video microscopy, Rink and colleagues have demonstrated that Rab5-marked early endosomes undergoes conversion to Rab7-marked late endosomes; and the conversion involves the HOPS complex, which has GEF activity towards Rab7 [24]. Overall, these studies demonstrate that Rab proteins and their exchange factors serve pivotal functions in the highly choreographed series of trafficking events in eukaryotic cells.

4. SEC4P GAPS

A rate-limiting step in the Rab cycle is the hydrolysis of GTP to GDP. This step directly translates into switching a Rab protein from its active to the inactive form. Because the intrinsic GTPase activity of Rab proteins is low, GTPase activating proteins (GAPs) are necessary to accelerate this process [27, 28] (Fig. **1**).

The first Rab GAP to be cloned was a yeast protein called Gyp6p [29]. It specifically activates the GTPase activity of Ypt6p and Ypt7p. Based on a BLAST database search, Du and colleagues identified Gyp1p as a novel protein with sequence homology to Gyp6p and Gyp7p [30]. GAP activity assays showed that Gyp1p possesses GAP activity for Sec4p, Ypt1p, Ypt7p and Ypt51p *in vitro*. However, it is unclear whether Gyp1p is a GAP for Sec4p *in vivo*.

Msb3p was previously described to have sequence homology with Gyp6p/Gyp7p [31]. Albert and coworkers [32] demonstrated that the homologous region lies within the catalytic domain. Msb3p accelerates the intrinsic GTPase activity of Sec4p, Ypt6p, Ypt51p and Ypt31p. Msb3p has a relatively low affinity for Sec4p; however its catalytic activity is very high: it can accelerate the intrinsic GTP hydrolysis rate of Sec4p (0.0016/min) by a factor of 5×10^5. Msb3p has a closely related homolog, Msb4p. Bi and colleagues [33] showed that both Msb3p and Msb4p localize to the bud tip, sites of polarized growth. They could further show that Msb3p and Msb4p function downstream of Cdc42p, a master regulator of polarized growth. Deletion of Msb3p and Msb4p led to the accumulation post-Golgi secretory vesicles, suggesting that they are the GAPs for Sec4p *in vivo* [34].

5. EFFECTORS OF SEC4P

5.1. The Exocyst

The exocyst is an octameric complex consisting of Sec3p, Sec5p, Sec6p, Sec8p, Sec10p, Sec15p, Exo70p, and Exo84p [35, 36]. The exocyst was the first multisubunit tethering complex identified as a Rab effector [37]. The first clue of Sec4p's interaction with the exocyst came from genetic studies showing that duplication of *SEC4* gene can suppress some exocyst mutants [38]. Subsequent yeast 2-hybrid assay and biochemical studies revealed the exocyst subunit Sec15p directly binds to Sec4p-GTP [37]. This interaction is specific to Sec4p but not to other closely related yeast Rab proteins such as Ypt1p and Ypt51p. Strikingly, a Sec4p-Ypt1p chimera with the effector domain of Sec4p replaced by that of Ypt1p abolished the interaction [37].

Sec15p binds to the exocyst component Sec10p, thus bridging Sec4p with the other exocyst components. On the plasma membrane, two other exocyst components, Sec3p and Exo70p bind to PI(4, 5)P$_2$ and interact with the Rho family of small GTPases such as Cdc42p, Rho1p, and Rho3p [39-46]. The interaction of Sec15p with Sec4p eventually links secretory vesicles containing most exocyst subunits with the exocyst Sec3p and Exo70p on the plasma membrane, allowing vesicles to be tethered to sites of polarized secretion [37, 45, 47] (Fig. **2**). Consistent with this model, a significant amount of Sec3 fails to assemble into the exocyst holo-complex in the *sec4-8* mutant [37].

Sec15p is also part of a positive feedback loop on the vesicle surface that keeps Sec4p active. This loop consists of interactions between Sec4p, its effector Sec15p, and its GEF, Sec2p. This model came from findings that Sec15p binds to Sec2p [48], while as mentioned in the previous section Sec2p binds to Ypt32p, a Rab GTPase controlling vesicle exit from the trans-Golgi network (TGN) [17]. In the model Ypt32p-GTP binds to Sec2p and localizes it to the vesicle surface. Sec2p activates Sec4p present on the vesicle, which in turn recruits Sec15p. Sec15p binds to Sec2p, displaces Ypt32p, and stabilizes Sec2p to the vesicle surface [48, 49]. This combination of GEF and effector is thought to establish a microdomain of active Sec4p, similar to Rab5/Rabex5/Rabaptin5 proteins on endosomes [50]. This type of effector-GEF interactions centered on a Rab protein may serve to generate a positive-feedback loop that activates certain steps of trafficking.

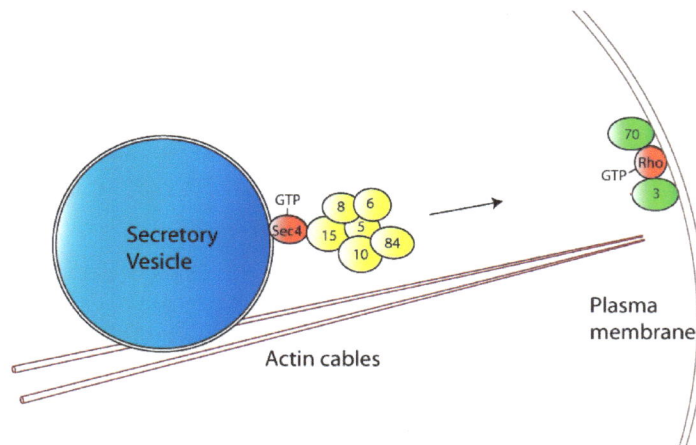

Figure 2: Polarized exocytosis in budding yeast. Sec4p is localized to the surface of secretory vesicles. In its GTP-bound form, Sec4p directly interacts with Sec15p, a member of the exocyst complex. The secretory vesicle carrying Sec15p and several members of the exocyst (marked in yellow) is transported along actin cables to the daughter cell plasma membrane. On the plasma membrane, two other exocyst components, Sec3p and Exo70p (marked in green), associate with the PI(4, 5)P$_2$ and the Rho GTPases. After vesicles are delivered to the plasma membrane, the exocyst components assemble into the octameric complex. This brings the vesicle in close contact with the plasma membrane ("tethering"), which facilitates subsequent SNARE assembly and membrane fusion.

Golgi-associated phosphatidylinositol 4-phosphate (PI4P) was recently found to play a role in directing the interactions between Sec4p, Ypt32p-GTP and Sec15p during vesicle transport to the plasma membrane [19]. Higher concentrations of PI4P closer to the TGN seem to inhibit Sec2p-Sec15p interaction and allow the localization of Sec2p to the vesicle surface through interaction with Ypt32p-GTP. Decreasing levels of PI4P as the vesicle nears the secretory site allow Sec15p to bind to Sec2p. This phosphatidylinositide gradient correlates with the flow of the cargo along the secretory pathway.

Mammalian Sec15 was identified as an effector of Rab11 [51, 52]. Structural study of the fly homologue mapped the Rab interaction to the C-terminus of Sec15 [53]. In addition to Rab11, it was found that other exocytic Rab proteins (*i.e.* Rab3, Rab8, Rab27) also interact with Sec15 in their GTP-bound form. These results suggest that the Rab-Sec15 interaction is evolutionarily conserved. However, in mammalian cells, further studies are needed to analyze the physiological implications of these interactions.

5.2. Sro7/77p

Sro7p and its paralog Sro77p are the yeast homologues of the *lethal giant larvae* (*lgl*) family of tumor suppressor proteins in *Drosophila melanogaster*. *lgl* proteins are important for maintaining cell polarization, and mutations in these proteins lead to development of malignant tumors, which are thought to be a result of loss of epithelial polarization. Sro7p was identified as an effector of Sec4p through an affinity purification [54]. *In vitro* binding assays and co-immunoprecipitation experiments later confirmed that Sro7p preferentially binds to the GTP-bound form of Sec4p [54]. Double deletions of Sro7p and its homologue Sro77p show a clear post-Golgi secretion block and the characteristic intracellular accumulation of vesicles similar to the late *sec* mutants [55]. Sro7p binds to the t-SNARE protein Sec9p, placing it in direct connection with the membrane fusion machinery. Furthermore, GTP–Sec4p, Sro7p, and Sec9p can form a ternary complex [54], suggesting that Sec4p regulates SNARE function through Sro7p. Structure studies have revealed that Sro7p approximates a twisted open clamshell followed by a tail of 60 amino acids. Deletion of this tail increases the binding affinity of Sro7p to the SNARE domain of Sec9p, suggesting that it serves to block Sro7p in an auto-inhibitory conformation [56]. It is possible that the interaction of GTP-bound Sec4p with Sro7p relieves the auto-inhibition, allowing the Sro7p-Sec9p binding. It is also interesting to note that Sro7p interacts with the exocyst component Exo84p [57] as well as Myo2p, the myosin type V motor involved in the polarized delivery of secretory vesicles [58]. Genetic data demonstrate that Sro7p interacts with the exocyst downstream of Sec4p [54, 57]. This suggests that Sec4p's downstream effectors may coordinate with each other in their action during polarized exocytosis.

6. COUPLING SECRETORY VESICLES TO THE ACTIN CYTOSKELETON

Secretory vesicles are transported along actin cables toward the daughter cell membrane for polarized exocytosis (Fig. **2**). Mutations in actin or tropomyosin result in depolarization of secretory vesicle transport [59, 60], signifying the importance of actin cables in polarized growth. Mutation in the Class V myosin protein Myo2p depolarizes cells, suggesting that it is required for the transport of post-Golgi vesicles to the bud [16, 61-64]. However, it is unclear how Myo2p is physically attached to the vesicles.

In *sec4* or *sec2* mutant cells, the directional transport of secretory vesicles is impaired; vesicles are randomly distributed in both the mother and daughter cells [16]. This indicates that activation of Sec4p is important for the engagement of vesicles to actin cyctoskeleton. Wagner and colleagues reported that Myo2p forms a complex with Sec4p and the myosin light chain Mlc1p [65]. However, Sec4p is not required for Myo1p and Mlc1p interaction, and Sec4p does not directly interact with Myo2p [66, 67]. Myo2p directly interacts with the GTP-bound form of Ypt32p, and this interaction is required for polarized secretion [66-68]. Ypt31/32p thus may coordinate the formation of vesicles, activation of Sec4p, and Myo2p-dependent transport.

7. CONCLUSIONS AND PERSPECTIVES

Past works in the field have demonstrated that Sec4p is a master regulator of post-Golgi secretion. Through its downstream effectors, Sec4p orchestrates various aspects of membrane trafficking including transport, tethering and docking of secretory vesicles to the plasma membrane. As good research would open a lot more new inquiries, the field now has a number of questions to address in the future. For example, phosphatidylinositides have been shown to function with the Rab to recruit proteins to the membrane [21]. It will be interesting to investigate how PI(4)P and PI(4, 5)P$_2$ function with Ypt32p and Sec4p to recruit proteins to mediate trafficking from Golgi to the plasma membrane. As mentioned above, in *sec4* and *sec2* mutant cells, the directional transport of secretory vesicles is impaired. Then through what proteins the activated Sec4p affects the polarized transport of secretory vesicles along the actin cables? Phosphorylation of GEF has been implicated in the regulation of Rab protein in response to signaling events. What are the kinases that mediate phosphorylation and how does phosphorylation affect Sec4p and the function of its downstream effectors? Future study of Sec4p will continue to provide molecular insights to the general principles of Rab function and regulation in eukaryotic cells. Furthermore, in the context of asymmetric growth of budding yeast, the study of Sec4p will help us better understand how exocytosis machinery coordinates with cytoskeleton and signaling molecules in many physiological processes such as cell morphogenesis and tissue organization.

ACKNOWLEDGEMENTS

Works from W.G. lab are supported by National Institutes of Health, American Heart Association, and the Pew Scholars Program in Biomedical Sciences. A.K. is supported by a postdoctoral fellowship from the Deutsche Forschungsgemeinschaft (DFG).

CONFLICT OF INTEREST

There is no conflict of interest from any of the authors.

REFERENCES

[1]   Novick P, Field C, Schekman R. Identification of 23 complementation groups required for post-translational events in the yeast secretory pathway. Cell 1980 Aug; 21(1):205-15.

[2]   Salminen A, Novick PJ. A ras-like protein is required for a post-Golgi event in yeast secretion. Cell 1987 May 22;49(4):527-38.

[3]   Goud B, Salminen A, Walworth NC, *et al.* A GTP-binding protein required for secretion rapidly associates with secretory vesicles and the plasma membrane in yeast. Cell 1988 Jun 3;53(5):753-68.

[4]   Kabcenell AK, Goud B, Northup JK, *et al.* Binding and hydrolysis of guanine nucleotides by Sec4p, a yeast protein involved in the regulation of vesicular traffic. J Biol Chem 1990 Jun 5;265(16):9366-72.

[5]   Garrett MD, Zahner JE, Cheney CM, *et al.* GDI1 encodes a GDP dissociation inhibitor that plays an essential role in the yeast secretory pathway. EMBO J 1994 Apr 1;13(7):1718-28.

[6]   Ullrich O, Horiuchi H, Bucci C, *et al.* Membrane association of Rab5 mediated by GDP-dissociation inhibitor and accompanied by GDPIGTP exchange. Nature 1994 Mar 10;368(6467):157-60.

[7]   Soldati T, Shapiro AD, Svejstrup AB, *et al.* Membrane targeting of the small GTPase Rab9 is accompanied by nucleotide exchange. Nature 1994 May 5;369(6475):76-8.

[8]   Seabra MC, Brown MS, Slaughter CA, *et al.* Purification of component A of Rab geranylgeranyl transferase: possible identity with the choroideremia gene product. Cell 1992 Sep 18;70(6):1049-57.

[9]   Andres DA, Seabra MC, Brown MS, *et al.* cDNA cloning of component A of Rab geranylgeranyl transferase and demonstration of its role as a Rab escort protein. Cell 1993 Jun 18;73(6):1091-9.

[10]   Calero M, Chen CZ, Zhu W, *et al.* Dual prenylation is required for Rab protein localization and function. Mol Biol Cell 2003 May;14(5):1852-67.

[11]   Calero M, Winand NJ, Collins RN. Identification of the novel proteins Yip4p and Yip5p as Rab GTPase interacting factors. FEBS Lett 2002 Mar 27;515(1-3):89-98.

[12]   Sivars U, Aivazian D, Pfeffer SR. Yip3 catalyses the dissociation of endosomal Rab-GDI complexes. Nature 2003 Oct 23;425(6960):856-9.

[13]   Chen CZ, Calero M, DeRegis CJ, *et al.* Genetic analysis of yeast Yip1p function reveals a requirement for Golgi-localized rab proteins and rab-Guanine nucleotide dissociation inhibitor. Genetics 2004 Dec;168(4):1827-41.

[14]   Elkind NB, Walch-Solimena C, Novick PJ. The role of the COOH terminus of Sec2p in the transport of post-Golgi vesicles. J Cell Biol 2000 Apr 3;149(1):95-110.

[15]   Dong G, Medkova M, Novick P, *et al.* A catalytic coiled coil: structural insights into the activation of the Rab GTPase Sec4p by Sec2p. Mol Cell 2007 Feb 9;25(3):455-62.

[16]   Walch-Solimena C, Collins RN, Novick PJ. Sec2p mediates nucleotide exchange on Sec4p and is involved in polarized delivery of post-Golgi vesicles. J Cell Biol 1997 Jun 30;137(7):1495-509.

[17]   Ortiz D, Medkova M, Walch-Solimena C, *et al.* Ypt32 recruits the Sec4p guanine nucleotide exchange factor, Sec2p, to secretory vesicles; evidence for a Rab cascade in yeast. J Cell Biol 2002 Jun 10;157(6):1005-15.

[18]   Nair J, Muller H, Peterson M, *et al.* Sec2 protein contains a coiled-coil domain essential for vesicular transport and a dispensable carboxy terminal domain. J Cell Biol 1990 Jun;110(6):1897-909.

[19]   Mizuno-Yamasaki E, Medkova M, Coleman J, *et al.* Phosphatidylinositol 4-phosphate controls both membrane recruitment and a regulatory switch of the Rab GEF Sec2p. Dev Cell 2010 May 18;18(5):828-40.

[20]   Jedd G, Mulholland J, Segev N. Two new Ypt GTPases are required for exit from the yeast trans-Golgi compartment. J Cell Biol 1997 May 5;137(3):563-80.

[21]   Stenmark H. Rab GTPases as coordinators of vesicle traffic. Nat Rev Mol Cell Biol 2009 Aug;10(8):513-25.

[22]   Knödler A, Feng S, Zhang J, *et al.* Coordination of Rab8 and Rab11 in primary ciliogenesis. Proc Natl Acad Sci U S A 2010 Apr 6;107(14):6346-51.

[23]   Wang W, Ferro-Novick S. A Ypt32p exchange factor is a putative effector of Ypt1p. Mol Biol Cell 2002 Sep;13(9):3336-43.

[24]   Rink J, Ghigo E, Kalaidzidis Y, *et al.* Rab conversion as a mechanism of progression from early to late endosomes. Cell 2005 Sep 9;122(5):735-49.

[25]   Morozova N, Liang Y, Tokarev AA, *et al.* TRAPPII subunits are required for the specificity switch of a Ypt-Rab GEF. Nat Cell Biol 2006 Nov;8(11):1263-9.

[26]   Poteryaev D, Datta S, Ackema K, *et al.* Identification of the switch in early-to-late endosome transition. Cell 2010 Apr 30;141(3):497-508.

[27]   Trahey M, McCormick F. A cytoplasmic protein stimulates normal N-ras p21 GTPase, but does not affect oncogenic mutants. Science 1987 Oct 23;238(4826):542-5.

[28]   Boguski MS, McCormick F. Proteins regulating Ras and its relatives. Nature 1993 Dec 16;366(6456):643-54.

[29]   Strom M, Vollmer P, Tan TJ, *et al.* A yeast GTPase-activating protein that interacts specifically with a member of the Ypt/Rab family. Nature 1993 Feb 25;361(6414):736-9.

[30]   Du LL, Collins RN, Novick PJ. Identification of a Sec4p GTPase-activating protein (GAP) as a novel member of a Rab GAP family. J Biol Chem 1998 Feb 6;273(6):3253-6.

[31]   Neuwald AF. A shared domain between a spindle assembly checkpoint protein and Ypt/Rab-specific GTPase-activators. Trends Biochem Sci 1997 Jul;22(7):243-4.

[32]   Albert S, Gallwitz D. Two new members of a family of Ypt/Rab GTPase activating proteins. Promiscuity of substrate recognition. J Biol Chem 1999 Nov 19;274(47):33186-9.

[33]   Bi E, Chiavetta JB, Chen H, *et al.* Identification of Novel, Evolutionarily Conserved Cdc42pinteracting Proteins and of Redundant Pathways Linking Cdc24p and Cdc42p to Actin Polarization in Yeast. Mol Biol Cell 2000 Feb;11(2):773-93.

[34]   Gao XD, Albert S, Tcheperegine SE, *et al.* The GAP activity of Msb3p and Msb4p for the Rab GTPase Sec4p is required for efficient exocytosis and actin organization. J Cell Biol 2003 Aug 18;162(4):635-46.

[35]   TerBush DR, Novick P. Sec6, Sec8, and Sec15 are components of a multisubunit complex which localizes to small bud tips in Saccharomyces cerevisiae. J Cell Biol 1995 Jul;130(2):299-312.

[36]   TerBush DR, Maurice T, Roth D, *et al.* The Exocyst is a multiprotein complex required for exocytosis in Saccharomyces cerevisiae. EMBO J 1996 Dec 2;15(23):6483-94.

[37]   Guo W, Roth D, Walch-Solimena C, *et al.* The exocyst is an effector for Sec4p, targeting secretory vesicles to sites of exocytosis. EMBO J 1999 Feb 15;18(4):1071-80.

[38]   Salminen A, Novick PJ. The Sec15 protein responds to the function of the GTP binding protein, Sec4p, to control vesicular traffic in yeast. J Cell Biol 1989 Sep;109(3):1023-36.

[39]   Finger FP, Hughes TE, Novick P. Sec3p is a spatial landmark for polarized secretion in budding yeast. Cell 1998 Feb 20;92(4):559-71.

[40]   Robinson NG, Guo L, Imai J, *et al.* Rho3 of Saccharomyces cerevisiae, which regulates the actin cytoskeleton and exocytosis, is a GTPase which interacts with Myo2 and Exo70. Mol Cell Biol 1999 May;19(5):3580-7.

[41]   Adamo JE, Rossi G, Brennwald P. The Rho GTPase Rho3 has a direct role in exocytosis that is distinct from its role in actin polarity. Mol Biol Cell 1999 Dec;10(12):4121-33.

[42]   Guo W, Tamanoi F, Novick, P. Spatial regulation of the exocyst complex by Rho1 GTPase. Nat Cell Biol 2001 Apr;3(4):353-60.

[43]   Zhang X, Bi E, Novick P, *et al.* Cdc42 interacts with the exocyst and regulates polarized secretion. J Biol Chem 2001 Dec 14;276(50):46745-50.

[44]   He B, Xi F, Zhang X, *et al.* Exo70 interacts with phospholipids and mediates the targeting of the exocyst to the plasma membrane. EMBO J 2007 Sep 19;26(18):4053-65.

[45]   Zhang X, Orlando K, He B, *et al.* Membrane association and functional regulation of Sec3 by phospholipids and Cdc42. J Cell Biol 2008 Jan 14;180(1):145-58.

[46]   Baek K, Knödler A, Lee SH, *et al.* Structure-function study of the N-terminal domain of exocyst subunit Sec3. J Biol Chem 2010 Apr 2;285(14):10424-33.

[47]   Boyd C, Hughes T, Pypaert M, *et al.* Vesicles carry most exocyst subunits to exocytic sites marked by the remaining two subunits, Sec3p and Exo70p. J Cell Biol 2004 Dec 6;167(5):889-901.

[48]   Medkova M, France YE, Coleman J, *et al.* The rab exchange factor Sec2p reversibly associates with the exocyst. Mol Biol Cell 2006 Jun;17(6):2757-69.

[49]   Novick P, Medkova M, Dong G, *et al.* Interactions between Rabs, tethers, SNAREs and their regulators in exocytosis. Biochem Soc Trans 2006 Nov;34(Pt 5):683-6.

[50]   Horiuchi H, Lippé R, McBride HM, *et al.* A novel Rab5 GDP/GTP exchange factor complexed to Rabaptin-5 links nucleotide exchange to effector recruitment and function. Cell 1997 Sep 19;90(6):1149-59.

[51]   Zhang XM, Ellis S, Sriratana A, *et al.* Sec15 is an effector for the Rab11 GTPase in mammalian cells. J Biol Chem 2004 Oct 8;279(41):43027-34.

[52]   Oztan A, Silvis M, Weisz OA, *et al.* Exocyst requirement for endocytic traffic directed toward the apical and basolateral poles of polarized MDCK cells. Mol Biol Cell 2007 Oct;18(10):3978-92.

[53]   Wu S, Mehta SQ, Pichaud F, *et al.* Sec15 interacts with Rab11 *via* a novel domain and affects Rab11 localization *in vivo*. Nat Struct Mol Biol 2005 Oct;12(10):879-85.

[54]   Grosshans BL, Andreeva A, Gangar A, *et al.* The yeast lgl family member Sro7p is an effector of the secretory Rab GTPase Sec4p. J Cell Biol 2006 Jan 2;172(1):55-66.

[55]   Lehman K, Rossi G, Adamo JE, *et al.* Yeast homologues of tomosyn and lethal giant larvae function in exocytosis and are associated with the plasma membrane SNARE, Sec9. J Cell Biol 1999 Jul 12;146(1):125-40

[56]   Hattendorf DA, Andreeva A, Gangar A *et al.* Structure of the yeast polarity protein Sro7 reveals a SNARE regulatory mechanism. Nature 2007 Mar 29;446(7135):567-71.

[57]   Zhang X, Wang P, Gangar A *et al.* Lethal giant larvae proteins interact with the exocyst complex and are involved in polarized exocytosis. J Cell Biol 2005 Jul 18;170(2):273-83.

[58]   Gangar A, Rossi G, Andreeva A, *et al.* Structurally conserved interaction of Lgl family with SNAREs is critical to their cellular function. Curr Biol 2005 Jun 21;15(12):1136-42.

[59]   Novick P, Botstein D. Phenotypic analysis of temperature-sensitive yeast actin mutants. Cell 1985 Feb;40(2):405-16.

[60]   Pruyne DW, Schott DH, Bretscher A. Tropomyosin-containing actin cables direct the Myo2p-dependent polarized delivery of secretory vesicles in budding yeast. J Cell Biol 1998 Dec 28;143(7):1931-45.

[61]   Johnston GC, Prendergast JA, Singer RA. The Saccharomyces cerevisiae MYO2 gene encodes an essential myosin for vectorial transport of vesicles. J Cell Biol 1991 May;113(3):539-51.

[62]   Govindan B, Bowser R, Novick P. The role of Myo2, a yeast class V myosin, in vesicular transport. J Cell Biol 1995 Mar;128(6):1055-68.

[63]   Schott D, Ho J, Pruyne D, *et al.* The COOH-terminal domain of Myo2p, a yeast myosin V, has a direct role in secretory vesicle targeting. J Cell Biol 1999 Nov 15;147(4):791-808.

[64]   Reck-Peterson SL, Novick PJ, Mooseker MS. The tail of a yeast class V myosin, myo2p, functions as a localization domain. Mol Biol Cell 1999 Apr;10(4):1001-17.

[65]   Wagner W, Bielli P, Wacha S, *et al.* Mlc1p promotes septum closure during cytokinesis *via* the IQ motifs of the vesicle motor Myo2p. EMBO J. 2002 Dec 2;21(23):6397-408.

[66]   Bielli P, Casavola EC, Biroccio A, *et al.* GTP drives myosin light chain 1 interaction with the class V myosin Myo2 IQ motifs *via* a Sec2 RabGEF-mediated pathway. Mol Microbiol 2006 Mar;59(5):1576-90.

[67]   Casavola EC, Catucci A, Bielli P, *et al.* Ypt32p and Mlc1p bind within the vesicle binding region of the class V myosin Myo2p globular tail domain. Mol Microbiol 2008 Mar;67(5):1051-66.

[68]   Lipatova Z, Tokarev AA, Jin Y, *et al.* Direct interaction between a myosin V motor and the Rab GTPases Ypt31/32 is required for polarized secretion. Mol Biol Cell 2008 Oct;19(10):4177-87.

CHAPTER 5

Masterclass with Rab3 and Rab27: Orchestrating Regulated Secretion

François Darchen[*] and Claire Desnos

CNRS/Université Paris Descartes, France

Abstract: A subset of Rab GTPases have instrumental roles in the biogenesis, trafficking, docking and exocytosis of secretory granules, secretory lysosomes and synaptic vesicles. The four Rab3 isoforms and the two Rab27 isoforms are the main members of this family of "secretory Rabs." Redundancy between the isoforms and between the Rab3 and Rab27 proteins has made the functional characterization of these proteins difficult. Data collected from different cell types suggest that the main role of Rab27 is to promote the recruitment of secretory vesicles at the release sites, while that of Rab3 is to control the number of ready-to-fuse vesicles. However, many observations that cannot be incorporated into this simplified scheme suggest that Rab3 and Rab27 have overlapping functions at different stages of the "life cycle" of secretory vesicles. Consistent with this, while some effector molecules are specific for Rab3 or Rab27, several interact with both of them.

Keywords: Rab3, Rab27, Exocytosis, Secretory Vesicle, Membrane Traffic.

1. INTRODUCTION

All eukaryotic cells have developed a secretory pathway that targets proteins synthesized in the endoplasmic reticulum and shuttles them through the Golgi apparatus to the plasma membrane (PM) or external milieu. In neurons and cells of the endocrine and exocrine glands, this secretory pathway has evolved to achieve the spatially and temporally controlled release of secretory products. In "regulated secretion", a multi-step process, the release of secretory products is triggered by a signal, generally calcium entry into the cell. In endocrine cells, for instance, water-soluble hormones are packaged into secretory granules (SGs) in the *trans*-Golgi network (TGN). SGs are transported along the microtubules to the cell periphery and interact with the actin-rich cortex until they attach to the PM (docking). Then, in a process called priming, Soluble NSF Attachment Protein Receptor (SNARE) complexes are formed between the fusing membranes and makes SGs ready to fuse with the PM. The process of regulated secretion is arrested at this stage until exocytosis is triggered by an elevation in calcium levels (Fig. **1A**) [1-3]. Synaptic vesicles (SVs) are formed at the synapse and undergo cycles of exocytosis and endocytosis (Fig. **1B**).

The secretory process highlights several points at which regulation is required. Specifically, how can a vesicle with a selective set of components be created? How is a vesicle transported to its release site? Finally, how is the vesicle attached to its target membrane, and how is exocytosis coupled with the stimulus? The Rab GTPases appear to be essential for these processes. By sequentially recruiting diverse effectors, they drive secretory vesicles from their biogenesis to their final exocytosis, while other Rabs are involved in the vesicle recycling steps.

Rab GTPases constitute a large family of membrane trafficking regulators (with approximately 60 isoforms in humans). Within this family, a subset of phylogenetically related Rab proteins have been linked to regulated secretion [4]. These "secretory Rabs" include Rab3a, b, c and d; Rab27a and b; Rab4a; Rab11b; Rab26; and Rab37. In this chapter, we review the data on Rab3 and Rab27, secretory Rabs that are widely distributed and that have been well characterized.

*Address correspondence to François Darchen: CNRS/Université Paris Descartes UMR 8192, Centre Universitaire des Saints-Peres, 47 rue des Saints-Peres, 75006 Paris, France; Tel: +33-1-70-64-99-16; Fax: +33-1-70-64-99-13 ; E-mail: francois.darchen@parisdescartes.fr

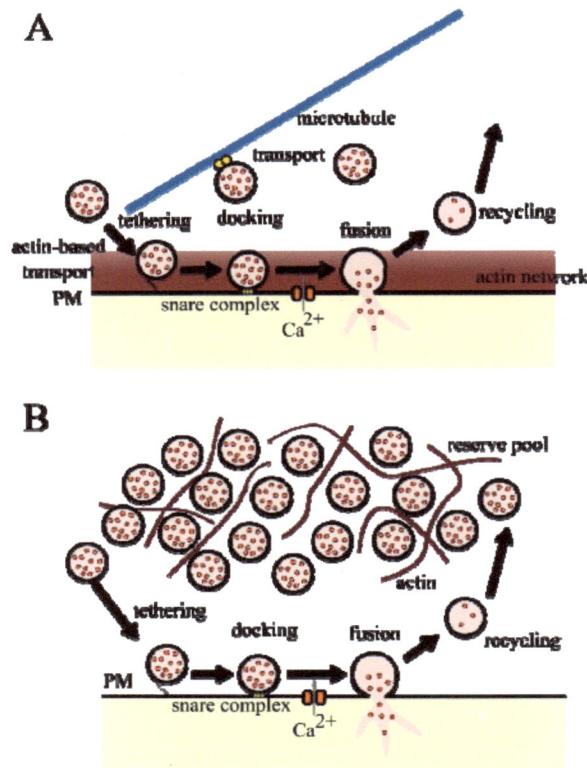

Figure 1: The final steps of the regulated secretory process. (A) In endocrine cells, secretory granules (SGs) are transported along microtubules from the TGN to the cell periphery. The subplasmalemmal actin meshwork is depicted in burgundy. Actin-binding proteins, including myosin Va, promote the dissociation of the SGs from the microtubules and the transport of the SGs to the plasma membrane (PM). Tethering factors capture SGs at the PM before they are stably docked. A priming reaction, during which SNAREs assemble, makes SGs ready to fuse. Exocytosis is triggered by stimulus-dependent calcium entry. **(B)** At the synapse, synaptic vesicles (SVs) undergo local cycling through exocytosis and endocytosis. At the active zone, clusters of SVs are stabilized *via* interactions with actin filaments and synapsins and form a reserve pool of vesicles. For clarity, an SV cluster is depicted at a non-proportional distance from the PM. The SVs undergo reactions similar to those of the SGs before fusing with the presynaptic membrane to release neurotransmitters. Tight temporal coupling between the stimulus and the neurotransmitter release is mediated by the close apposition of the SVs to the calcium channels.

2. RAB3 AND RAB27 ARE IMPLICATED IN REGULATED SECRETION

2.1. Multiple Rabs are Present on Secretory Vesicles

Four highly homologous *Rab3* genes are found in mammals [5]. Rab3a and c are mainly expressed in the brain and neuroendocrine cells. Rab3b is present at low levels in neurons but is abundant in the pituitary gland and epithelial cells [6-8]. Rab3d is expressed at very low levels in the brain but is enriched in adipocytes, muscle, lungs and in the exocrine pancreas [9, 10]. Despite these differences in tissue distribution, several isoforms of Rab3 are often simultaneously expressed in single cells [7, 11].

Rab27a is also expressed in a broad range of secretory cells of endocrine and exocrine glands, in immune cells and in melanocytes [4]. Using mice expressing *LacZ* under the control of the *Rab27b* promoter, Gomi and collaborators showed that the Rab27b expression pattern differs from that of Rab27a [12] but is still restricted to specialized secretory cells. For instance, in the pancreas, the expression of Rab27a is dominant in the islets and that of Rab27b is dominant in acinar cells. Notably, Rab27b, but not Rab27a, is expressed in neurons from many regions of the brain [13]. Rab27b is also expressed in cells (such as those of the skin, esophageal epithelium and bladder) that increase their surface area, most likely by exocytosis, in response to mechanical stress. The reason for this diversity is unknown, and this diversity is surprising because knockout studies have revealed that Rab isoforms can generally compensate for the lack of the other isoforms (see below).

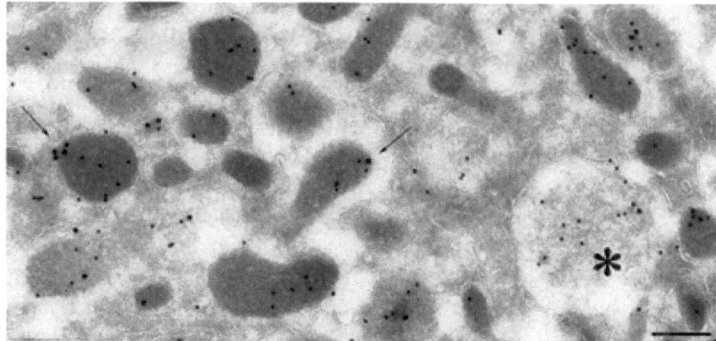

Figure 2: The association of Rab27a with secretory granules. Ultrathin cryosections of adrenal chromaffin cells were double immunogold-labeled for Rab27a (15-nm gold particles) and chromogranin A/B (10-nm gold particles), a component of the granule matrix. Rab27A localizes to the chromogranin-positive dense core granules (arrows) and to immature granules (star). Bar, 200 nm. Originally published in [19].

Many studies have identified the localization of Rab3 and Rab27 to secretory vesicles (Fig. **2**). Rab3a and Rab3c are associated with synaptic vesicles in neurons [14-16], while Rab3a, Rab3b, Rab3c, Rab27a and Rab27b are found on SGs in neuroendocrine cells [13, 17-22]. Rab3d is associated with the granules of exocrine cells [10, 23-25]. Rab27 is also found on secretory organelles such as melanosomes [26], secretory lysosomes [27, 28] multi-vesicular bodies [29] and Weibel-Palade bodies [30]. In *Caenorhabditis elegans*, Rab27 colocalizes with Rab3 in neurons. Rab27 is mislocalized in *C. elegans aex-3* mutants (which are defective in the exchange factor for Rab3 and Rab27) and in animals deficient in the kinesin UNC104, suggesting that Rab27 is associated with SVs [31]. The association of Rab27b with synaptic vesicles has been recently demonstrated [32, 33]. Fukuda and colleagues fused a collection of Rabs to GFP and expressed them in neuroendocrine PC12 cells. Only Rab3 (a-d), Rab27 (a, b) and Rab37 showed a specific association with SGs [34].

2.2. Rab3 and Rab27 Regulate Secretion

2.2.1. Rab3

Many studies using antisense oligonucleotides or overexpressed Rab3 constructs have implicated Rab3 in regulated exocytosis [6, 7, 21, 22, 35-41]. Positive and negative roles for Rab3 in secretion have been reported, and no consensus on its role has emerged from these studies. Redundancy between Rabs is certainly one of the reasons for the difficulty in solving the "Rab3 issue". Another reason is that these studies were performed in different cell types using different methods, and each study highlighted one aspect of Rab3 function.

Knockout (KO) mouse lines for each of the four *RAB3* genes have been generated. The single Rab3 KO mice are viable and fertile [42, 43], and only mild phenotypes were found in these mice. The circadian period of locomotor activity is shortened in Rab3a KO mice and in *earlybird* mice, which contain mutations in *RAB3a* [44]. Rab3a KO mice perform well in memory tasks and display only mildly increased locomotor activity, suggesting a role for this gene in behavioral stability [45]. In the hippocampal neurons of Rab3a KO mice, the amplitude of the excitatory postsynaptic potential (EPSP, a measure of the postsynaptic response to released neurotransmitters) is similar to [46, 47] or even greater than [48] that found in control animals. These data suggest that, despite the status of Rab3a as the most abundant Rab protein on SVs, Rab3a is not essential for synaptic vesicle exocytosis but rather is a modulator of the release process. Consistent with this, several forms of short-term or long-term synaptic plasticity are affected by Rab3a deletion: paired-pulse facilitation is increased in cultured hippocampal neurons [48], there is a modest synaptic depression of hippocampal CA1 synapses during the repetitive stimulation of Schaffer collaterals [47], and long-term potentiation (LTP) is abolished in hippocampal CA3-mossy fiber synapses [46].

The inactivation of a single *RAB3* gene is well tolerated. However, quadruple Rab3 KO mice die immediately after birth because they cannot breathe [43]. Viability is restored by a single Rab3a allele or by two Rab3b or Rab3c alleles and is partially restored by two Rab3d alleles. These observations clearly indicate that the Rab3 isoforms can compensate, at least partially, for the lack of the others. Strikingly, only mild synaptic defects have been observed in Rab3 abcd quadruple KO mouse embryos. The amplitude of the evoked response was reduced by only 30% in neuronal cultures, and spontaneous release events (miniatures) were not significantly modified [42, 43, 48]. In *C. elegans*, in which a single *RAB3* gene exists, *RAB3* deletion also causes relatively minor synaptic defects [31, 49].

Although Rab3 is not mandatory for exocytosis in some cells, it seems to be essential in others. For example, reducing Rab3b expression in anterior pituitary cells [6] or Rab3a in intermediate pituitary cells [7] with antisense oligonucleotides severely inhibited secretion. Furthermore, the inability of the quadruple Rab3 KO mice to breathe [43] suggests that exocytosis is severely impaired at the neuromuscular junction. The simplest explanation would be that another Rab GTPase has redundant activity that can compensate, at least partially, for the lack of Rab3. The following findings support this possibility: (i) mutations in *GDI1*, which encodes αGDI, a factor that delivers Rabs:GDP to the membranes and extract them for recycling (Box 1), are more severe than the deletion of Rab3 [50-52]; and (ii) deletion of the Rab3 exchange factor, Rab3-GEF (Box 1), causes a marked reduction in the evoked neurotransmitter release in mice [53, 54] and *C. elegans* [55]. These findings suggest that these Rab3 exchange factors activate another Rab. Several lines of research indicate that this other Rab is Rab27.

Box 1: The life cycles of Rab3 and Rab27

Rabs oscillate between GDP- and GTP-bound forms. GDP-bound Rabs are mostly found in the cytosol in complex with Rab-GDI (guanine nucleotide dissociation inhibitor), whereas Rab-GTP complexes are anchored in membranes *via* a geranylgeranyl moiety. Rab-GDI is instrumental in the "life cycle" of Rabs: it can extract or deliver Rab:GDP to membranes [175, 176]. Mutations in *GDI1*, which encodes GDIα, the most abundant form of GDI in the brain, cause X-linked nonspecific mental retardation [50]. *GDI1* KO mice have impaired short-term memory and display altered social behavior [51]. The relatively severe synaptic defects caused by GDIα deletion may be related to improper recycling or delivery of Rab3 and Rab27 or may be due to the inhibition of endocytic Rabs, as suggested by the reduced numbers of synaptic vesicles found in nerve terminals [177]. Park and colleagues [178] found that calmodulin (CaM) is able to extract Rab3 from synaptosomal membranes in a Ca^{2+}-dependent manner, but the physiological consequences of this interaction are unclear. The delivery of Rabs to membranes is facilitated by GDI displacement factors (GDFs) [175]. In the case of Rab3, the only known molecule with such activity is synapsin, which competes with Rab-GDI for Rab3 binding [127, 128].

The equilibrium between Rab:GDP and Rab:GTP depends on the intrinsic GTPase activity of the Rab proteins, which is low in the case of Rab3 and Rab27 [179, 180], and on the respective activities of Rab-GEF (which catalyses the exchange of GTP for GDP) and Rab-GAP (which catalyses the hydrolysis of GTP). Rab3-GEF (also called DENN/MADD) is also active against Rab27, but not against other Rabs [181, 182]. Rab3A, but not Rab27A, cycles rapidly between the granule membrane and the cytosol [74] and readily dissociates from membranes during active secretion [176, 183]. Rab27 may be stably bound to GTP and rapidly reactivated by a GEF or more resistant to GDI-mediated extraction [184]. In an attempt to correlate the nucleotide status of Rab3a in synaptosomes with the secretory process, Stahl *et al.* observed a marked increase in the GDP/GTP ratio on Rab3a after strong stimulation of exocytosis [124]. These results indicate that GTP hydrolysis by Rab3 and Rab27 is somehow coupled to the secretory activity.

Rab3- and Rab27-specific GTPase activating proteins (Rab3-GAP and EPI64/Rab27-GAP) have been described [180, 185, 186]. Rab3-GAP consists of two subunits, the catalytic subunit p130 and the noncatalytic subunit p150. Mutations in the gene encoding p130 cause Warburg Micro syndrome, which is characterized by ocular and neurodevelopmental defects and severe mental retardation [187]. P130 null mice are viable and fertile [188]. Although no obvious defect in basal neurotransmission has been reported, glutamate release from synaptosomes is severely impaired in these animals. The data indicate that GTP hydrolysis by Rab3 is rate-limiting at some stage in the secretory process. How Rab3-GAP activity is regulated during secretion is not known. Calcium ions do not change the activity of p130 [179]. The Rab3-GAP binding site involves the switch I region of Rab3 and overlaps with the rabphilin-binding domain. Dissociation of the Rab3-effector interaction upon calcium entry may thus allow Rab3-GAP to trigger GTP hydrolysis.

Although GEFs and GAPs are thought to function by regulating the nucleotide cycle of the Rabs, they may exert other functions. For instance, Rab3-GEF interacts with the kinesins Kif1Bb and Kif1A, regulates axonal transport of synaptic vesicle precursors [189] and is involved in neuroprotection [190].

2.2.2. Rab27

RAB27a was identified as the gene mutated in Griscelli syndrome type II [27, 56] and in the coat-color mutant *ashen* [57]. Griscelli syndrome (GS; MIM 214450) type II is an often fatal disease combining albinism and hemophagocytic syndrome, an uncontrolled activation of lymphocytes and macrophages [58]. At the cellular level, Rab27a deletion induces defects in melanosome transport and lysosome exocytosis in cytotoxic T lymphocytes [27, 28, 59] (see also the chapter by J. Hammer in this eBook). Dendritic cells from *ashen* mice demonstrate a deficiency in antigen cross-presentation caused by the premature acidification of the phagosomes [60]. The impaired delivery of lysosome-related organelles containing the NADPH oxidase Nox2 to the phagosome is responsible for this defect. In addition to these defects, moderate glucose intolerance has been found in *ashen* mice, which correlates with decreased glucose-induced insulin secretion [61]. Despite the expression of Rab27b in many secretory cells and neurons [12], the only phenotype caused by Rab27b deletion is a hemorrhagic tendency, which has been attributed to the reduced density and exocytosis of the dense granules in platelets [62]. The phenotype caused by *RAB27A* and *RAB27B* inactivation simply combines the defects found in the single KOs. Again, redundancy between the Rabs and especially between Rab3 and Rab27 may explain the mild phenotype of *RAB27* deletion. Nevertheless, interfering with Rab27a or Rab27b function impairs stimulus-dependent secretion in neuroendocrine cells [12, 13, 19, 34, 63, 64], platelets [62, 65], pancreatic acinar cells [66] and neutrophils [67-69]. RNA silencing of Rab27a or Rab27b also impairs the secretion of exosomes, the luminal vesicles contained in multi-vesicular bodies [29]. Conversely, overexpression of Rab27a promotes secretion in endocrine cells [13, 20] and the secretion of prostate-specific markers [70].

2.2.3. Synergy between Rab3 and Rab27

Rab27b is expressed in neurons [12] and is associated with synaptic vesicles [32, 33]. In agreement with a role for Rab27 in SV exocytosis, it was found that an anti-Rab27 antibody inhibited neurotransmitter release at the squid giant synapse and impaired the recovery of the releasable pool of SVs after the stimulus-induced exhaustion of this pool [71]. In *C. elegans*, the inactivation of *RAB27* in the *aex-6* mutant causes minor synaptic defects, as do mutations in *RAB3* [49]. However, both spontaneous and evoked neurotransmission are more severely impaired in the *RAB3*; *RAB27* double mutant than in the *RAB3* or *RAB27* single mutants [31]. The authors of this study also found that the exchange factor aex-3 activates both Rab3 and Rab27 and that inactivating *aex-3* causes a severe phenotype similar to that of the *RAB3*; *RAB27* double mutant. These data suggest that Rab3 and Rab27 have overlapping functions in secretion, and one may compensate for the lack of the other. Thus, it will be interesting to cross Rab3- and Rab27-deficient mice to analyze the resulting effects in neurons and other secretory cells.

3. REGULATION OF SECRETION STEPS BY RAB3, RAB27 AND THEIR EFFECTORS

3.1. Biogenesis and Maturation of Secretory Vesicles

Secretory granules are formed in the *trans*-Golgi network as immature SGs. They are converted to mature SGs through a process that combines membrane fusion and fission, matrix condensation and the retrieval of various components [72]. A 50% reduction in the number of SGs has been observed in the adrenal chromaffin cells of the quadruple Rab3 KO mice [73], suggesting a role for Rab3 in granule biogenesis. However, an increased homotypic fusion of immature SGs or uncontrolled constitutive exocytosis may also account for the observed reduction in the number of SGs, and there is no direct evidence that Rab3 controls the budding of immature

SGs from the *trans*-Golgi network. Both Rab3 and Rab27 are found on immature SGs [19, 74]. However, live-cell imaging has revealed that GFP-Rab3a and GFP-Rab27a associate with immature SGs only after a lag period of approximately 20 min after the release of these proteins from the *trans*-Golgi network [74], arguing against their involvement in granule formation. Rab3 and Rab27 may nevertheless contribute to granule maturation. Indeed, Rab3d deletion causes an increase in the size of SGs in pancreatic acinar cells and in the parotid gland, with the volume being doubled [75]. Similar results were obtained in PC12 cells upon expression of dominant negative Rab3a or Rab3d mutants [76] and in Rab27b-deficient platelets [62]. The size of the multivesicular endosomes was also increased in HeLa B6HA cells upon RNA silencing of Rab27a or its effector, granuphilin [29], perhaps as a result of increased homotypic fusion. Myosin Va, probably recruited by Rab27 and its effector MyRIP (myosin- and Rab-interacting protein; also called Slac2c), participates in SG maturation [77]. The proposed mechanism is the facilitation, by the tethering of the vesicle membrane to actin, of the retrieval of components pulled by microtubule-based motors.

3.2. Recruitment of Secretory Vesicles at Release Sites

3.2.1. A Role for Rab3 and Rab27 in Targeting Secretory Vesicles to Release Sites

In endocrine cells, SGs are transported along microtubules from the *trans*-Golgi network (TGN) toward the cell periphery [78], where they accumulate in the actin-rich cortex [79]. To become available for release, SGs must cross the actin cortex and interact with the plasma membrane (Fig. 1A). The actin-rich cortex has long been viewed as a barrier to fusion [80], but it is also important for vesicle recruitment at release sites [81]. Indeed, impairing the interaction of SGs with actin (by interfering with Rab27, MyRIP or myosin Va; see below) induces a redistribution of SGs from the actin-rich cell cortex to the perinuclear area [82] (Fig. 3). Most likely, this is due to the fact that without the help of actin-binding proteins and actin-based motor molecules such as myosin Va, minus-end directed microtubule-based motors have an advantage over plus-end directed ones and drive SGs away from the PM because of the polarity of the microtubule network [81]. Presynaptic terminals also contain an actin-rich cytomatrix that controls the mobility of SVs. SVs interact with F-actin *via* synapsin, and these interactions are thought to mediate the formation of a reserve pool of SVs and its availability for release (Fig. 1B). Indeed, deletion of synapsin reduces the number of vesicles found at the synapse [83, 84].

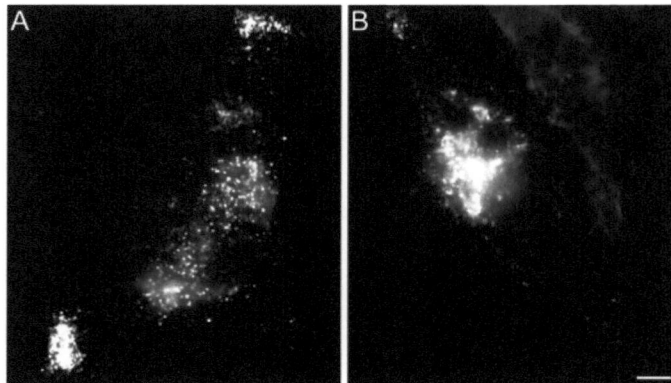

Figure 3: MyRIP RNA silencing impairs secretory granule distribution. Enterochromaffin BON cells were transfected with a control siRNA duplex (**A**) or with siRNAs direted to MyRIP (**B**) and with GFP-tagged neuropeptide-Y, a secretory granule marker. In control cells, SGs display a marked enrichment at the cell periphery, particularly in cell extensions. However, in cells treated with MyRIP siRNAs, SGs accumulate in the perinuclear region. Bar, 3 μM.

Rab3, Rab27 and several of their effectors have been implicated in the recruitment of secretory vesicles at the cell periphery [34, 49, 69, 82]. Notably, several of these effectors (MyRIP, rabphilin and synapsin) interact with actin or actin-binding proteins. Controlling the association of vesicles with actin filaments to maintain them in the vicinity of the release sites, to control their availability for release or to power vesicle movement along actin tracks may be a common theme of those effectors.

Attaching vesicles to the plasma membrane, referred to as docking, is another important step in the secretory process. Secretory vesicle docking is impaired in Rab27-deficient cytotoxic T lymphocytes [28, 85], pancreatic β-cells [64], pituitary cells [12] and in squid synapses loaded with anti-Rab27 antibodies [71]. The docking of multivesicular endosomes is also impaired by the knockdown of the Rab27a or Rab27b genes [29]. There is also evidence that Rab3 is involved in secretory vesicle docking. SG docking is increased upon Rab3 overexpression [76, 86] and decreased in PC12 cells expressing Rab3a or Rab3d dominant negative mutants [76]. Docking is unchanged, however, in the chromaffin cells of quadruple Rab3 KO mice [73], which may be accounted for by the remaining expression of Rab27. Although SV docking is not modified in hippocampal neurons from Rab3a null mice, the increase in SV docking that normally follows calcium elevation is abolished in these cells [87]. Moreover, SV docking is reduced in the neuromuscular junctions of Rab3a null mice [88]. Taken together, the data suggest that Rab3 facilitates SV attachment to the PM. The direct interaction between Rab3 and munc18-1 has been proposed to mediate this effect, but the association of Rab3 with wild-type munc18-1 is weak and independent of the nucleotide-bound status of Rab3 [89].

Recent evidence suggests that a loose mode of vesicle attachment, referred to as tethering, precedes docking. The transition between the two states has been visualized by total internal reflection fluorescence microscopy (TIRFM) as a 20-nm move toward the PM that occurs a few seconds before exocytosis [90]. The SNARE proteins syntaxin 1 and SNAP-25 are anchored in the PM and, with the help of Munc18 and Munc13, can assemble with synaptotagmin-1 on the vesicle [91-93]. This complex is likely to mediate docking (rather than tethering), given the small size of SNAREs. Consistently, SVs are not docked in Munc13-deficient neurons but remain attached to the PM by small filaments [94]. It is unknown whether Rab3 and Rab27 are involved in SNARE-mediated docking or in tethering. Large proteins act as tethering factors in the secretory and endocytic pathways [95]. The Rab effectors MyRIP, rabphilin and granuphilin (see Box 2) may have a similar function in SG tethering.

Box 2: Rab3 and Rab27 Partnership

The family of Rab3 and Rab27 effectors is comprised of a dozen members. With the exception of Munc13-4, the effectors contain a helical Rab-binding domain. Some of them contain, in addition, a zinc finger and an aromatic motif (SGAWFF in rabphilin) that contribute to the interaction. Most of the effectors interact with Rab27, but only a subset bind to Rab3. The C2 domains, which differ between the different partners, and other regions allow the effectors to interact with different molecules. The main features are summarized in the table below without distinction between the Rab isoforms (see [4, 191]).

Name	Rab3 binding	Rab27 binding	Helical RBD	Zinc finger	C2 domains	Interactions	References
Rabphilin	‡	‡	+	+	+	Ca2+, phospholipids, SNAP25, a-actinin, rabaptin 5	[47, 121, 122]
Rim	+		+	+	+	calcium channels, liprins, RimBP, Munc13, CAST, piccolo, ELKS, synaptotagmin, SNAP-25, Scrapper	[156, 192, 193]
Slp1/JFC1		‡	+		+	NADPH oxidase, PI(3,4,5)P3	[67]
Slp2-a		‡	+		+	PS, PIP2	[120, 194, 195]
Slp3-a		+	+	+	+		[196]
Slp4/granuphilin	+	‡	+	+	+	syntaxin-1a, Munc18, PS, PIP2	[13, 20, 116]

Box: 2 cont….

Slp5	+	+	+	+		[197]
Slac2-a/melanophilin	‡	+	+		myosin Va, actin, EB1	[97-99, 198]
Slac2-b	+	+				[199]
MyRIP	‡	+	+		myosin Va, myosin VIIa, actin, sec6, sec8, AKAP*	[104, 105, 107]
Noc2	+	‡	+	+		[142, 144, 192]
Munc13-4		+		+		[65, 148, 200]
Synapsin	‡				Actin	[127, 128]
Munc18-1	+				syntaxin-1, granuphilin	[89]

* A-kinase anchoring proteins; ‡ Interaction demonstrated with endogenous proteins

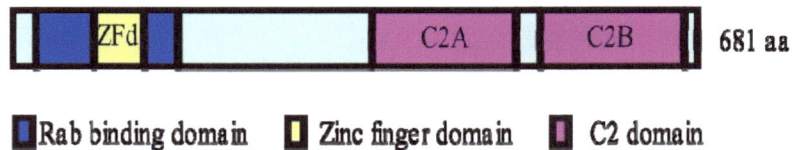

A

■ **Rab binding domain**   □ **Zinc finger domain**   ■ **C2 domain**

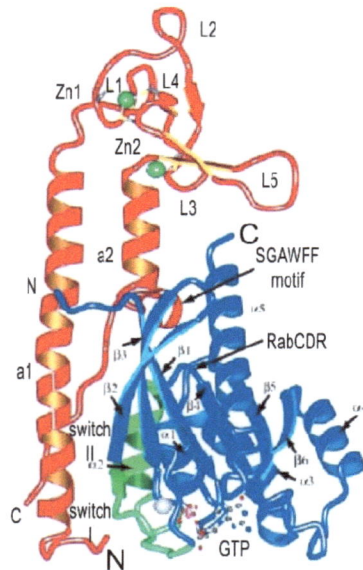

B

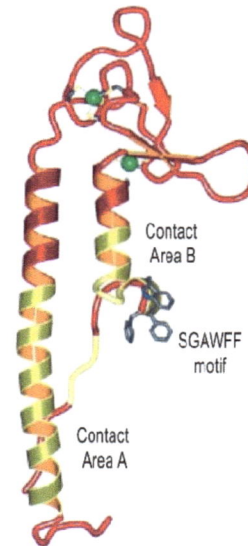

C

Rabphilin is the prototypical Rab3 and Rab27 effector. It was the first Rab3 effector to be identified [121], and it contains several features that are also found in many other effectors. Rabphilin consists of an N-terminal helical Rab-binding domain (RBD), which contacts Rab3 or Rab27, a zinc finger motif, a proline-rich linker region, and two C2 domains (Panel A). The C2B domain binds calcium ions [201] and phospholipids (phosphatidylserine and phosphatidylinositol 4,5-bisphosphate) with high Ca^{2+} affinity [134, 202]. It also associates with the SNARE protein SNAP-25 in a calcium-insensitive manner [134, 135, 137, 202]. The structure of rabphilin complexed with Rab3a has been characterized at a 2.6-Å resolution [203]. In panel B (reprinted from

Box: 2 cont….

Ostermeier and Brunger [203] with permission), residues 40 to 170 of rabphilin are shown in red, and Rab3a is shown in blue. The Rab3a:GTP structure is similar to the catalytic core of other G proteins, with a central six-stranded ß sheet surrounded by α–helices. The switch I and switch II regions of Rab3a (whose conformations change upon GTP hydrolysis; in green in panel B) span residues 49 to 57 and 80 to 95, respectively. The N-terminal domain of rabphilin consists of two antiparallel α–helices separated by a subdomain with two Zn2+-binding sites and several interspersed loops. The second, shorter α-helix ends in a turn composed of a SGAWFF motif (conserved in the sequences of rabphilin, Noc2, Rim and MyRIP) and is followed by an extended polypeptide segment. The structure reveals two distinct interfaces between Rab3a and rabphilin (in yellow in panel C, also reprinted from Ref. [203]). Rabphilin contacts the switch regions of Rab3a through part of its long α-helix α1 and its C-terminal extended segment. The second contact area involves the C-terminal end of rabphilin α–helix α2, the adjacent SGAWFF motif, and regions of Rab3a called RabCDRs (complementarity determining regions), which consist of residues 19 to 22, 94 to 96 (the α2β4 loop), 124 to 128 (the α3β5 loop), and 182 to 193 (the C-terminal half of helix α5). In contrast to the switch regions, RabCDRs exhibit a high degree of sequence variability within the Rab family. RabCDRs could be primarily involved in providing binding specificity, whereas the switch regions might function as a Rab conformation sensor.

3.2.2. Rab Effectors in the Recruitment of Secretory Vesicles to Release Sites

MyRIP. Rab27 can recruit myosin Va *via* melanophilin, and this tripartite complex mediates the interaction of melanosomes with actin, the retention of melanosomes in actin-rich areas of the cell and their transfer to keratinocytes [96-99]. Disruption of this complex is responsible for the partial albinism observed in Griscelli syndrome and in *ashen*, *leaden* and *dilute* coat-color mutants [27, 57, 100-103]. *Via* MyRIP, which has strong similarity to melanophilin, Rab27 can also recruit actin-based motors onto SGs, retinal melanosomes and Weibel-Palade bodies [19, 63, 82, 104-106]. MyRIP interacts with Rab27:GTP *via* an N-terminal helix and with myosin VIIa or myosin Va *via* a central region [19, 104, 105, 107-110]. In addition, MyRIP interacts directly with actin *via* its C-terminal region. This region restricts SG mobility within the actin cortex [19] and may contribute to SG docking (CD and FD, unpublished data). The interactions of MyRIP with sec6 and sec8, two components of the exocyst complex, and the A-kinase-anchoring protein (AKAP) have also been reported [111].

Impairing the function of MyRIP or myosin Va inhibits secretion [19, 82, 112-114, 204]. This inhibition is largely due to a reduction in SG recruitment to release sites (Fig. **3**). Indeed, lowering the levels of Rab27, MyRIP or myosin Va reduces the density of SGs near the PM as observed by TIRFM [34, 69, 82, 204]. Different mechanisms may contribute to this effect. (i) MyRIP and myosin Va promote the capture of SGs in the actin-rich cortex by competing with microtubule-based motors [82, 110]. (ii) MyRIP and myosin Va may power the directed motion of SGs along actin tracks toward the PM. Although there is no direct evidence for the myosin Va-driven movement of SGs through the actin cortex, myosin V motors have been shown to propel organelles along actin cables [81]. (iii) MyRIP and myosin Va promote SG tethering at the PM. Analysis of the trajectories of single SGs imaged by TIRFM revealed that impairing myosin Va or MyRIP activity reduced the occurrence of long-lasting (>10 s) immobilization events, which likely reflects the attachment of SGs to the PM [82, 204]. Strikingly, only a MyRIP knockdown reduces the characteristic time of immobilization. Taken together, these data suggest that (i) myosin Va promotes the retention of SGs in the actin cortex and powers their motion toward the PM, where they can find a docking platform, and that (ii) MyRIP promotes docking not only by recruiting myosin Va but also by stabilizing the attachment of SGs to the PM.

Granuphilin. Granuphilin is preferentially expressed in pancreatic β-cells [115] and pituitary cells [12] but is not expressed in neurons. It interacts with Rab27a and Rab3 [13, 20, 64, 116]. In addition, it directly binds to the PM-anchored SNARE protein syntaxin-1 [117, 118] and to Munc18-1 [116], which participate in SG docking [91-93, 116]. Overexpression of granuphilin redistributes insulin granules to the cell periphery [117], whereas deletion of granuphilin severely reduces the number of docked SGs, as measured by electron microscopy [12, 64].

Despite the severe reduction in SG docking caused by granuphilin deletion, there is no defect in SG secretion [12, 64, 117, 119]; rather, SG secretion is increased. Consistent with these results, overexpressed granuphilin inhibits secretion in several cell lines [19, 20, 116, 118]. Similar observations have been made for the Rab27 effector Slp2-a in pancreatic α-cells [120]. In addition, spontaneous SG secretion is increased by the deletion of granuphilin [119]. Granuphilin thus increases the threshold for secretion and tightens the coupling between stimulation and secretion. Notably, granuphilin interacts with the closed conformation of syntaxin, which cannot pair with other SNAREs [118]. By preventing the assembly of a productive SNARE complex, granuphilin may function as a fusion clamp. However, granuphilin also exerts positive effects on secretion. For instance, in HeLa B6H4 cells, a granuphilin gene knockdown inhibits exosome secretion [29]. How the positive and negative effects of granuphilin are coordinated is unclear. One possibility is that fusion-incompetent granuphilin/syntaxin complexes recruit vesicles to the PM and are replaced by fusion-competent SNARE complexes upon stimulation of the secretory activity.

Rabphilin. Rabphilin (see Box 2) binds to Rab3:GTP [121] and Rab27:GTP [31, 122], and both Rab3 and Rab27 recruit rabphilin onto secretory vesicles [31, 122-124]. Overexpression of rabphilin in neuroendocrine PC12 cells increases the density of SGs detected in the vicinity of the plasma membrane by TIRFM [34], suggesting that rabphilin is involved in vesicle docking at the plasma membrane. However, secretory vesicles that are not physically attached to the PM have been observed by TIRFM [19, 34, 82]. Therefore, the role of rabphilin in docking must be further tested. Alternatively, rabphilin may promote the retention of vesicles in the actin-rich cell cortex. In support of this possibility, rabphilin interacts with α-actinin and ß-adducin, promotes the actin-bundling activity of α-actinin and stimulates the association of SGs with F-actin in the presence of α-actinin [125, 126].

Synapsin. An interaction between Rab3 and synapsin has been reported [127, 128]. Rab3 interferes with several properties of synapsin: its ability to bind actin filaments and induce their bundling as well as its ability to aggregate phospholipid vesicles. The interaction between synapsin and Rab3 likely occurs *in vivo* because the amount of Rab3a associated with SVs is reduced in synapsin I, synapsin II and synapsin I/II KO mice [127]. By reducing the attachment of SVs to actin-bound clusters, Rab3 may increase the availability of SVs for exocytosis. The functional consequences of the Rab3-synapsin interaction have not yet been formally tested. The decrease in the activity-dependent recruitment of SVs to the releasable pool observed in the Rab3a null cells might be accounted for by the Rab3-synapsin interaction [47]. However, an analysis of neuromuscular junctions from *Rab3a* and *synapsin II* double KO mice has suggested that Rab3a controls SV docking, whereas synapsin II regulates the size of the reserve pool of SVs [83].

3.3. Accumulation of a Pool of Ready-to-Fuse Vesicles

3.3.1. Rab3 and Rab27 in Vesicle Priming

At the plasma membrane, vesicles undergo a priming reaction that makes them ready to fuse upon the elevation of calcium levels. In molecular terms, priming most likely corresponds to the assembly of SNARE complexes, the core of the fusion machinery [2, 129]. The number of primed (ready-to-fuse) vesicles can be measured by combining the UV photolysis of caged calcium and time-resolved membrane capacitance measurements [130]. Upon UV delivery, the calcium concentration rises almost instantaneously from resting to saturation levels (Fig. **4A**), and the different kinetic components of the secretory response can be resolved. An exocytotic burst that corresponds to the fusion of primed vesicles and lasts approximately one second precedes a slower, sustained phase of release that corresponds to the sequential priming and fusion of the vesicles that were not primed at the time the stimulus was given.

Using this approach, it was found that Rab3 controls the number of primed vesicles: in adrenal chromaffin cells from quadruple Rab3 KO mice, the fast exocytotic burst was severely reduced, whereas the sustained phase was unaffected [73] (Fig. **4A**). Similar results were obtained in Rab3a null pancreatic β-cells [131]. The lack of change in the sustained component of the release demonstrates that Rab3 is not essential for exocytosis *per se*. In the absence of Rab3, SGs can reach the plasma membrane, undergo priming and exocytose, provided that calcium is elevated. If priming proceeds normally, the severe reduction of the exocytotic burst suggests that SGs cannot remain in the primed state. Therefore, Rab3 would be needed to stabilize priming. This represents a specific and important function, especially in neurons or neuroendocrine cells, in which the stimulus is transient and can operate only on primed vesicles to trigger exocytosis in less than a millisecond.

Other possibilities can be envisioned for Rab3 function. (i) Priming may proceed normally at high but not low calcium concentrations in the absence of Rab3. This possibility is unlikely because lowering Rab3a levels using antisense oligonucleotides increased rather than decreased priming at low calcium levels [37] (Fig. **4B**). (ii) A Rab3 deficiency may increase spontaneous release, leading to the continuous consumption of the pool of primed vesicles. This possibility is also unlikely because the frequency of miniatures was unchanged in quadruple Rab3 KO mice [43] and because the time constant of the fusion reaction was not modified in cells lacking Rab3 [73]. However, it would be worth testing this possibility carefully.

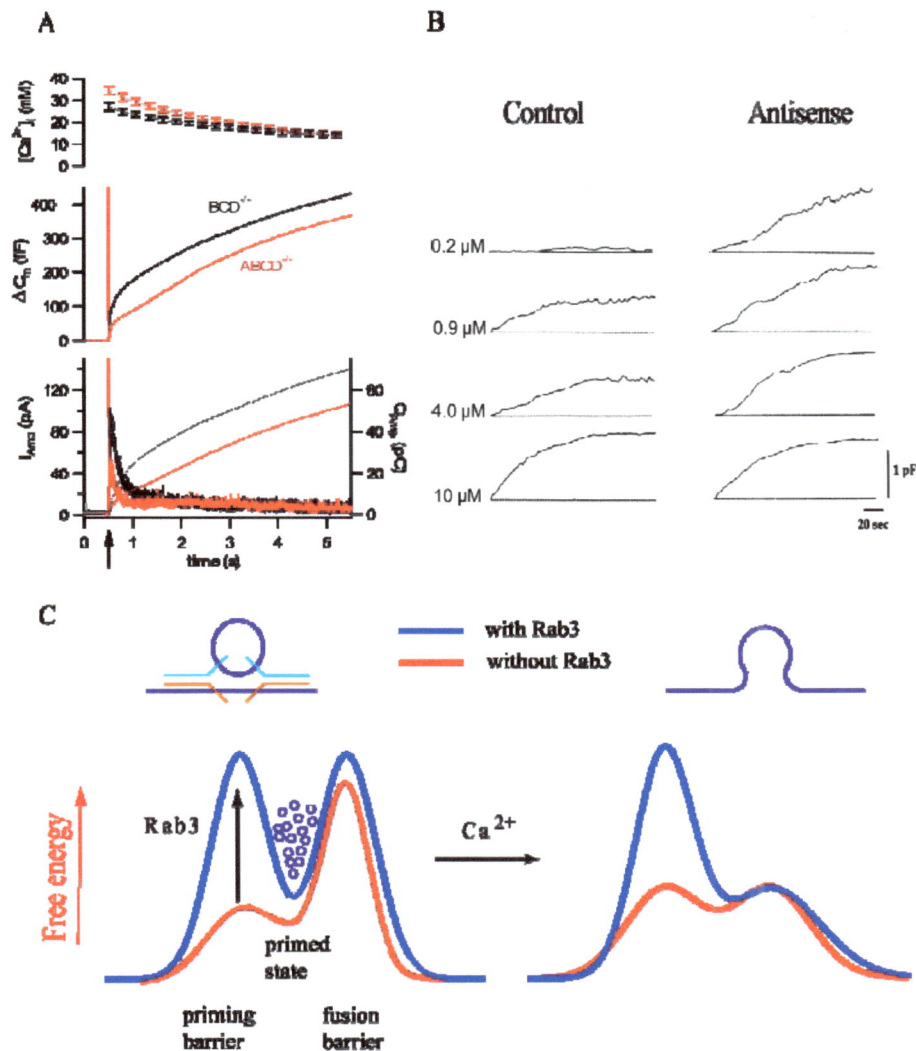

Figure 4: Rab3 controls the number of ready-to-fuse vesicles. (A) Deletion of Rab3 reduces the exocytotic burst but does not affect the sustained phase of release. Secretion was triggered in adrenal chromaffin cells by UV-flash-induced Ca^{2+} uncaging (at 500 ms; arrow). The secretory response was monitored by membrane capacitance recording (middle panel), and catecholamine oxidation was measured by amperometry (the lowest panel shows the amperometric current and the integrated trace). Intracellular Ca^{2+} levels were monitored using ratiometric fluorescence measurement of Fura dyes (upper panel). Analysis of the responses from control cells ($A^{+/+}BCD^{-/-}$ or $A^{+/-}BCD^{-/-}$, denoted here as $BCD^{-/-}$ and shown in black) and quadruple Rab3 KO cells ($ABCD^{-/-}$, shown in red) following Ca^{2+} uncaging shows that the size of the burst phase is strongly compromised in the quadruple KO, whereas the sustained phase is not dramatically modified. Reprinted from [73] with permission. (B) Lowering Rab3a levels increases the secretory response of adrenal chromaffin cells monitored by measuring changes in membrane capacitance (Cm). Exocytosis was elicited by dialyzing the cells through the patch pipette with solutions buffered at various Ca^{2+} concentrations (shown on the left). The Cm

traces obtained after injection into chromaffin cells of control (left) or antisense (right) oligonucleotides directed to Rab3a mRNA are shown. Whole-cell recordings were performed 5 days after microinjection. Reprinted from [37]. (**C**) The energy landscape of exocytosis. Two energy barriers must be overcome before exocytosis can proceed (blue line). The first barrier corresponds to the priming reaction and the second to membrane fusion. Primed vesicles are contained between these two peaks in a metastable state and are endowed with assembled SNARE complexes (depicted in the upper left panel). Calcium elevation lowers the energy barrier for fusion (downward arrow in the right panel), allowing primed vesicles to fuse (upper right drawing). Deletion of Rab3 (red line) is proposed to lower the priming energy barrier. This mechanism would account for the reduced exocytotic burst (panel A) due to the accumulation of fewer vesicles in the primed state in the absence of Rab3 because they undergo unpriming before the arrival of the stimulus. This mechanism would also explain the increase in secretory activity observed during sustained calcium elevation (panel B); in that case, vesicles could fuse once they were primed, and there would be no need for stable priming.

In the exocytosis energy landscape depicted in Fig. **4C**, the primed vesicles sit between the two peaks corresponding to the priming and fusion reactions. The barrier to fusion is lowered by the entry of calcium ions, which are thought to act on synaptotagmins [1-3, 129]. The increased unpriming rate found in Rab3 null mice suggests that the priming barrier is reduced in the absence of Rab3. A similar phenotype (unstable priming) was found in snapin KO mice [132]. Snapin binds to the SNARE protein SNAP-25 and may help to recruit synaptotagmin to the SNARE complex. Rab3 may thus cooperate with snapin to stabilize the fusion machinery. Priming would be facilitated if the priming barrier were reduced. Consistent with this, inhibiting Rab3a expression in chromaffin cells increased the responses to long-lasting calcium elevation (calcium dialysis *via* use of a patch pipette, a procedure that primarily probes the priming reaction) and repetitive stimulation [37, 38] (Fig. **4B**). Conversely, increased Rab3 levels should inhibit priming. Indeed, overexpression of Rab3 or GTPase-deficient Rab3 constructs inhibited secretion in several endocrine cell types and in neurons [22, 35-38, 40]. The strong inhibitory effect of GTPase-deficient Rab3 suggests that GTP hydrolysis by Rab3 may be rate limiting in the priming reaction and may be mandatory to terminate the priming reaction. Accordingly, GTP hydrolysis has been proposed to occur downstream of SNARE protein assembly because toxin-insensitive SNARE complexes accumulated in Aplysia neurons injected with GTPase-deficient Rab3 proteins [39].

The data supporting a role for Rab27 in priming are less compelling. Glucose-dependent insulin secretion was reduced in Rab27a-deficient *ashen* β-cells [61, 119], but the pool of ready-to-fuse insulin granules was not affected. Additionally, the initial response to a depolarization-induced Ca^{2+} influx was increased in Rab27a null cells, arguing against a role for Rab27 in priming [131]. The fact that the replenishment of the readily releasable pool (RRP) of SGs was slower in Rab27a-deficient cells than in control ones, may rather suggest a role for Rab27 in the priming reaction [131]. But because Rab27 deficiency impairs the recruitment and docking of insulin granules at the plasma membrane [119], priming may be reduced simply by mass action.

3.3.2. Rab Effectors in Secretory Vesicle Priming

Rabphilin. Rab3 and Rab27 have many effectors (see Box 2). However, it is currently unclear which of these effectors can mediate their priming effect. In endocrine cells, secretion is increased upon overexpression of rabphilin, whereas it is inhibited when rabphilin is impaired [133-135]. Rabphilin-deficient nematodes are lethargic, indicating a mild locomotor defect [136]. However, the rabphilin phenotype is more severe in the Rab3 null background, indicating that Rab27 and rabphilin function synergistically with Rab3 at the synapse [31]. There is also a synergistic interaction with SNARE proteins containing hypomorphic mutations, which suggests that the physical coupling of rabphilin and SNAP-25 is physiologically relevant [135, 137]. Inactivation of rabphilin in mice does not induce any detectable phenotype [138]. Basal neurotransmission, short-term synaptic plasticity and long-term potentiation are unchanged in the hippocampal synapses of rabphilin-deficient mice [138]. However, in cultured neurons from rabphilin null mice, an increase in the recovery rate of synaptic responses after depletion of the releasable pool of vesicles has been observed [137]. Altogether, the data suggest a role for rabphilin in recruiting vesicles to the releasable pool, but the molecular mechanisms of this recruitment are still unclear. Other studies have suggested a role for rabphilin in the coupling between exocytosis and endocytosis [139], which is eventually mediated by the interaction between rabphilin and the Rab5 effector Rabaptin-5 [140, 141].

Noc2. Like rabphilin, Noc2 binds to both Rab3 and Rab27, but it lacks the C2 domains (hence its name; see Box 2) [122]. Overexpression studies have demonstrated both positive and negative effects of Noc2 on exocytosis [142-144]. Inactivation of the *NOC2* gene in mice causes glucose intolerance, but only in animals stressed by water immersion [145]. This defect in insulin secretion is apparently due to an upregulation of Go or Gi signaling and cannot be taken as evidence for Noc2 being involved in the release mechanism. Noc2 deficiency has more profound effects in exocrine glands, causing an accumulation of SGs and a defect in amylase secretion, but the step at which secretion is arrested is not known [146]. An *in vitro* interaction between Noc2 and Munc13 has been reported [142], which suggests that Noc2 may function in priming because Munc13 is an important priming factor [147].

Munc13. *Munc13-4* mutations cause familial hemophagocytic lymphohistiocytosis type 3 [85]. As in Griscelli syndrome, an uncontrolled activation of macrophages and T cells results from a defect in secretory lysosome exocytosis. However, in contrast to *Rab27a* mutations, *Munc13-4* mutations do not impair cytolytic granule docking at the plasma membrane [85]. Munc13-4 thus functions at a post-docking stage, most likely in priming, similar to other members of the Munc13 family [147]. This effect on priming is probably dependent on Rab27a, which directly interacts with Munc13-4 [65, 148] and recruits it onto secretory lysosomes [149]. Munc13-4 also has a Rab27-independent role in the maturation of cytolytic granules [150]. The function of Munc13-4 is not restricted to T cells; it also promotes the secretion of secretory lysosomes in platelets, mast cells and neutrophils [65, 148, 151, 152]. Concerning Rab3, an interaction with Munc13-1 *via* Rim have been described [153] and this complex promotes the priming of synaptic vesicles within the active zone [154].

3.4. Triggering Fusion: Coupling Synaptic Vesicles to Calcium Channels

At the active zone, some SVs are attached to the plasma membrane, and a subset of these docked vesicles are physically coupled to Ca^{2+} channels. This spatial organization of the nerve terminals contributes to the extremely fast kinetics and efficacy of neurotransmitter release. Ca^{2+} gradients are sharp, and vesicles can experience high amounts of Ca^{2+} (approximately 100 μM) near the mouth of open channels and a submicromolar Ca^{2+} concentration a few microns away. Therefore, the ability to fuse is not sufficient to ensure a rapid neurotransmitter release. Wadel and colleagues [155] demonstrated that rapid release can be triggered by calcium uncaging after depletion of the pool of vesicles responsible for the synchronous release triggered by action potentials. These authors concluded that the recruitment of synaptic vesicles to sites where Ca^{2+} channels cluster is a rate-limiting step for rapid neurotransmitter release.

At neuromuscular junctions in Rab3a knockout mice, the defect in evoked release is more pronounced at low calcium concentrations than at physiological calcium concentrations, suggesting reduced calcium sensitivity [88]. Conversely, calcium elevation partially rescues Rab3a deficiency [156]. Because synchronous, but not asynchronous, release was found to be Rab3a-dependent, the data suggest that Rab3 mediates the recruitment of SVs to calcium channel clusters. The regulation of Rab3 effects by neuronal activity was also observed in the cholinergic neurons of *Aplysia californica* [35]. In these cells, the inhibitory effect of a GTPase-deficient Rab3 protein was potentiated by the prior injection of a Ca^{2+} chelator but reduced by a train of electrical stimuli, a treatment known to increase the Ca^{2+} concentration in the nerve terminal. Facilitation was also increased in cultured autaptic hippocampal neurons from quadruple Rab3 KO mice [42], which suggests that calcium elevation overcomes Rab3 deficiency. These studies are consistent with a role for Rab3 in targeting synaptic vesicles to calcium channels.

Rim. The Rab3 effector Rim [157] interacts with calcium channels and could thus mediate the Rab3-dependent tethering of synaptic vesicles to calcium channels. Rim interacts with Rab3 [158] but not with Rab27. The Rim family includes seven members encoded by four genes (*RIM1* through *4*). The extensive alternative splicing of Rim1 and Rim2 produces many Rim variants, but little is known about their function. Only Rim1α and Rim2α include the zinc finger and the SGAWFF motif that confer Rab3 binding activity (see Box 2). The other Rim domains consist of a central PDZ domain and two C-terminal atypical C2 domains that do not associate with phospholipids [159].

Rim is concentrated at the active zone (the region of the presynaptic compartment where synaptic vesicles are docked) and interacts with other components of the active zone protein network such as CAST, Munc13, Piccolo, Bassoon and ELKS [157, 160-162]. Rab3 and Munc13 can bind simultaneously to Rim [163] (see also [162] for further discussion). Calcium-dependent associations of the Rim C2A domain with the SNARE protein SNAP-25, synaptotagmin [164] (see also [165] for contrasting evidence) and the alpha subunits of voltage-gated calcium channels [164] have also been described. The Rim C2B domains interact with Liprin-α [166], synaptotagmin 1 [164, 166], the ubiquitin ligase Scrapper [167] and the β-subunits of voltage-gated calcium channels [168]. Finally, Rim binds to Rim-binding proteins (RBP 1 and 2) *via* a proline-rich region located between the two C2 domains [169]. RBP1 and 2 are connected to the α-subunits of calcium channels and thus can also physically link Rim to calcium channels.

Inactivation of the *RIM* gene in *C. elegans* [170] provokes an uncoordinated phenotype more severe than the *RAB3* phenotype. Both spontaneous neurotransmitter release and evoked responses are greatly reduced. Using high pressure freezing to maintain the architecture of the synapse, Weimer *et al.* [171] highlighted a selective decrease in the number of SVs close to dense projections (specialized areas enriched in calcium channels that contain Liprin-α proteins), while the overall number of SVs in the synaptic boutons remained unaffected.

In mammals, functional studies of Rim function have been complicated by the existence of several Rim genes. Rim1α null mice are viable and fertile. They have difficulties in spatial learning and fear conditioning and have a defect in maternal behavior. Their basal neurotransmission is not modified, but their short-term and long-term synaptic plasticity are affected [166, 172]. Mice lacking both Rim1α and Rim2α die immediately after birth because they cannot breathe. At the neuromuscular junction, spontaneous release is normal, indicating that the membrane fusion mechanism is not impaired, but evoked responses are severely reduced [173]. Altogether, the data suggest that the main function of Rim is to position SVs near calcium channels where the fusion machinery can ideally sense changes in calcium concentration upon the arrival of a stimulus. This positioning may be mediated by the direct interaction of Rim with calcium channels or interaction *via* liprins or RBP. In the absence of Rim1α, the levels of Munc13 are reduced. Because Munc13 has an important role in vesicle priming, some of the effects of Rim inactivation may be accounted for by this Munc13 effect.

4. CONCLUSIONS AND PERSPECTIVES

A large number of studies have been devoted to Rab proteins since their discovery in 1987 [174]. Even when study of these proteins is limited to their roles in secretion, Rab functions are complex. One source of complexity is the redundancy between the Rab isoforms and between Rab3 and Rab27. Clearly, Rab3 and Rab27 have specific roles and effectors, but there is also some overlap between their functions. Another source of complexity comes from the time courses of regulated secretion in various types of secretory cells. In neurons, research emphasis is put on time-resolved techniques because the relevant unit of time for neurotransmission is the millisecond. Hence, much importance has been placed on the priming reaction, which is largely controlled by Rab3. In endocrine or exocrine cells, given the slower kinetics of the secretory responses, the emphasis is put on techniques that measure the sustained components of release and on the recruitment of vesicles at release sites, a process largely controlled by Rab27.

By recruiting multiple effector molecules, Rab3 and Rab27 orchestrate multiple events that proceed from the biogenesis of a secretory vesicle to its exocytosis. Further work is needed to precisely delineate the functions of these effectors and to determine the chronology of the above events. How the recruitment of different effectors is regulated in time and space and coordinated with the progress of the vesicle along the secretory process is an intriguing issue.

ACKNOWLEDGEMENTS

We thank O. Jouannot and I. Fanget for their help in preparing the figures and S. O'Regan for revising the manuscript. This work was supported by the Centre National de la Recherche Scientifique and the Agence Nationale de la Recherche. FD is supported by INSERM.

CONFLICT OF INTEREST

There is no conflict of interest from any of the authors.

REFERENCES

[1]　Rizo J, Rosenmund C. Synaptic vesicle fusion. Nat Struct Mol Biol 2008 Jul;15(7):665-74.

[2]　Sorensen JB. Conflicting views on the membrane fusion machinery and the fusion pore. Annual Rev Cell Dev Biol 2009;25:513-37.

[3]　Südhof TC. Neurotransmitter release. Handbook Exp Pharm 2008(184):1-21.

[4]　Fukuda M. Regulation of secretory vesicle traffic by Rab small GTPases. Cell Mol Life Sci 2008 Sep;65(18):2801-13.

[5]　Darchen F, Goud B. Multiple aspects of Rab protein action in the secretory pathway: focus on Rab3 and Rab6. Biochimie 2000 Apr;82(4):375-84.

[6]　Lledo PM, Vernier P, Vincent JD, Mason WT, Zorec R. Inhibition of Rab3B expression attenuates Ca(2+)-dependent exocytosis in rat anterior pituitary cells. Nature 1993;364(6437):540-4.

[7]　Rupnik M, Kreft M, Nothias F, *et al*. Distinct role of Rab3A and Rab3B in secretory activity of rat melanotrophs. Am J Physiol Cell Physiol 2007 Jan;292(1):C98-105.

[8]　Weber E, Berta G, Tousson A, *et al*. Expression and polarized targeting of a rab3 isoform in epithelial cells. J Cell Biol 1994;125(3):583-94.

[9]　Baldini G, Hohl T, Lin HY, Lodish HF. Cloning of a Rab3 isotype predominantly expressed in adipocytes. Proc Natl Acad Sci USA 1992;89(11):5049-52.

[10]　Valentijn JA, Gumkowski FD, Jamieson JD. The expression pattern of rab3D in the developing rat exocrine pancreas coincides with the acquisition of regulated exocytosis. Eur J Cell Biol 1996;71(2):129-36.

[11]　Stettler O, Nothias F, Tavitian B, Vernier P. Double *in situ* hybridization reveals overlapping neuronal populations expressing the low molecular weight GTPases Rab3a and Rab3b in Rat brain. Eur J Neurosci 1995;7(4):702-13.

[12]　Gomi H, Mori K, Itohara S, Izumi T. Rab27b is expressed in a wide range of exocytic cells and involved in the delivery of secretory granules near the plasma membrane. Mol Biol Cell 2007 Nov;18(11):4377-86.

[13]　Yi Z, Yokota H, Torii S, *et al*. The Rab27a/granuphilin complex regulates the exocytosis of insulin-containing dense-core granules. Mol Cell Biol 2002 Mar;22(6):1858-67.

[14]　Fischer von Mollard G, Mignery GA, Baumert M, *et al*. rab3 is a small GTP-binding protein exclusively localized to synaptic vesicles. Proc Natl Acad Sci USA 1990;87(5):1988-92.

[15]　Fischer von Mollard G, Stahl B, Khokhlatchev A, Sudhof TC, Jahn R. Rab3C is a synaptic vesicle protein that dissociates from synaptic vesicles after stimulation of exocytosis. J Biol Chem 1994;269(15):10971-4.

[16]　Takamori S, Rhee JS, Rosenmund C, Jahn R. Identification of a vesicular glutamate transporter that defines a glutamatergic phenotype in neurons. Nature 2000;407(6801):189-94.

[17]　Darchen F, Senyshyn J, Brondyk WH, *et al*. The GTPase Rab3a is associated with large dense core vesicles in bovine chromaffin cells and rat PC12 cells. J Cell Sci 1995;108(Pt 4):1639-49.

[18]　Darchen F, Zahraoui A, Hammel F, Monteils MP, Tavitian A, Scherman D. Association of the GTP-binding protein Rab3A with bovine adrenal chromaffin granules. Proc Natl Acad Sci USA 1990;87(15):5692-6.

[19]　Desnos C, Schonn JS, Huet S, *et al*. Rab27A and its effector MyRIP link secretory granules to F-actin and control their motion towards release sites. J Cell Biol 2003 Nov 10;163(3):559-70.

[20]　Fukuda M, Kanno E, Saegusa C, Ogata Y, Kuroda TS. Slp4-a/Granuphilin-a Regulates Dense-core Vesicle Exocytosis in PC12 Cells. J Biol Chem 2002;277(42):39673-8.

[21]　Iezzi M, Escher G, Meda P, *et al*. Subcellular distribution and function of Rab3A, B, C, and D isoforms in insulin-secreting cells. Mol Endocrinol 1999;13(2):202-12.

[22]　Regazzi R, Ravazzola M, Iezzi M, *et al*. Expression, localization and functional role of small GTPases of the Rab3 family in insulin-secreting cells. J Cell Sci 1996;109(Pt 9):2265-73.

[23]　Baldini G, Wang G, Weber M, *et al*. Expression of Rab3D N135I inhibits regulated secretion of ACTH in AtT-20 cells. J Cell Biol 1998;140(2):305-13.

[24]　Ohnishi H, Ernst SA, Wys N, McNiven M, Williams JA. Rab3D localizes to zymogen granules in rat pancreatic acini and other exocrine glands. Am J Physiol 1996;271(3 Pt 1):G531-8.

[25]　Tang LH, Gumkowski FD, Sengupta D, Modlin IM, Jamieson JD. rab3D protein is a specific marker for zymogen granules in gastric chief cells of rats and rabbits. Gastroenterol 1996;110(3):809-20.

[26]   Seabra MC, Coudrier E. Rab GTPases and myosin motors in organelle motility. Traffic 2004 Jun;5(6):393-9.

[27]   Menasche G, Pastural E, Feldmann J, *et al.* Mutations in RAB27A cause Griscelli syndrome associated with haemophagocytic syndrome. Nat Genet 2000 Jun;25(2):173-6.

[28]   Stinchcombe JC, Barral DC, Mules EH, *et al.* Rab27a is required for regulated secretion in cytotoxic T lymphocytes. J Cell Biol 2001 Feb 19;152(4):825-34.

[29]   Ostrowski M, Carmo NB, Krumeich S, *et al.* Rab27a and Rab27b control different steps of the exosome secretion pathway. Nat Cell Biol 2010 Jan;12(1):19-30; sup pp 1-13.

[30]   Hannah MJ, Hume AN, Arribas M, *et al.* Weibel-Palade bodies recruit Rab27 by a content-driven, maturation-dependent mechanism that is independent of cell type. J Cell Sci 2003 Oct 1;116(Pt 19):3939-48.

[31]   Mahoney TR, Liu Q, Itoh T, *et al.* Regulation of synaptic transmission by RAB-3 and RAB-27 in Caenorhabditis elegans. Mol Biol Cell 2006 Jun;17(6):2617-25.

[32]   Pavlos NJ, Gronborg M, Riedel D, *et al.* Quantitative analysis of synaptic vesicle Rabs uncovers distinct yet overlapping roles for Rab3a and Rab27b in Ca2+-triggered exocytosis. J Neurosci 2010 Oct 6;30(40):13441-53.

[33]   Takamori S, Holt M, Stenius K, *et al.* Molecular anatomy of a trafficking organelle. Cell 2006 Nov 17;127(4):831-46.

[34]   Tsuboi T, Fukuda M. Rab3A and Rab27A cooperatively regulate the docking step of dense-core vesicle exocytosis in PC12 cells. J Cell Sci 2006 Jun 1;119(Pt 11):2196-203.

[35]   Doussau F, Clabecq A, Henry JP, Darchen F, Poulain B. Calcium-dependent regulation of rab3 in short-term plasticity. J Neurosci 1998;18(9):3147-57.

[36]   Holz RW, Brondyk WH, Senter RA, Kuizon L, Macara IG. Evidence for the involvement of Rab3A in Ca(2+)-dependent exocytosis from adrenal chromaffin cells. J Biol Chem 1994;269(14):10229-34.

[37]   Johannes L, Lledo PM, Chameau P, Vincent JD, Henry JP, Darchen F. Regulation of the Ca2+ sensitivity of exocytosis by Rab3a. J Neurochem 1998;71(3):1127-33.

[38]   Johannes L, Lledo PM, Roa M, Vincent JD, Henry JP, Darchen F. The GTPase Rab3a negatively controls calcium-dependent exocytosis in neuroendocrine cells. EMBO J 1994;13(9):2029-37.

[39]   Johannes L, Perez F, Laran-Chich MP, Henry JP, Darchen F. Characterization of the interaction of the monomeric GTP-binding protein Rab3a with geranylgeranyl transferase II. Eur J Biochem 1996;239(2):362-8.

[40]   Thiagarajan R, Tewolde T, Li Y, Becker PL, Rich MM, Engisch KL. Rab3A negatively regulates activity-dependent modulation of exocytosis in bovine adrenal chromaffin cells. J Physiol 2004 Mar 1;555(Pt 2):439-57.

[41]   Yunes R, Tomes C, Michaut M, *et al.* Rab3A and calmodulin regulate acrosomal exocytosis by mechanisms that do not require a direct interaction. FEBS Lett 2002 Aug 14;525(1-3):126-30.

[42]   Schlüter OM, Basu J, Sudhof TC, Rosenmund C. Rab3 superprimes synaptic vesicles for release: implications for short-term synaptic plasticity. J Neurosci 2006 Jan 25;26(4):1239-46.

[43]   Schlüter OM, Schmitz F, Jahn R, Rosenmund C, Sudhof TC. A complete genetic analysis of neuronal Rab3 function. J Neurosci 2004 Jul 21;24(29):6629-37.

[44]   Kapfhamer D, Valladares O, Sun Y, *et al.* Mutations in Rab3a alter circadian period and homeostatic response to sleep loss in the mouse. Nat Genet 2002;32(2):290-5.

[45]   D'Adamo P, Wolfer DP, Kopp C, Tobler I, Toniolo D, Lipp HP. Mice deficient for the synaptic vesicle protein Rab3a show impaired spatial reversal learning and increased explorative activity but none of the behavioral changes shown by mice deficient for the Rab3a regulator Gdi1. Eur J Neurosci. 2004 Apr;19(7):1895-905.

[46]   Castillo PE, Janz R, Südhof TC, Tzounopoulos T, Malenka RC, Nicoll RA. Rab3A is essential for mossy fibre long-term potentiation in the hippocampus. Nature 1997;388(6642):590-3.

[47]   Geppert M, Bolshakov VY, Siegelbaum SA, *et al.* The role of Rab3A in neurotransmitter release. Nature 1994 Jun 9;369(6480):493-7.

[48]   Geppert M, Goda Y, Stevens CF, Südhof TC. The small GTP-binding protein Rab3A regulates a late step in synaptic vesicle fusion. Nature 1997;387(6635):810-4.

[49]   Nonet ML, Staunton JE, Kilgard MP, *et al.* Caenorhabditis elegans rab-3 mutant synapses exhibit impaired function and are partially depleted of vesicles. J Neurosci 1997;17(21):8061-73.

[50]   D'Adamo P, Menegon A, Lo Nigro C, *et al.* Mutations in GDI1 are responsible for X-linked non-specific mental retardation [see comments] [published erratum appears in Nat Genet 1998 Jul;19(3):303]. Nat Genet. 1998;19(2):134-9.

[51]   D'Adamo P, Welzl H, Papadimitriou S, *et al.* Deletion of the mental retardation gene Gdi1 impairs associative memory and alters social behavior in mice. Hum Mol Genet 2002 Oct 1;11(21):2567-80.

[52]   Ishizaki H, Miyoshi J, Kamiya H, *et al.* Role of rab GDP dissociation inhibitor alpha in regulating plasticity of hippocampal neurotransmission. Proc Natl Acad Sci USA 2000;97(21):11587-92.

[53]   Tanaka M, Miyoshi J, Ishizaki H, *et al.* Role of rab3 gdp/gtp exchange protein in synaptic vesicle trafficking at the mouse neuromuscular junction. Mol Biol Cell 2001;12(5):1421-30.

[54]   Yamaguchi K, Tanaka M, Mizoguchi A, *et al.* A GDP/GTP exchange protein for the Rab3 small G protein family up-regulates a postdocking step of synaptic exocytosis in central synapses. Proc Natl Acad Sci USA 2002;99(22):14536-41.

[55]   Iwasaki K, Staunton J, Saifee O, Nonet M, Thomas JH. aex-3 encodes a novel regulator of presynaptic activity in C. elegans. Neuron 1997;18(4):613-22.

[56]   Griscelli C, Durandy A, Guy-Grand D, Daguillard F, Herzog C, Prunieras M. A syndrome associating partial albinism and immunodeficiency. Am J Med 1978 Oct;65(4):691-702.

[57]   Wilson SM, Yip R, Swing DA, *et al.* A mutation in Rab27a causes the vesicle transport defects observed in ashen mice. Proc Natl Acad Sci USA 2000;97(14):7933-8.

[58]   Stinchcombe J, Bossi G, Griffiths GM. Linking albinism and immunity: the secrets of secretory lysosomes. Science 2004 Jul 2;305(5680):55-9.

[59]   Haddad EK, Wu X, Hammer JA, 3rd, Henkart PA. Defective granule exocytosis in Rab27a-deficient lymphocytes from Ashen mice. J Cell Biol 2001 Feb 19;152(4):835-42.

[60]   Jancic C, Savina A, Wasmeier C, *et al.* Rab27a regulates phagosomal pH and NADPH oxidase recruitment to dendritic cell phagosomes. Nat Cell Biol 2007 Apr;9(4):367-78.

[61]   Kasai K, Ohara-Imaizumi M, Takahashi N, *et al.* Rab27a mediates the tight docking of insulin granules onto the plasma membrane during glucose stimulation. J Clin Invest 2005 Feb;115(2):388-96.

[62]   Tolmachova T, Abrink M, Futter CE, Authi KS, Seabra MC. Rab27b regulates number and secretion of platelet dense granules. Proc Natl Acad Sci USA 2007 Apr 3;104(14):5872-7.

[63]   Waselle L, Coppola T, Fukuda M, *et al.* Involvement of the Rab27 binding protein Slac2c/MyRIP in insulin exocytosis. Mol Biol Cell 2003 Oct;14(10):4103-13.

[64]   Gomi H, Mizutani S, Kasai K, Itohara S, Izumi T. Granuphilin molecularly docks insulin granules to the fusion machinery. J Cell Biol 2005 Oct 10;171(1):99-109.

[65]   Shirakawa R, Higashi T, Tabuchi A, *et al.* Munc13-4 is a GTP-Rab27-binding protein regulating dense core granule secretion in platelets. J Biol Chem 2004 Mar 12;279(11):10730-7.

[66]   Chen X, Li C, Izumi T, Ernst SA, Andrews PC, Williams JA. Rab27b localizes to zymogen granules and regulates pancreatic acinar exocytosis. Biochem Biophys Res Commun 2004 Oct 29;323(4):1157-62.

[67]   Munafo DB, Johnson JL, Ellis BA, Rutschmann S, Beutler B, Catz SD. Rab27a is a key component of the secretory machinery of azurophilic granules in granulocytes. Biochem J 2007 Mar 1;402(2):229-39.

[68]   Herrero-Turrion MJ, Calafat J, Janssen H, Fukuda M, Mollinedo F. Rab27a regulates exocytosis of tertiary and specific granules in human neutrophils. J Immunol 2008 Sep 15;181(6):3793-803.

[69]   Johnson JL, Brzezinska AA, Tolmachova T, *et al.* Rab27a and Rab27b regulate neutrophil azurophilic granule exocytosis and NADPH oxidase activity by independent mechanisms. Traffic 2010 Apr;11(4):533-47.

[70]   Johnson JL, Ellis BA, Noack D, Seabra MC, Catz SD. The Rab27a-binding protein, JFC1, regulates androgen-dependent secretion of prostate-specific antigen and prostatic-specific acid phosphatase. Biochem J 2005 Nov 1;391(Pt 3):699-710.

[71]   Yu E, Kanno E, Choi S, *et al.* Role of Rab27 in synaptic transmission at the squid giant synapse. Proc Natl Acad Sci USA 2008 Oct 14;105(41):16003-8.

[72]   Morvan J, Tooze SA. Discovery and progress in our understanding of the regulated secretory pathway in neuroendocrine cells. Histochem Cell Biol 2008 Mar;129(3):243-52.

[73]   Schonn J-S, van Weering JRT, Mohrmann R, *et al.* Rab3 proteins involved in vesicle biogenesis and priming in embryonic mouse chromaffin cells. Traffic 2010;11:1415-8.

[74]   Handley MT, Haynes LP, Burgoyne RD. Differential dynamics of Rab3A and Rab27A on secretory granules. J Cell Sci 2007 Mar 15;120(Pt 6):973-84.

[75]   Riedel D, Antonin W, Fernandez-Chacon R, *et al.* Rab3D is not required for exocrine exocytosis but for maintenance of normally sized secretory granules. Mol Cell Biol 2002;22(18):6487-97.

[76]   Martelli AM, Baldini G, Tabellini G, Koticha D, Bareggi R. Rab3A and Rab3D control the total granule number and the fraction of granules docked at the plasma membrane in PC12 cells. Traffic 2000;1(12):976-86.

[77]   Kogel T, Rudolf R, Hodneland E, *et al.* Distinct roles of myosin Va in membrane remodeling and exocytosis of secretory granules. Traffic 2010 May;11(5):637-50.

[78]   Varadi A, Tsuboi T, Johnson-Cadwell LI, Allan VJ, Rutter GA. Kinesin I and cytoplasmic dynein orchestrate glucose-stimulated insulin-containing vesicle movements in clonal MIN6 beta-cells. Biochem Biophys Res Commun 2003 Nov 14;311(2):272-82.

[79]   Rudolf R, Kogel T, Kuznetsov SA, *et al.* Myosin Va facilitates the distribution of secretory granules in the F-actin rich cortex of PC12 cells. J Cell Sci 2003 Apr 1;116(Pt 7):1339-48.

[80]   Malacombe M, Bader MF, Gasman S. Exocytosis in neuroendocrine cells: new tasks for actin. Biochimica et biophysica acta 2006 Nov;1763(11):1175-83.

[81]   Desnos C, Huet S, Darchen F. 'Should I stay or should I go?': myosin V function in organelle trafficking. Biol Cell 2007 Aug;99(8):411-23.

[82]   Desnos C, Huet S, Fanget I, *et al.* Myosin va mediates docking of secretory granules at the plasma membrane. J Neurosci 2007 Sep 26;27(39):10636-45.

[83]   Fdez E, Hilfiker S. Vesicle pools and synapsins: new insights into old enigmas. Brain Cell Biol 2006 Jun;35(2-3):107-15.

[84]   Fornasiero EF, Bonanomi D, Benfenati F, Valtorta F. The role of synapsins in neuronal development. Cell Mol Life Sci 2010 May;67(9):1383-96.

[85]   Feldmann J, Callebaut I, Raposo G, *et al.* Munc13-4 is essential for cytolytic granules fusion and is mutated in a form of familial hemophagocytic lymphohistiocytosis (FHL3). Cell 2003 Nov 14;115(4):461-73.

[86]   van Weering JR, Toonen RF, Verhage M. The role of Rab3a in secretory vesicle docking requires association/dissociation of guanidine phosphates and Munc18-1. PLoS One 2007;2(7):e616.

[87]   Leenders AG, Lopes da Silva FH, Ghijsen WE, Verhage M. Rab3a is involved in transport of synaptic vesicles to the active zone in mouse brain nerve terminals. Mol Biol Cell 2001 Oct;12(10):3095-102.

[88]   Coleman WL, Bill CA, Bykhovskaia M. Rab3a deletion reduces vesicle docking and transmitter release at the mouse diaphragm synapse. Neurosci 2007 Aug 10;148(1):1-6.

[89]   Graham ME, Handley MT, Barclay JW, *et al.* A gain-of-function mutant of Munc18-1 stimulates secretory granule recruitment and exocytosis and reveals a direct interaction of Munc18-1 with Rab3. Biochem J 2008 Jan 15;409(2):407-16.

[90]   Karatekin E, Tran VS, Huet S, Fanget I, Cribier S, Henry JP. A 20-nm step toward the cell membrane preceding exocytosis may correspond to docking of tethered granules. Biophys J 2008 Apr 1;94(7):2891-905.

[91]   de Wit H, Walter AM, Milosevic I, *et al.* Synaptotagmin-1 docks secretory vesicles to syntaxin-1/SNAP-25 acceptor complexes. Cell 2009 Sep 4;138(5):935-46.

[92]   Toonen RF, Kochubey O, de Wit H, *et al.* Dissecting docking and tethering of secretory vesicles at the target membrane. EMBO J 2006 Aug 23;25(16):3725-37.

[93]   Verhage M, Sorensen JB. Vesicle docking in regulated exocytosis. Traffic 2008 Sep;9(9):1414-24.

[94]   Siksou L, Varoqueaux F, Pascual O, Triller A, Brose N, Marty S. A common molecular basis for membrane docking and functional priming of synaptic vesicles. Eur J Neurosci 2009 Jul;30(1):49-56.

[95]   Pfeffer S. Vesicle tethering factors united. Mol Cell 2001 Oct;8(4):729-30.

[96]   Fukuda M. Synaptotagmin-like Protein (Slp) Homology Domain 1 of Slac2- a/Melanophilin Is a Critical Determinant of GTP-dependent Specific Binding to Rab27A. J Biol Chem 2002;277(42):40118-24.

[97]   Hume AN, Collinson LM, Hopkins CR, *et al.* The leaden gene product is required with Rab27a to recruit myosin Va to melanosomes in melanocytes. Traffic 2002;3(3):193-202.

[98]   Provance DW, James TL, Mercer JA. Melanophilin, the product of the leaden locus, is required for targeting of myosin-Va to melanosomes. Traffic 2002;3(2):124-32.

[99]   Wu XS, Rao K, Zhang H, *et al.* Identification of an organelle receptor for myosin-Va. Nat Cell Biol 2002;4(4):271-8.

[100]   Matesic LE, Yip R, Reuss AE, *et al.* Mutations in Mlph, encoding a member of the Rab effector family, cause the melanosome transport defects observed in leaden mice. Proc Natl Acad Sci USA 2001 Aug 28;98(18):10238-43.

[101]   Menasche G, Ho CH, Sanal O, *et al.* Griscelli syndrome restricted to hypopigmentation results from a melanophilin defect (GS3) or a MYO5A F-exon deletion (GS1). J Clin Invest 2003 Aug;112(3):450-6.

[102]   Mercer JA, Seperack PK, Strobel MC, Copeland NG, Jenkins NA. Novel myosin heavy chain encoded by murine dilute coat colour locus. Nature 1991 Feb 21;349(6311):709-13.

[103]   Pastural E, Barrat FJ, Dufourcq-Lagelouse R, *et al.* Griscelli disease maps to chromosome 15q21 and is associated with mutations in the myosin-Va gene. Nat Genet 1997 Jul;16(3):289-92.

[104]   El-Amraoui A, Schonn JS, Kussel-Andermann P, *et al.* MyRIP, a novel Rab effector, enables myosin VIIa recruitment to retinal melanosomes. EMBO Rep 2002 May;3(5):463-70.

[105]   Fukuda M, Kuroda TS. Slac2-c (synaptotagmin-like protein homologue lacking C2 domains-c), a novel linker protein that interacts with Rab27, myosin Va/VIIa, and actin. J Biol Chem 2002;277(45):43096-103.

[106]   Nightingale TD, Pattni K, Hume AN, Seabra MC, Cutler DF. Rab27a and MyRIP regulate the amount and multimeric state of VWF released from endothelial cells. Blood 2009 May 14;113(20):5010-8.

[107] Imai A, Yoshie S, Nashida T, Shimomura H, Fukuda M. The small GTPase Rab27B regulates amylase release from rat parotid acinar cells. J Cell Sci 2004 Apr 15;117(Pt 10):1945-53.

[108] Kuroda TS, Fukuda M. Functional analysis of Slac2-c/MyRIP as a linker protein between melanosomes and myosin VIIa. J Biol Chem 2005 Jul 29;280(30):28015-22.

[109] Klomp AE, Teofilo K, Legacki E, Williams DS. Analysis of the linkage of MYRIP and MYO7A to melanosomes by RAB27A in retinal pigment epithelial cells. Cell Motil Cytoskeleton 2007 Jun;64(6):474-87.

[110] Lopes VS, Ramalho JS, Owen DM, *et al.* The ternary Rab27a-Myrip-Myosin VIIa complex regulates melanosome motility in the retinal pigment epithelium. Traffic 2007 May;8(5):486-99.

[111] Goehring AS, Pedroja BS, Hinke SA, Langeberg LK, Scott JD. MyRIP anchors protein kinase A to the exocyst complex. J Biol Chem 2007 Nov 9;282(45):33155-67.

[112] Ivarsson R, Jing X, Waselle L, Regazzi R, Renstrom E. Myosin 5a controls insulin granule recruitment during late-phase secretion. Traffic 2005 Nov;6(11):1027-35.

[113] Rose SD, Lejen T, Casaletti L, Larson RE, Pene TD, Trifaro JM. Myosins II and V in chromaffin cells: myosin V is a chromaffin vesicle molecular motor involved in secretion. J Neurochem 2003 Apr;85(2):287-98.

[114] Varadi A, Tsuboi T, Rutter GA. Myosin Va transports dense core secretory vesicles in pancreatic MIN6 beta-cells. Mol Biol Cell 2005 Jun;16(6):2670-80.

[115] Wang J, Takeuchi T, Yokota H, Izumi T. Novel rabphilin-3-like protein associates with insulin-containing granules in pancreatic beta cells. J Biol Chem 1999 Oct 1;274(40):28542-8.

[116] Coppola T, Frantz C, Perret-Menoud V, Gattesco S, Hirling H, Regazzi R. Pancreatic beta-cell protein granuphilin binds Rab3 and Munc-18 and controls exocytosis. Mol Biol Cell 2002 Jun;13(6):1906-15.

[117] Torii S, Takeuchi T, Nagamatsu S, Izumi T. Rab27 effector granuphilin promotes the plasma membrane targeting of insulin granules *via* interaction with syntaxin 1a. J Biol Chem 2004 May 21;279(21):22532-8.

[118] Torii S, Zhao S, Yi Z, Takeuchi T, Izumi T. Granuphilin modulates the exocytosis of secretory granules through interaction with syntaxin 1a. Mol Cell Biol 2002 Aug;22(15):5518-26.

[119] Kasai K, Fujita T, Gomi H, Izumi T. Docking is not a prerequisite but a temporal constraint for fusion of secretory granules. Traffic 2008 Jul;9(7):1191-203.

[120] Yu M, Kasai K, Nagashima K, *et al.* Exophilin4/Slp2-a targets glucagon granules to the plasma membrane through unique Ca2+-inhibitory phospholipid-binding activity of the C2A domain. Mol Biol Cell 2007 Feb;18(2):688-96.

[121] Shirataki H, Kaibuchi K, Sakoda T, *et al.* Rabphilin-3A, a putative target protein for smg p25A/rab3A p25 small GTP-binding protein related to synaptotagmin. Mol Cell Biol 1993 Apr;13(4):2061-8.

[122] Fukuda M, Kanno E, Yamamoto A. Rabphilin and Noc2 are recruited to dense-core vesicles through specific interaction with Rab27A in PC12 cells. J Biol Chem 2004 Mar 26;279(13):13065-75.

[123] Li C, Takei K, Geppert M, *et al.* Synaptic targeting of rabphilin-3A, a synaptic vesicle Ca2+/phospholipid-binding protein, depends on rab3A/3C. Neuron 1994;13(4):885-98.

[124] Stahl B, von Mollard GF, Walch-Solimena C, Jahn R. GTP cleavage by the small GTP-binding protein Rab3A is associated with exocytosis of synaptic vesicles induced by alpha-latrotoxin. J Biol Chem 1994;269(40):24770-6.

[125] Baldini G, Martelli AM, Tabellini G, *et al.* Rabphilin localizes with the cell actin cytoskeleton and stimulates association of granules with F-actin cross-linked by {alpha}-actinin. J Biol Chem 2005 Oct 14;280(41):34974-84.

[126] Kato M, Sasaki T, Ohya T, *et al.* Physical and functional interaction of rabphilin-3A with alpha-actinin. J Biol Chem 1996;271(50):31775-8.

[127] Giovedi S, Darchen F, Valtorta F, Greengard P, Benfenati F. Synapsin is a novel Rab3 effector protein on small synaptic vesicles. II. Functional effects of the Rab3A-synapsin I interaction. J Biol Chem 2004 Oct 15;279(42):43769-79.

[128] Giovedi S, Vaccaro P, Valtorta F, *et al.* Synapsin is a novel Rab3 effector protein on small synaptic vesicles. I. Identification and characterization of the synapsin I-Rab3 interactions *in vitro* and in intact nerve terminals. J Biol Chem 2004 Oct 15;279(42):43760-8.

[129] Wojcik SM, Brose N. Regulation of membrane fusion in synaptic excitation-secretion coupling: speed and accuracy matter. Neuron 2007 Jul 5;55(1):11-24.

[130] Rettig J, Neher E. Emerging roles of presynaptic proteins in Ca++-triggered exocytosis. Science 2002 Oct 25;298(5594):781-5.

[131] Merrins MJ, Stuenkel EL. Kinetics of Rab27a-dependent actions on vesicle docking and priming in pancreatic beta-cells. J Physiol 2008 Nov 15;586(Pt 22):5367-81.

[132] Tian JH, Wu ZX, Unzicker M, *et al.* The role of Snapin in neurosecretion: snapin knock-out mice exhibit impaired calcium-dependent exocytosis of large dense-core vesicles in chromaffin cells. J Neurosci 2005 Nov 9;25(45):10546-55.

[133] Arribas M, Regazzi R, Garcia E, Wollheim CB, Decamilli P. The Stimulatory Effect of Rabphilin 3A on Regulated Exocytosis from Insulin-Secreting Cells Does Not Require an Association-Dissociation Cycle with Membranes Mediated by Rab-3. Eur J Cell Biol 1997;74(3):209-16.

[134] Chung SH, Takai Y, Holz RW. Evidence that the Rab3a-binding protein, rabphilin3a, enhances regulated secretion. Studies in adrenal chromaffin cells. J Biol Chem 1995;270(28):16714-8.

[135] Tsuboi T, Fukuda M. The C2B domain of rabphilin directly interacts with SNAP-25 and regulates the docking step of dense core vesicle exocytosis in PC12 cells. J Biol Chem 2005 Nov 25;280(47):39253-9.

[136] Staunton J, Ganetzky B, Nonet ML. Rabphilin potentiates soluble N-ethylmaleimide sensitive factor attachment protein receptor function independently of rab3. J Neurosci 2001 Dec 1;21(23):9255-64.

[137] Deak F, Shin OH, Tang J, et al. Rabphilin regulates SNARE-dependent re-priming of synaptic vesicles for fusion. EMBO J 2006 Jun 21;25(12):2856-66.

[138] Schlüter OM, Schnell E, Verhage M, et al. Rabphilin knock-out mice reveal that rabphilin is not required for rab3 function in regulating neurotransmitter release. J Neurosci 1999 Jul 15;19(14):5834-46.

[139] Burns ME, Sasaki T, Takai Y, Augustine GJ. Rabphilin-3A - A Multifunctional Regulator of Synaptic Vesicle Traffic. J Gen Physiol 1998;111(2):243-55.

[140] Ohya T, Sasaki T, Kato M, Takai Y. Involvement of Rabphilin3 in endocytosis through interaction with Rabaptin5. J Biol Chem 1998 Jan 2;273(1):613-7.

[141] Coppola T, Hirling H, Perret-Menoud V, et al. Rabphilin dissociated from Rab3 promotes endocytosis through interaction with Rabaptin-5. J Cell Sci 2001 May;114(Pt 9):1757-64.

[142] Cheviet S, Coppola T, Haynes LP, Burgoyne RD, Regazzi R. The Rab-binding protein Noc2 is associated with insulin-containing secretory granules and is essential for pancreatic beta-cell exocytosis. Mol Endocrinol 2004 Jan;18(1):117-26.

[143] Haynes LP, Evans GJ, Morgan A, Burgoyne RD. A direct inhibitory role for the rab3-specific effector, noc2, in ca2+-regulated exocytosis in neuroendocrine cells. J Biol Chem 2001;276(13):9726-32.

[144] Kotake K, Ozaki N, Mizuta M, Sekiya S, Inagaki N, Seino S. Noc2, a putative zinc finger protein involved in exocytosis in endocrine cells. The J Biol Chem 1997;272(47):29407-10.

[145] Matsumoto M, Miki T, Shibasaki T, et al. Noc2 is essential in normal regulation of exocytosis in endocrine and exocrine cells. Proc Natl Acad Sci USA 2004 Jun 1;101(22):8313-8.

[146] Imai A, Yoshie S, Nashida T, Shimomura H, Fukuda M. Functional involvement of Noc2, a Rab27 effector, in rat parotid acinar cells. Arch Biochem Biophys 2006 Nov 15;455(2):127-35.

[147] Brose N, Rosenmund C, Rettig J. Regulation of transmitter release by Unc-13 and its homologues. Curr Opin Neurobiol 2000 Jun;10(3):303-11.

[148] Neeft M, Wieffer M, de Jong AS, et al. Munc13-4 is an effector of rab27a and controls secretion of lysosomes in hematopoietic cells. Mol Biol Cell 2005 Feb;16(2):731-41.

[149] Wood SM, Meeths M, Chiang SC, et al. Different NK cell-activating receptors preferentially recruit Rab27a or Munc13-4 to perforin-containing granules for cytotoxicity. Blood 2009 Nov 5;114(19):4117-27.

[150] Menager MM, Menasche G, Romao M, et al. Secretory cytotoxic granule maturation and exocytosis require the effector protein hMunc13-4. Nat Immunol 2007 Mar;8(3):257-67.

[151] Brzezinska AA, Johnson JL, Munafo DB, et al. The Rab27a effectors JFC1/Slp1 and Munc13-4 regulate exocytosis of neutrophil granules. Traffic 2008 Dec;9(12):2151-64.

[152] Pivot-Pajot C, Varoqueaux F, de Saint Basile G, Bourgoin SG. Munc13-4 regulates granule secretion in human neutrophils. J Immunol 2008 May 15;180(10):6786-97.

[153] Dulubova I, Lou X, Lu J, et al. A Munc13/RIM/Rab3 tripartite complex: from priming to plasticity? EMBO J 2005 Aug 17;24(16):2839-50.

[154] Betz A, Thakur P, Junge HJ, et al. Functional interaction of the active zone proteins Munc13-1 and RIM1 in synaptic vesicle priming. Neuron 2001;30(1):183-96.

[155] Wadel K, Neher E, Sakaba T. The coupling between synaptic vesicles and Ca2+ channels determines fast neurotransmitter release. Neuron 2007 Feb 15;53(4):563-75.

[156] Coleman WL, Bykhovskaia M. Rab3a-mediated vesicle recruitment regulates short-term plasticity at the mouse diaphragm synapse. Mol Cell Neurosci 2009 Jun;41(2):286-96.

[157] Mittelstaedt T, Alvarez-Baron E, Schoch S. RIM proteins and their role in synapse function. Biol Chem 2010 Jun;391(6):599-606.

[158] Wang Y, Okamoto M, Schmitz F, Hofmann K, Südhof TC. Rim Is a Putative Rab3 Effector in Regulating Synaptic-Vesicle Fusion. Nature 1997;388(6642):593-8.

[159] Wang Y, Sugita S, Sudhof TC. The RIM/NIM family of neuronal C2 domain proteins. Interactions with Rab3 and a new class of Src homology 3 domain proteins. J Biol Chem 2000 Jun 30;275(26):20033-44.

[160] Takao-Rikitsu E, Mochida S, Inoue E, *et al.* Physical and functional interaction of the active zone proteins, CAST, RIM1, and Bassoon, in neurotransmitter release. J Cell Biol 2004 Jan 19;164(2):301-11.

[161] Lu J, Li H, Wang Y, Sudhof TC, Rizo J. Solution structure of the RIM1alpha PDZ domain in complex with an ELKS1b C-terminal peptide. J Mol Biol 2005 Sep 16;352(2):455-66.

[162] Betz A, Thakur P, Junge HJ, *et al.* Functional interaction of the active zone proteins Munc13-1 and RIM1 in synaptic vesicle priming. Neuron 2001;30(1):183-96.

[163] Dulubova I, Lou X, Lu J, *et al.* A Munc13/RIM/Rab3 tripartite complex: from priming to plasticity? EMBO J 2005 Aug 17;24(16):2839-50.

[164] Coppola T, Magnin-Luthi S, Perret-Menoud V, Gattesco S, Schiavo G, Regazzi R. Direct interaction of the Rab3 effector RIM with Ca2+ channels, SNAP-25, and synaptotagmin. J Biol Chem 2001 Aug 31;276(35):32756-62.

[165] Dai H, Tomchick DR, Garcia J, Sudhof TC, Machius M, Rizo J. Crystal structure of the RIM2 C2A-domain at 1.4 A resolution. Biochemistry 2005 Oct 18;44(41):13533-42.

[166] Schoch S, Castillo PE, Jo T, *et al.* RIM1alpha forms a protein scaffold for regulating neurotransmitter release at the active zone. Nature 2002 Jan 17;415(6869):321-6.

[167] Yao I, Takagi H, Ageta H, *et al.* SCRAPPER-dependent ubiquitination of active zone protein RIM1 regulates synaptic vesicle release. Cell 2007 Sep 7;130(5):943-57.

[168] Kiyonaka S, Wakamori M, Miki T, *et al.* RIM1 confers sustained activity and neurotransmitter vesicle anchoring to presynaptic Ca2+ channels. Nat Neurosci 2007 Jun;10(6):691-701.

[169] Hibino H, Pironkova R, Onwumere O, Vologodskaia M, Hudspeth AJ, Lesage F. RIM binding proteins (RBPs) couple Rab3-interacting molecules (RIMs) to voltage-gated Ca(2+) channels. Neuron 2002 Apr 25;34(3):411-23.

[170] Koushika SP, Richmond JE, Hadwiger G, Weimer RM, Jorgensen EM, Nonet ML. A post-docking role for active zone protein Rim. Nat Neurosci 2001 Oct;4(10):997-1005.

[171] Weimer RM, Gracheva EO, Meyrignac O, Miller KG, Richmond JE, Bessereau JL. UNC-13 and UNC-10/rim localize synaptic vesicles to specific membrane domains. J Neurosci 2006 Aug 2;26(31):8040-7.

[172] Castillo PE, Schoch S, Schmitz F, Sudhof TC, Malenka RC. RIM1alpha is required for presynaptic long-term potentiation. Nature 2002 Jan 17;415(6869):327-30.

[173] Schoch S, Mittelstaedt T, Kaeser PS, *et al.* Redundant functions of RIM1alpha and RIM2alpha in Ca(2+)-triggered neurotransmitter release. EMBO J 2006 Dec 13;25(24):5852-63.

[174] Touchot N, Chardin P, Tavitian A. Four additional members of the ras gene superfamily isolated by an oligonucleotide strategy: molecular cloning of YPT-related cDNAs from a rat brain library. Proc Natl Acad Sci USA 1987 Dec;84(23):8210-4.

[175] Pfeffer S. Membrane domains in the secretory and endocytic pathways. Cell 2003 Feb 21;112(4):507-17.

[176] Sakisaka T, Meerlo T, Matteson J, Plutner H, Balch WE. Rab-alphaGDI activity is regulated by a Hsp90 chaperone complex. EMBO J 2002 Nov 15;21(22):6125-35.

[177] Bianchi V, Farisello P, Baldelli P, *et al.* Cognitive impairment in Gdi1-deficient mice is associated with altered synaptic vesicle pools and short-term synaptic plasticity, and can be corrected by appropriate learning training. Hum Mol Genet 2009 Jan 1;18(1):105-17.

[178] Park JB, Farnsworth CC, Glomset JA. Ca2+/calmodulin causes Rab3A to dissociate from synaptic membranes. J Biol Chem 1997;272(33):20857-65.

[179] Clabecq A, Henry JP, Darchen F. Biochemical characterization of Rab3-GTPase-activating protein reveals a mechanism similar to that of Ras-GAP. J Biol Chem 2000 Oct 13;275(41):31786-91.

[180] Itoh T, Fukuda M. Identification of EPI64 as a GTPase-activating protein specific for Rab27A. J Biol Chem 2006 Oct 20;281(42):31823-31.

[181] Figueiredo AC, Wasmeier C, Tarafder AK, Ramalho JS, Baron RA, Seabra MC. Rab3GEP is the non-redundant guanine nucleotide exchange factor for Rab27a in melanocytes. J Biol Chem 2008 Aug 22;283(34):23209-16.

[182] Yoshimura S, Gerondopoulos A, Linford A, Rigden DJ, Barr FA. Family-wide characterization of the DENN domain Rab GDP-GTP exchange factors. J Cell Biol Oct 18;191(2):367-81.

[183] Fischer von Mollard G, Südhof TC, Jahn R. A small GTP-binding protein dissociates from synaptic vesicles during exocytosis. Nature 1991;349(6304):79-81.

[184] Kondo H, Shirakawa R, Higashi T, *et al.* Constitutive GDP/GTP exchange and secretion-dependent GTP hydrolysis activity for Rab27 in platelets. J Biol Chem 2006 Sep 29;281(39):28657-65.

[185] Fukui K, Sasaki T, Imazumi K, Matsuura Y, Nakanishi H, Takai Y. Isolation and characterization of a GTPase activating protein specific for the Rab3 subfamily of small G proteins. J Biol Chem 1997;272(8):4655-8.

[186] Nagano F, Sasaki T, Fukui K, Asakura T, Imazumi K, Takai Y. Molecular cloning and characterization of the noncatalytic subunit of the Rab3 subfamily-specific GTPase-activating protein. J Biol Chem 1998;273(38):24781-5.

[187] Aligianis IA, Johnson CA, Gissen P, *et al*. Mutations of the catalytic subunit of RAB3GAP cause Warburg Micro syndrome. Nat Genet 2005 Mar;37(3):221-3.

[188] Sakane A, Manabe S, Ishizaki H, *et al*. Rab3 GTPase-activating protein regulates synaptic transmission and plasticity through the inactivation of Rab3. Proc Natl Acad Sci USA 2006 Jun 27;103(26):10029-34.

[189] Niwa S, Tanaka Y, Hirokawa N. KIF1Bbeta- and KIF1A-mediated axonal transport of presynaptic regulator Rab3 occurs in a GTP-dependent manner through DENN/MADD. Nat Cell Biol 2008 Nov;10(11):1269-79.

[190] Miyoshi J, Takai Y. Dual role of DENN/MADD (Rab3GEP) in neurotransmission and neuroprotection. Trends Mol Med 2004 Oct;10(10):476-80.

[191] Izumi T. Physiological roles of Rab27 effectors in regulated exocytosis. Endocr J 2007 Dec;54(5):649-57.

[192] Fukuda M. Distinct Rab binding specificity of Rim1, Rim2, rabphilin, and Noc2. Identification of a critical determinant of Rab3A/Rab27A recognition by Rim2. J Biol Chem 2003 Apr 25;278(17):15373-80.

[193] Sun L, Bittner MA, Holz RW. Rab3a binding and secretion-enhancing domains in Rim1 are separate and unique. Studies in adrenal chromaffin cells. J Biol Chem 2001 Apr 20;276(16):12911-7.

[194] Holt O, Kanno E, Bossi G, *et al*. Slp1 and Slp2-a localize to the plasma membrane of CTL and contribute to secretion from the immunological synapse. Traffic 2008 Apr;9(4):446-57.

[195] Kuroda TS, Fukuda M. Rab27A-binding protein Slp2-a is required for peripheral melanosome distribution and elongated cell shape in melanocytes. Nat Cell Biol 2004 Dec;6(12):1195-203.

[196] Fukuda M. The C2A domain of synaptotagmin-like protein 3 (Slp3) is an atypical calcium-dependent phospholipid-binding machine: comparison with the C2A domain of synaptotagmin I. Biochem J 2002 Sep 1;366(Pt 2):681-7.

[197] Kuroda TS, Fukuda M, Ariga H, Mikoshiba K. Synaptotagmin-like protein 5: a novel Rab27A effector with C-terminal tandem C2 domains. Biochem Biophys Res Commun 2002 May 10;293(3):899-906.

[198] Fukuda M. The synaptotagmin-like protein (Slp) homology domain 1 of Slac2- a/melanophilin is a critical determinant of GTP-dependent, specific binding of Rab27A. J Biol Chem 2002;19:19.

[199] Kuroda TS, Fukuda M, Ariga H, Mikoshiba K. The Slp homology domain of synaptotagmin-like proteins 1-4 and Slac2 functions as a novel Rab27A binding domain. J Biol Chem 2002 Mar 15;277(11):9212-8.

[200] Goishi K, Mizuno K, Nakanishi H, Sasaki T. Involvement of Rab27 in antigen-induced histamine release from rat basophilic leukemia 2H3 cells. Biochem Biophys Res Commun 2004 Nov 5;324(1):294-301.

[201] Ubach J, Garcia J, Nittler MP, Südhof TC, Rizo J. Structure of the janus-faced C2B domain of rabphilin. Nat Cell Biol 1999;1(2):106-12.

[202] Yamaguchi T, Shirataki H, Kishida S, *et al*. Two functionally different domains of rabphilin-3A, Rab3A p25/smg p25A-binding and phospholipid- and Ca(2+)-binding domains. J Biol Chem 1993 Dec 25;268(36):27164-70.

[203] Ostermeier C, Brunger AT. Structural basis of Rab effector specificity: crystal structure of the small G protein Rab3A complexed with the effector domain of rabphilin- 3A. Cell 1999;96(3):363-74.

[204] Huet S, Fanget I, Jouannot O, *et al*. Myrip couples the capture of secretory granules by the actin-rich cell cortex and their attachment to the plasma membrane. J Neurosci 2012;32(7):2564-77.

CHAPTER 6

Functions of Rab27a in Melanocytes and Cytotoxic T Lymphocytes

John A. Hammer III[*] and Xufeng Wu

National Institutes of Health, USA

Abstract: Melanocytes and cytotoxic T lymphocytes are the two cell types where the functional importance of the Rab GTPases Rab27a was first convincingly shown. Indeed, Rab27a is a resident Rab in the signature organelles present in these two cell types - the melanosome in the melanocyte and the lytic granule in the T cell. For both cell types, loss of Rab27a results in a striking defect in the function of their signature organelle. In the case of the melanocyte, Rab27a in the melanosome membrane serves to connect the organelle to the actin cytoskeleton by binding to its effector protein melanophilin, which in turn binds to the actin-based motor protein myosin Va. This myosin-based interaction of the melanosome with the actin cytoskeleton in the periphery of the melanocyte serves, together with long-range, bidirectional, microtubule-dependent movement of the organelle, to drive the proper intracellular distribution of melanosomes required for normal pigmentation in mammals. Delineation of this mechanism was aided enormously by the positional cloning of the mouse coat color genes encoding Rab27a (*ashen*), melanophilin (*leaden*) and myosin Va (*dilute*). Interestingly, a related tripartite complex containing Rab27a, its effector protein MyRip/Slac2-c, and myosin VIIa connects melanosomes to the actin cytoskeleton in retinal pigmented epithelial cells. In the case of the T cell, Rab27a in the membrane of the lytic granule is required for the docking and fusion of the granule at the immunological synapse, which in turn is required for target cell killing. Intense efforts are underway to define the downstream effectors of Rab27a in T cells that mediate this essential function. These efforts have been aided to a very large extent by the identification of several genes mutated in a family of closely-related lymphocyte diseases referred to as Familial Hemophagocytic Lymphohistiocytosis. Those studies have identified Munc13-4, a member of a family of proteins involved in controlling SNARE pairing, as one key downstream effector of Rab27a in T cells. Other studies have suggested that the Rab27a effector proteins Slp1 and Slp2a may facilitate the docking of lytic granules *via* their C2 domains.

Keywords: Rab27a, Lymphocyte, Melanocyte, Melanosome, Lytic Granule.

1. Rab27A FUNCTION IN MAMMALIAN MELANOCYTES

1.1. The Tripartite Complex of Rab27a, Melanophilin, and Myosin Va Connects Melanosomes to the Actin Cytoskeleton in Skin Melanocytes

Melanocytes in mammalian skin are neural crest-derived cells that reside at the base of the epidermis and the bottom of the hair follicle [1, 2]. Their principle function is to produce pigment within a specialized lysosome called the melanosome, and to donate this pigment to keratinocytes, which comprise most of the hair and skin (the same organ, just arranged differently). Each epidermal melanocyte provides pigment to many keratinocytes by virtue of its extensive dendritic arbor, which allows it to contact upwards of 35 keratinocytes. Melanosomes form in the melanocyte's central cytoplasm, fill with pigment, and then translocate out the cell's dendrites, accumulating at dendritic tips. The transfer of pigment to keratinocytes is thought to occur at these tips, although the mechanism of transfer is unclear [3]. In mammals (as opposed to fish and amphibians), this intercellular transfer of pigment is required for visible pigmentation, as the dissemination of pigment within keratinocytes is required to make the animal visibly pigmented. The mechanism responsible for positioning melanosomes at the melanocyte's dendritic tips is thought to involve cooperation between long-range, bidirectional, microtubule-dependent transport, which is responsible for moving the organelles out the cell's dendritic extensions, and local myosin Va-based movement, which is responsible for capturing the organelles at actin-rich dendritic tips (the Cooperative Capture model) [4].

*Address correspondence to John A. Hammer III:** Laboratory of Cell Biology, National heart, Lung and Blood Institute, National Institutes of Health, Bethesda, MD 20824, USA; Tel: 301-496-8960; Fax: 301-402-1519; E-mail: hammer@nhlbi.nih.gov

The central role of Rab27a and its effector protein melanophilin/Slac2a in positioning melanosomes within mouse skin melanocytes by virtue of their ability as a complex to recruit myosin Va on to the surface of the melanosome has already been the subject of numerous reviews [5-11]. The basic mechanism, gleaned from studies performed in parallel in several labs [12-16], was propelled to a large extent by the positional cloning and sequencing of the three mouse coat color mutants *dilute* [17], *ashen* [18] and *leaden* [19], which encode the heavy chain of myosin Va, Rab27a, and melanophilin, respectively. Numerous approaches were used to elucidate the mechanism by which Rab27a governs melanosome distribution [12-15]. These include (i) the characterization of the common defect in the intracellular distribution of melanosomes in primary melanocytes cultured from these mutant mice (the organelles accumulate in the central cytoplasm rather than at the cell's dendritic tips), (ii) the complementation of this distribution defect in mutant melanocytes, (iii) the localization of myosin Va, Rab27a, and melanophilin on melanosomes, and (iv) the demonstration of the hierarchy of physical interactions between these three proteins. Put most simply, Rab27a in the melanosome membrane recruits melanophilin in a GTP-dependent fashion, and melanophilin then recruits myosin Va (Fig. **1**). Moreover, the interaction between melanophilin and myosin Va requires Exon F, a 27-residue exon that is inserted into the central stalk domain of the melanocyte-spliced isoform of myosin Va (but not, for example, into the brain-spliced isoform) [12]. The bridging of the indirect interaction between Rab27a and myosin Va by melanophilin appears to be obligatory *in vivo* (at least as regards melanosome distribution), despite the fact some class V myosins have been reported to interact directly with certain Rab GTPases *in vitro* (*e.g.,* myosin Vb and Rab11a) [20]. Importantly, the assembly of this tripartite complex of Rab27a, melanophilin and myosin Va on the melanosome surface enables the organelle to interact with the actin cytoskeleton (and, presumably, the subsequent myosin Va-dependent movement of the organelle along actin filaments- see below). Moreover, this interaction serves to drive the switching of melanosomes from the long-range, bidirectional, microtubule-dependent transport that moves the organelles to the ends of dendrites, but cannot on its own (*i.e.* in the absence of a functioning tripartite complex) drive their accumulation there (because of its inherent bi-directionality), to local movement on F-actin, thereby causing their accumulation at dendritic tips [4]. This mechanism is entirely consistent with the genetics of these three mouse coat color mutants. Specifically, they give identical phenotypes on a black background and their effects are not additive, as removal of any one of these three proteins totally abrogates the link between the melanosome and the actin cytoskeleton (Fig. **1**). Together, these results represented the first example at the molecular level of an organelle receptor for an actin-based motor. Moreover, they provided an example where the alternative splicing of a motor protein is used to specify one of its cargos. Finally, these studies expanded the functional repertoire of Rab GTPases to include the regulation of vesicle transport, in addition to its known roles in vesicle formation and the fusion of the vesicle at its target membrane [12-15].

While the above studies demonstrated that Rab27 and melanophilin are both required to recruit myosin Va onto the melanosome surface *in vivo*, they did not demonstrate that they are sufficient. Evidence that they are probably sufficient came subsequently when the tripartite complex was reconstituted *in vitro* from pure proteins and shown to move at ~1 um/sec along actin filaments [21, 22].This result, and the fact that myosin Va contains several design features that should make it a superb organelle motor *in vivo* (*e.g.,* highly processive, a large step size that spans the half helical repeat of F-actin, allowing the myosin to walk straight along the helical actin filament rather than having to spiral around it) [23], raises the question of whether myosin Va actually moves melanosomes *in vivo*. Such movement was in fact not central to the Cooperative Capture model described above. In this model, the minimal/essential role of myosin Va is simply to capture melanosomes at actin-rich dendritic tips following their delivery there from the central cytoplasm by the long range, bi-directional, microtubule-dependent transport that operates constitutively along the cell's dendrites [4]. That said, the same study demonstrated the existence of a class of melanosome movements that are myosin Va-dependent, microtubule-independent, relatively slow (~0.15 um/sec), and exhibit little directional persistence [4]. This last feature would be consistent with the fact that the organization of F-actin in the melanocyte's cortex, as in all vertebrate cell types, is largely isotropic. That said, filopodia have been argued to serve as a conduit for melanosome transfer to keratinocytes [24]. This is interesting, as filopodia represent one of the few environments within vertebrate cells where F-actin is uniformly polarized (its barbed ends are oriented uniformly towards the tip of the filopodia). This organization would allow myosin Va, a barbed end-directed motor, to in principle move melanosomes efficiently to the tip of filopodia. Slow, myosin V-dependent melanosome movements that exhibit little directional persistence have also been identified in fish and frog melanophores (although the melanosomes in these cells have not been shown to use the Rab27a/melanophilin "receptor" to recruit myosin V, this is generally assumed to be the case) [8, 25, 26]. Unlike the myosin Va-dependent movements of melanosomes in mouse melanocytes (4), these movements in frog melanophores appear capable of driving significant melanosome spreading in the cytoplasm over time [8, 27]. Normally, these movements cooperate with the hormonally-controlled, synchronous, kinesin-dependent, microtubule plus end-directed, centrifugal movement of melanosomes in order to create the even cytoplasmic distribution of the organelles required to darken the animal. Finally, using a variation of a technique that can pinpoint the position of fluorescence molecules with nanometer resolution (FIONA), melanosomes within the melanophore's cytoplasm were sometimes seen to move in ~35 nm increments, a value that corresponds to myosin Va's step size [28].

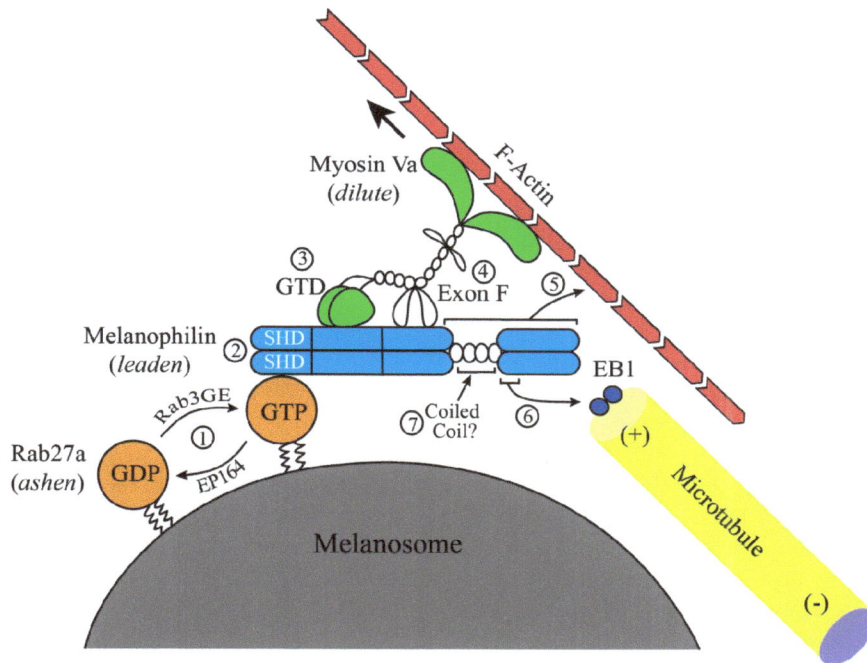

Figure 1: Organization and regulation of the melanosome receptor for myosin Va. Shown is the overall organization of the melanosome receptor for myosin Va. Put most simply, Rab27a attached to the melanosome membrane by virtue of its geranylgeranyl groups binds melanophilin in a GTP-dependent fashion, at which point melanophilin then binds myosin Va in an Exon-F-dependent fashion. This tripartite complex serves to connect the melanosome to the cortical actin cytoskeleton following its long range, microtubule-dependent transport to the distal end of the melanocyte's dendrites. This cooperative: capture mechanism of melanosome transport serves to drive the normal concentration of melanosomes at the tips of the melanocyte's dendrites, the site of the organelle's transfer to keratinocytes. Seven sites where this tripartite complex can be regulated or interact with additional components are highlighted. Site 1 high lights the regulation of the nucleotide state of Rab27a by a Rab27a-specific GEF and GAP, where only the GTP-bound form of Rab27a binds appreciably to melanophilin. Site 2 denotes the presence in melanophilin of the conserved helix-zinc finger-helix motif termed the SHD domain, which spans residues ~1-150 in melanophilin, and which binds specifically to Rab27a in melanophilin and in other SHD domain-containing Rab27a effector proteins. Site 3 refers to the low affinity binding site in melanophilin (residues ~150-240) that binds the globular tail domain (GTD) of myosin Va. The GTD is required but not sufficient for the physiological interaction between myosin Va and melanophilin. Site 4 marks the presence in myosin Va of the alternatively spliced, 27-residue exon, Exon F, which is present in the melanocyte-spliced isoform of myosin Va, which interacts with high affinity with residues ~320-410 in melanophilin, and which is required (but also not sufficient) for the physiological interaction between myosin Va and melanophilin. Site 5 high lights the presence in the C-terminal domain of melanophilin (residues ~400-590) of a binding site for F-actin. This interaction may facilitate melanosome capture in the actin-rich periphery and/or increase the processivity of myosin Va when it is bound to melanophilin. Site 6 marks the presence in melanophilin's C-terminal domain of two "SKIP" motifs that mediate melanophilin's interaction with the core microtubule plus end tracking protein EB1, through which melanophilin associates transiently with the plus end of growing microtubules. This interaction may influence the frequency of melanosome movement on microtubules and/or

the switching of melanosomes from microtubule-dependent movement to movement on F-actin. Finally, site 7 denotes the position in melanophilin of a putative coiled coil domain (residues 440-483), which is required for melanophilin to function in the receptor complex for myosin Va. Please see the main text for the references pertaining to the points made in this figure.

1.2. The Interaction of Melanophilin with F-Actin

In addition to the binding sites in melanophilin for Rab27a and myosin Va, which are essential for melanophilin to bridge the indirect interaction between the rab and the motor (see [12, 15, 29-34] for details as regards these binding sites, and [35] for why the Rab27a-binding "SHD" domain of melanophilin is Rab27a-specific), melanophilin also contains an apparent binding site for F-actin in its C-terminal ~190 residues [33] (Fig. **1**). The *in vitro* binding data supporting this interaction is not particularly rigorous, as no Kd, stochiometry, salt sensitivity, or fine mapping have been reported. That said, this C-terminal fragment, as well as full length melanophilin, clearly targets dramatically to F-actin when over expressed in cells, and it appears on high expression to increase the cell's F-actin content [33]. We do note, however, that endogenous melanophilin does not appear to co localize appreciably with cortical actin [12]. That said, cAMP may promote the interaction of melanophilin with F-actin *in vivo* [36]. Interestingly, the over expression in wild type melanocytes of melanophilin lacking its C-terminal actin binding domain results, presumably *via* a dominant negative mechanism, in a *dilute/ashen/leaden* phenotype (*i.e.* melanosomes accumulated in the central cytoplasm) [33]. Based on this observation, Kuroda *et al.* [33] argued that, like melanophilin's recruitment of myosin Va to the melanosome surface *via* its simultaneous interactions with Rab27a and the myosin, the direct interaction of melanophlin with cortical actin *via* its actin binding domain is essential to drive the peripheral concentration of melanosomes (note that this study also identified function blocking point mutations in full length melanophilin that abrogate its interactions with Rab27a, myosin Va and F-actin *in vivo*, which have been very valuable to the field). In striking contrast, Hume *et al.* [29] showed that the expression in cultured *leaden* melanocytes of a version of melanophilin containing a function blocking point mutation in its actin binding domain fully rescues the defect in melanosome distribution. This result argues strongly that melanophilin's C-terminal actin binding site is not essential for generating the normal peripheral concentration of melanosomes. Structure/function relationships obtained by rescue like this are in principle more convincing than such relationships generated by the over expression of dominant negative constructs, arguing that melanophilin's actin binding domain is indeed not essential for normal melanosome distribution. That said, roles for this C-terminal actin binding site in melanosome movement and positioning might become apparent when tested in melanocytes *in situ*. This is because the large increase in the length and complexity of the dendritic arbor seen for melanocytes in tissue relative to melanoctyes in culture should place a much higher demand on the machinery driving the cooperative capture mechanism. Moreover, by analogy with other motors like dynein, where the ability of accessory factors like the dynactin complex to interact with the microtubule serves to increase dynein's processivity [37] (the duration of its movement following a productive encounter with a microtubule), melanophilin's actin binding domain may serve to increase myosin Va's processivity, enhancing the actin-based movement of melanosomes. While this role may be difficult to discern in end-point assays of cultured *leaden* melanocytes rescued by melanophilin with and without its actin binding domain, it should be apparent using *in vitro* motility assays of the tripartite complex like those performed by Wu *et al.* [21].

1.3. Melanophiln and Microtubule Plus end Tracking

Melanophiln also marks dynamically the growing end of the microtubule in a process called plus end tracking [38]. Melanophilin accomplishes this by binding to the plus end tracking protein EB1 [38, 39] (Fig. **1**), which is expressed by all vertebrate cell types and is one of the so-called "core" plus end tracking proteins or +TIPs [40, 41]. Melanophilin can also recruit myosin Va on to the microtubule plus end in an Exon F-dependent process (at least in cells over expressing fluorescently-tagged versions of melanophilin and myosin Va) [38]. Interestingly, a microtubule plus end complex analogous to this EB1/melanophilin/myosin Va plus end complex is also present in yeast, where the EB1 homolog Bim1 recruits the bridging protein Kar9, which then recruits the yeast type V myosin Myo2p [42, 43]. This yeast plus end complex plays an important role in pre-anaphase spindle positioning. The role of the

EB1/melanophilin/myosin Va plus end complex in melanocytes is unclear, however. This complex has been proposed to play a role in focusing at the microtubule plus end the switching of melanosomes from microtubule-based transport to actomyosin Va-based transport [38]. The ability of a version of melanophilin that cannot interact with EB1 (by virtue of point mutations in the sequence SRYP, which corresponds to melanophilin's version of the consensus EB1-binding "SKIP" motif present in many proteins that plus end track by hitch hiking on EB1) to rescue peripheral melanosome distribution in cultured *leaden* melanocytes argues against this proposal, however [39]. That said, as mentioned above, tests of the functional significance of melanophilin-EB1 interaction under conditions where melanocytes are much more highly polarized than in culture might reveal a significant role for this interaction in melanosome transport and positioning.

1.4. Melanophilin's Coiled Coil Domain

In terms of other functional domains in melanophilin, in addition to its well-characterized, N-terminal, Rab27a-binding SHD domain, and its two apparent myosin Va binding domains (a low affinity interaction site for myosin Va's globular tail domain located in the ~90 residues that immediately follow the SHD domain, and a high affinity Exon-F binding site located between residues 320 and 410 [29, 32, 33], a short region of sequence predicted to form coiled coil (residues 440 to 483) has also been shown to be required for function using rescue of *leaden* melanocytes [29, 39] (Fig. **1**). This observation is consistent with the fact that many adaptors proteins for motor protein are constitutive or regulated dimmers [20]. Whether or not melanophilin as isolated is dimeric (or in a monomer: dimer equilibrium) has not been established, however. In summary, the three properties of melanophilin known to be required for it to function in melanosome positioning in cultured melanocytes are (i) its ability to bind to Rab27a (in a GTP-dependent fashion), (ii) its ability to bind to myosin Va (in an Exon-F dependent fashion), and (iii) the presence of an apparent coiled coil domain. Indeed, the argument has been made that the other two easily recognizable interactions exhibited by melanophilin- binding to F-actin and binding to EB1- are not required [39]. In other words, it has been argued that melanophilin's interactions with Rab27a and myosin Va are not only required but also *sufficient* for melanosome positioning. That said, further tests of melanophilin function under conditions where melanocytes are much more highly polarized than in culture (*e.g., in situ*) may reveal roles for binding to F-actin and EB1 in terms of the efficiency and/or extent of melanosome movement and positioning.

1.5. Regulation of the Assembly and Function of the Tripartite Complex of Rab27a, Melanophilin, and Myosin Va

The nature of the tripartite complex linking melanosomes to actin through myosin Va provides several sites for possible regulation of assembly/disassembly/efficiency, a design feature that is probably common to most motor protein receptors [20]. Three of these sites have been demonstrated experimentally: regulation of receptor avidity [44, 45], regulation of motor: adaptor interaction [46], and regulation of motor mechanochemistry by receptor engagement [47, 49]. First, because Rab27a binds melanophilin only in its GTP-bound conformation [12-15, 30], the extent to which melanophilin (and hence myosin Va) is recruited to the melanosome surface should be regulated by GEFs that push Rab27a to its active, GTP-bound state and GAPs that push it to its inactive, GDP-bound state (Fig. **1**). Consistent with this prediction, knockdown of Rab3GEP, a non-redundant GEF for Rab27a in melanocytes, causes the clustering of melanosomes in the perinuclear area and the destabilization of the Rab27a effector melanophilin, just as in Rab27a-null *ashen* melanocytes [44]. Similarly, over-expression of EP164, a GAP specific for Rab27a, causes the loss of Rab27a in the melanosome membrane (presumably due to the extraction of GDP-Rab27a by GDI) and melanosome clustering, and this effect requires EP164's orphan TBC (GAP) domain [45]. Second, the cell cycle-dependent phosphorylation of myosin V at a single site in its cargo-binding globular tail domain by calmodulin-dependent protein kinase II results in the release of the myosin from the melanosome surface in frog pigment cells [46] (this does not appear to happen in mouse skin melanocytes, however). Indeed, the modulation of cargo binding by phosphorylation of the motor's cargo binding domain may be a general mechanism for regulating motor: organelle interactions [20]. Finally, as regards the Rab27a/melanophilin-dependent regulation of myosin Va's mechanochemical properties, *in vitro* studies have shown that the myosin assumes a folded, enzymatically and mechanically inert 14S conformation in EGTA (that appears as a compact, triangular shaped molecule in electron micrographs), and an extended, enzymatically active 11S conformation when the free calcium concentrations is above ~1 uM [47, 48]. Importantly, while micro molar calcium unfolds the myosin and activates its actin-activated ATPase *in vitro*, it also severely compromises the myosin's mechanical properties by causing the dissociation of calmodulin light chains from the myosin's lever arm (causing this arm to loose rigidity). The more likely trigger *in vivo* is the presence of cargo, which would serve to push allosterically or kinetically trap sterically the myosin in its unfolded, active state. Direct evidence for such a trigger has been found *in vitro* using melanophilin as a myosin Va cargo [49]. Such cargo-dependent regulation of a motor's mechanochemical activity may be general phenomena, as it makes good biological sense. Specifically, it avoids energy waste, prevents cargo-free motors from piling up at the ends of cytoskeletal tracks, and promotes motor recycling by diffusion [20].

1.6. Rab27a, Slp2a and Melanosome Positioning

The most recent major paper on the role of Rab27a in melanosome movement and positioning in skin melanocytes focused on another Rab27a effector protein, the C2 domain-containing protein Slp2a [50]. Specifically, Kuroda and Fukuda [50] presented evidence that Rab27a recruits Slp2a following the Rab27a/melanophilin/myosin Va-dependent switching of melanosomes from long-range, bidirectional, microtubule-dependent movement along dendrites to capture and possible local movement on actin in the cortex. Moreover, they presented evidence that Slp2a, by virtue of its membrane-binding C2 domains, drives the final step in melanosome positioning by attaching the organelle to the plasma membrane [50]. Evidence for this last step, which is argued to be essential for generating the normal peripheral distribution of melanosomes, came largely from knockdown of Slp2a by RNAi in melan-a melanocytes. Knockdown yielded two phenotypes: (1) a dramatic change in cell shape, characterized by a rounded rather than dendritic morphology and the presence in most cells of a broad, actin-rich lamella completely surrounding the cell, and (2) the displacement of melanosomes away from the plasma membrane. This phenotype was referred to as "peripheral dilution", to contrast it with the perinuclear accumulation phenotype exhibited by melanocytes lacking melanophilin. Rescue of the first defect required that Slp2a have just functional C2 domains, while rescue of the second defect required that Sp2a also have the ability to interact with Rab27a. One concern we have regarding this study is that the defect in melanosome distribution seen in their Slp2a knockdown cells, which involves a relative large displacement of melanosomes away from the plasma membrane, also involves the formation of broad, actin-rich lamella between the displaced melanosomes in the central cytoplasm and the plasma membrane. Actin-rich lamella, which are also seen in a subset of normal melanocytes in culture (see Fig. 7 in [51]), as well as in wild type melanocytes over expressing melanophilin's actin binding domain [33]), invariably exclude most organelles, including all melanosomes. Perhaps it is this physical barrier, rather than the inability of melanosomes to bind to phosphatidylserine in the plasma membrane, that is the basis for the defect seen in melanosome positioning in Slp2a KD cells. We also note that the Slp2a knockout mice generated by this lab several years ago have not been reported to have a *dilute*-like coat color defect, although this could be due to compensation in the mouse [52]. Regardless, the fact that endogenous Slp2a co localizes with melanosomes and that it is by far the most abundant Rab27a effector in melanocytes [15, 50] suggest that it should play a significant role in some aspect of melanosome biology if not the one argued by Kuroda and Fukuda [50].

1.7. The Tripartite Complex of Rab27a, MyRip/Slac2-c, and Myosin VIIa Connects Melanosomes to the Actin Cytoskeleton in Retinal Pigmented Epithelial Cells

In a clear variation on theme, Rab27a also plays an essential role in determining the intracellular distribution of melanosomes in retinal pigmented epithelial (RPE) cells. Specifically, Rab27a on the surface of RPE melanosomes recruits the effector protein MyRip/Slac2-c, which in turn recruits the class VII unconventional myosin myosin VIIa, thereby connecting the organelle to the actin cytoskeleton [30, 53-58]. The existence of this tripartite complex on RPE melanosomes was revealed by (i) localization studies, (ii) protein: protein interaction assays, and (iii) the defect in melanosome distribution (perinuclear accumulation rather than concentrated apically) exhibited by RPE cells lacking either Rab27a (RPE from *ashen* mice) or myosin VIIa (RPE from *shaker* mice and humans with Usher's syndrome). Consistent with its role in bridging Rab27a to a myosin, MyRip/Slac2-c possesses the same overall domain organization as melanophilin/Slac2-a (they are ~70% similar in sequence). Moreover, like melanophilin, MyRip/Slac2-c binds to Rab27a through an N-terminal SHD domain and to myosin VIIa *via* sequences in its C-terminal half [30, 53, 56, 57]. MyRip/Slac2-c can also bind to myosin Va *in vitro* (preferring the melanocyte spliced isoform over the brain spliced isoform), although its interaction with myosin VIIa is ~10 fold stronger [30, 56]. It should be noted, however, that MyRip cannot rescue melanosome distribution in *leaden* skin melanocytes [56], suggesting that its low affinity interaction with myosin Va may not be physiologically relevant (although see [59] for a different take on this and other points as regards the interactions between MyRip/Slac2-c and both myosin Va and myosin VIIa). Conversely, melanophilin does not bind appreciably to myosin VIIa [30]. Evidence that the tripartite complex of Rab27a/MyRip/myosin VIIa actually assembles on the surface of melanosomes in RPE cells (beyond that provided by the defect in melanosome distribution observed in RPE cells from *ashen* and *shaker* mice) came initially from the heterologous, *in vivo* reconstitution of the complex in skin melanocytes [56]. Specifically, by complementing *leaden* (melanophilin null) skin melanoctyes with myosin VIIa and MyRip (both wild type and mutated versions), Kuroda and Fukuda [56] provided evidence for the formation of a functional Rab27a/MyRip/myosin VIIa complex on the surface of the melanosome (see also [59]). Subsequently, Lopes *et al.* [57].provided evidence that this tripartite complex regulates melanosome movement in RPE cells. Specifically, the frequency of fast, microtubule-dependent movements of melanosomes in primary mouse RPE cells was shown to increase when any one of the three players in the complex was missing (obtained by analyzing RPE cells from *ashen* and *shaker-1* mice, and following knockdown of MyRIP/Slac2-c in wild type cells). These observations, and the fact that similar results were obtained by treatment of cells with cytochalasin D, argue that the complex of Rab27a/MyRip/myosin VIIa on the surface of RPE melanosomes serves to connect the organelles to the actin cytoskeleton, which in turn tends to damp the bidirectional, microtubule-dependent movements of the organelle. Similar damping was reported previously in skin melanocytes by comparing the speed and persistence of microtubule-dependent melanosome movements in melanocyte's that contain or lack myosin Va [4].These and complimentary data published by Klomp *et al.* [58], who also analyzed melanosome movements in RPE cells from *ashen* and *shaker-1* mice, argue that the mechanism responsible for generating the correct apical distribution of melanosomes in RPE cells is similar to the Cooperative Capture mechanism responsible for accumulating melanosomes at the dendritic tips of skin mekanocytes [4].

1.8. Other Melanosome-Associated Rab GTPases that may Cooperate with Rab27a

Rab27a may not be the only Rab GTPase that influences the distribution of melanosomes in skin melanocytes. Specifically, Jordens *et al.* [60] have presented evidence that Rab7, a Rab GTPase that normally resides on late endosomes and lysosomes, and that is involved in the recruitment of the microtubule minus end-directed motor dynein (through a complex involving the Rab7 effector protein RILP, as well as other proteins [61]), also resides on melanosomes. Moreover, their data suggests that immature melanosomes possess more Rab7 than Rab27a, while the reverse is the case for mature melanosomes. Such a maturation dependent switch in Rab content in the melanosome membrane could serve to keep immature melanosomes in the central cytoplasm by strongly recruiting dynein to the organelles surface, and to push mature melanosomes to the actin-rich periphery by strongly recruiting myosin Va. Such a Rab-dependent switch in melanosme distribution would serve to ensure that only fully mature organelles are transferred to keratinocytes, which makes perfect biological sense. Finally, based on the use of an *in vitro* motility assay for melanosomes, as well as on the over expression of a dominant active variant, it has been argued that Rab8 also plays a role in the movement of melanosomes on F-actin in mouse melanocytes [22]. It should be noted, however, that a Rab8-dependent recruitment of melanosomes to F-actin is not apparent in melanocytes that lack Rab27a.

2. RAB27A FUNCTION IN CYTOTOXIC T LYMPHOCYTES

2.1. Rab27a is Required for Lytic Granule Docking and Secretion in Cytotoxic T Lymphocytes

Cytotoxic T lymphocytes (CTLs) kill target cells (*e.g.,* virally-infected cells, tumorigenic cells) *via* the polarized secretion of lytic granules, which are secretory lysosomes that harbor pore-forming proteins (perforin) and enzymes (granzymes) that trigger apoptosis in the target cell [62-65]. CD8$^+$ CTLs recognize target cells through their T cell receptor (TCR), which recognizes in a highly specific fashion a cognate peptide presented on the surface of the target cell *via* a MHC class 1 receptor. TCR engagement leads to a complex array of signaling reactions and protein: protein interactions that drive the process of polarized secretion. This process involves a dramatic rearrangement of membrane proteins in the portion of the T cell's plasma membrane that is in contact with the target cell, resulting in the formation of the immunological synapse (IS). The mature IS is characterized by a central accumulation of TCRs at the "cSMAC" and a peripheral accumulation of the T cell integrin LFA-1 at the "pSMAC", which forms a sealing gasket around the cSMAC *via* interaction with ICAM in the target cell plasma membrane [66]. During IS maturation, the T cell's interphase microtubule array undergoes a dramatic and rapid reorientation in which the centrosome, to which all of the cells microtubules are attached *via* there minus ends, is pulled very close to the T cell's plasma membrane at the IS. This reorientation of the centrosome is thought to be driven by the action of the minus end-directed microtubule motor dynein, which reels the centrosome over because it is anchored at the IS [67-71]. Centrosome repositioning is followed by the microtubule-dependent, minus end-directed movement of lytic granules, which results in their accumulation at the centrosome adjacent to the plasma membrane at the IS [72, 73]. This movement appears to be driven by the action of dynein on the surface of the granules [72, 73]. The final steps of polarized lytic granule secretion involve the docking of the granules to the plasma membrane at the cSMAC [74], followed by their calcium- and SNARE-dependent fusion with the plasma membrane, leading to release of the granule's contents into the cleft between the two cells (see [75] for how lytic granule components are thought to enter the target cell and trigger programmed cell death) [74]. By analogy with the SNARE-dependent fusion of synaptic vesicles in neurons [76], a priming step in between the docking and fusion steps, in which the lytic granule is "primed" for fusion, may also occur in T cells [74]. Natural killer (NK) cells appear to share many aspects of the cytolytic pathway exhibited by T cells. That said, as members of the innate immune system (as opposed to the adaptive immune system, of which CTLs are a part), NK cells do not recognize target cells *via* a TCR. Rather, NK cells use a set of germ line-encoded receptors that recognize generic signals such as the constant portion of IgGs [77].

The fact that Rab27a plays an essential role in the secretion of lytic granules by cytotoxic T cells (CTLs) (as well as by NK cells) was revealed by back to back papers published in 2001 by Stinthcombe *et al.* [78] and Haddad *et al.* [79]. Together, these papers showed that T cells isolated from *ashen* mice, which are homozygous for a functional null allele at the Rab27a locus, are normal up to the point of lytic granule secretion. Specifically, *ashen* T cells recognize target cells, exhibit robust TCR signaling, create a normal IS, reorient their centrosome to the IS with normal dynamics, and readily move their lytic granules to the vicinity of the IS. Where *ashen* T cells exhibit a dramatic (>95%) deficit is in the secretion of these lytic granules and subsequent target cell killing. Electron micrographs revealed that in the absence of Rab27a the lytic granules do not exhibit morphological docking. Specifically, while they move to the vicinity of the plasma membrane at the IS, they do not get very close to the membrane as in wild type T cells. Such "docking" is presumably a prerequisite for the subsequent SNARE-dependent exocytosis of the granules [74, 76]. Similar observations were subsequently made in T cells from humans with Griscellis Syndrome Type 2, who harbor function blocking mutations in Rab27a [80]. Together, these studies suggested that Rab27a in T cells is playing the traditional role assigned to Rab GTPases by driving the docking of lytic granules to the plasma membrane at the IS, and, perhaps, by also facilitating their subsequent, SNARE-dependent membrane fusion.

2.2. Munc13-4 is a Critical Downstream Effector of Rab27a in T Cells

The initial results from analyses of *ashen* T cells described above raised the next major question- what effector protein (s) is downstream of Rab27a in T lymphocytes? For example, is it an asymmetric tethering molecule like those seen for other Rab GTPases. Clearly it is not the duos seen in pigment cells (melanophilin/myosin Va in skin melanocytes and MyRip/myosin VIIa in RPE cells), as T cells that lack myosin Va (*i.e.* T cells isolated from *dilute* mice [79, 81] and humans with Griscellis Syndrome Type 1 [82]), melanophilin (*i.e.* T cells from *leaden* mice and humans with Griscellis Syndrome Type 3 [83]), or myosin VIIa (*i.e.* T cells from *shaker-1* mice [81]) are normal. The breakthrough came from the positional cloning of the locus for FHL3, one of a family of related immunological diseases collectively known as Familial Hemophagocytic Lymphohistiocytosis (FHL), all of which are characterized by multisystemic inflammation and organ infiltration by CD8+ T cells and macrophages [84]. Previous studies had identified the pore-forming, lytic granule protein perforin as the protein mutated in FHL2, and the T-SNARE syntaxin-11 as the protein mutated FHL4 [85, 86]. Cloning of the FHL3 locus identified the C2 domain-containing protein Munc13-4 as the protein mutated in this disease. This was exciting, as Munc family proteins are known to play important roles in SNARE-dependent exocytosis in the nervous system [76]. Specifically, in neurons the Munc family member Munc18-1 is thought to prevent the formation of the trans SNARE complex that drives membrane fusion by locking the T SNARE syntaxin-1 in its closed conformation. Moreover, the Munc family member Munc13-1 is thought to drive a "priming" step by facilitating the dissociation of Munc18-1 from syntaxin-1, allowing the T SNARE to proceed in the fusion process. Importantly, Feldman *et al.* [84] showed that while T cells from FHL3 patients exhibit a profound defect in lytic granule secretion (and target cell killing), they do exhibit morphologically docking of their lytic granules at the IS. This observation clearly placed Munc13-4 function downstream from Rab27a-dependent docking. Importantly, two papers published at about the same time showed that (i) Rab27a interacts with Munc13-4 in a GTP-dependent fashion (although not *via* an SHD domain, as in most other Rab27a effector proteins) [87], and (ii) that this interaction appears to control the release of dense core granules from platelets and secretory lysosomes from mast cells [88]. Together, these results argued that Munc13-4 serves as a critical downstream effector of Rab27a in T cells following the Rab27a-dependent docking of the lytic granule at the IS, where it probably facilitates the SNARE-dependent fusion of the lytic granule with the plasma membrane by "priming" the granule for fusion. Consistently, GFP-tagged Munc13-4 was shown to localize to lytic granules at the IS in T cell-target cell conjugates (but only minimally to lytic granules in unengaged T cells) [84]. Importantly, these results did not rule out the existence of an intermediate step prior to the interaction between Rab27a and Munc13-4 that involves the recruitment by Rab27a of another effector protein that drives morphological docking by tethering the lytic granule to the plasma membrane.

2.3. The Formation of Mature, Fusion-Competent, Rab27a-Positive Lytic Granules may Involve the Fusion of Several Distinct Membrane Compartments

The picture up to this point, although not yet fully supported by experiments, was a fairly straightforward one in which Rab27a localized on the lytic granule facilitates the docking of the granule at the IS, perhaps *via* the recruitment of some kind of tethering effector protein. At this point, Rab27a would then recruit Munc13-4, whose role would be to facilitate the priming of the granule for subsequent SNARE-dependent fusion. Both of these processes argue that Rab27a should be present on lytic granules. Indeed, prior efforts to localize Rab27a in T cells at the light and EM levels provided evidence that it resides on a subset of lytic granules in unengaged and engaged T cells [79, 89] (although only [79] co-localized it with a bonafide granule marker, granzyme B). Similarly, Rab27a (and Munc13-4) were shown to localize on secretory lysosomes in mast cells [88]. This relatively simple, linear model for lytic granule assembly/function was shattered, however, by the study of Menager *et al.* [90] published in 2007. The first key observation they reported was that there is essentially no co localization of GFP-tagged Rab27a with core lytic granule markers like perforin in unengaged T cells. Using a host of tagged, expressed versions of Rab27a, Munc13-4, and markers for various membrane compartments, the authors then presented evidence for the existence of a much more complex process driving the formation of mature lytic granules (Fig. **2**). Specifically, the authors argued that Munc13-4 present on Rab11-positive recycling endosomes first fuses with a Rab27a-positive, late endosomal compartment (but not *via* Rab27a/Munc13-4 interaction) to create what they term exocytic vesicles. This step does not require T cell activation. Creation of the mature lytic granule, which they argue occurs (i) only in T cells that have engaged a target cell and become activated, (ii) only at the last moment before granule secretion, and (iii) only very close to the membrane at the IS, involves the fusion of these Rab27a/Munc13-4-positive exocytic vesicles with a third and final compartment, a secretory lysosome that contains all of the lytic granule mediators (*e.g.,* perforin, granzymes) but lacks both Rab27a and Munc13-4 (see also [91] for somewhat similar data in resting versus engaged NK cells). At this point in time, Munc13-4 on the mature lytic granule might then facilitate the granule priming step, presumably through interaction with Rab27a. The authors note that by separating the executors of cytotoxic function (perforin and granzymes) from the effectors of their regulated secretion (Rab27a and Munc13-4), unactivated T cells possess an important checkpoint for target cell killing. Nevertheless, as the authors themselves note, certain aspects of their model are quite unusual. For example, there are few if any examples in the literature of such a direct link between the recycling endosome pathway and the pathway for regulated secretion. Given this, and the overall complexity of the model for lytic granule maturation that they propose (which raises many issues, such as how the polarization of multiple compartments and their subsequent fusion are regulated) (Fig. **2**), we think it is important at some point to confirm their model by careful localization of the endogenous molecules (*e.g.,* Rab27a, Munc13-4). This is especially important given that their model is deduced largely from the over expression of GFP-tagged proteins, which, in principle, could be significantly mislocalized.

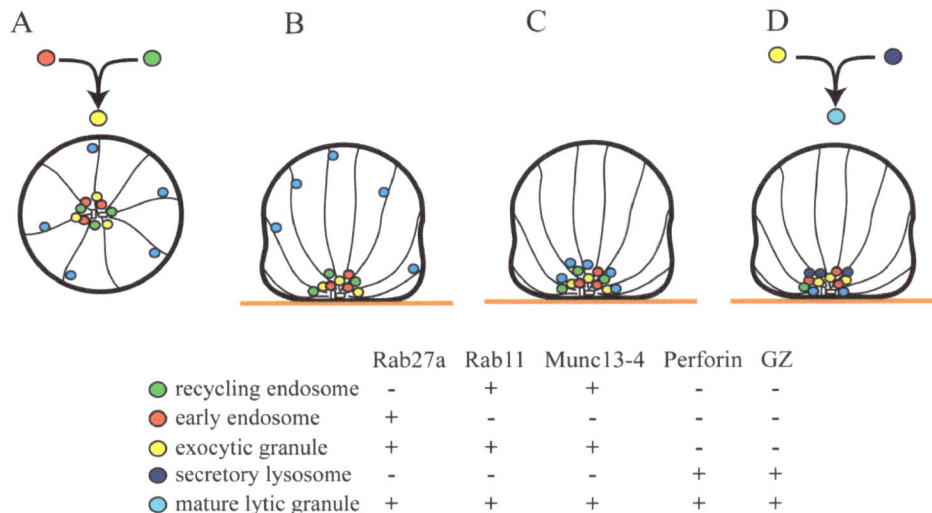

	Rab27a	Rab11	Munc13-4	Perforin	GZ
● recycling endosome	-	+	+	-	-
● early endosome	+	-	-	-	-
○ exocytic granule	+	+	+	-	-
● secretory lysosome	-	-	-	+	+
○ mature lytic granule	+	+	+	+	+

Figure 2: Pathway for maturation of lytic granules. Shown is the complex pathway for lytic granule maturation as proposed by Menager *et al.* [90]. These authors argue that Munc13-4- and Rab11-positive recycling endosomes (green) first fuse with a Rab27a-positive, late endosomal compartment (red) to create what they term exocytic vesicles (yellow) (this step does not require activation of the T cell by a target cell) (A). Engagement of a target cell (depicted here by the orange line) results in the repositioning of the microtubule cytoskeleton, such that the centrosome becomes closely opposed to the T cell membrane at the IS (B). Although not directly addressed by Menager *et al.* [90], our cartoon suggests that centrosome repositioning would serve to bring the three compartments described above near to the IS. Following centrosome repositioning, secretory lysosomes (dark blue) that contain all of the lytic granule mediators (*e.g.,* perforin, granzymes) but lack both Rab27a and Munc13-4 are moved, presumably by the microtubule minus end-directed motor dynein, to the minus ends of microtubules, which are anchored at the centrosome (C). Note, however, that in some cases (*e.g.,* NK cells) lytic granules may converge at the centrosome prior to it being repositioned adjacent to the IS. Creation of the mature lytic granule, which Menager *et al.* [90] argue occurs only at the last moment before granule secretion, and only very close to the plasma membrane at the IS, involves the fusion of the Rab27a-, Munc13-4-, and Rab11-positive exocytic vesicles (yellow) with the perforin- and granzyme-positive secretory lysosomes (dark blue) to create the final, mature lytic granule (teal) (D). Only at this moment is an organelle created that contains both the executors of cytotoxic function (perforin and granzymes) and the effectors of their regulated secretion (Rab27a and Munc13-4). Keeping these two functional classes of molecules separate until the last moment might provide T cells with an additional checkpoint for target cell killing. Please see the main text for the references pertaining to the points made in this figure.

2.4. The Rab27a Effector Proteins Slp1 and Slp2a may Facilitate the Docking of Lytic Granules *via* their C2 domains

As mentioned above, the information to date does not exclude the possibility that, prior to interaction with Munc13-4, Rab27a recruits some kind of tethering factor to mediate its role in docking lytic granules at the IS. Indeed, such an interaction would be in line with the biology of many other Rab GTPases. Two recent papers point to the known Rab27a effector proteins Slp1 and Slp2a as playing such a tethering role, although the acid test of this role has yet to come. In the first paper, Holt *et al.* [92] demonstrated that of the five known Slp family members (all of which are composed of an N-terminal Rab27a-binding SHD domain, a central linker domain, and two C-terminal C2 domains that can bind the plasma membrane), only Slp1 and Slp2a are expressed in T cells. While a GFP-tagged version of Slp2a localizes largely to the plasma membrane and exhibits no clear co localization with lytic granules in unengaged cells, it does appear to concentrate at the IS in T cell: target cell conjugates. This localization suggests that Slp2a might play a role there in the Rab27a-dependent docking of lytic granules *via* its ability to bind both Rab27a on the granule and the plasma membrane. Consistent with such an important functional link, Slp2a is dramatically destabilized in the absence of Rab27a [92]. Finally, while T cells from knock out mice lacking either Slp1 or Slp2a are normal as regards target cell killing (suggesting that they are functionally redundant), T cells over expressing Slp2a's SHD domain, which should act as a dominant negative for Rab27a's interaction with both Slp1 and Slp2a, do exhibit a significant inhibition in killing ability [92]. It is important to point out, however, that this dominant negative construct would almost certainly block the interaction of Rab27a with other Rab27a effectors, both known (Munc13-4) and unknown. This problem makes the interpretation of this dominant negative experiment problematic. Unfortunately, the critical test of Slp1/Slp2a function in T cells, which would come from the characterization of T cells isolated from Slp1/Slp2a double knockout mice, was not reported. That said, it should be noted that in at least one case, the deletion of a Rab27a effector protein involved in the tethering of a vesicle to its target membrane can give somewhat unexpected results. Specifically, while pancreatic B cells from Slp4/granulophilin knockout mice exhibit a significant reduction in the morphological docking of insulin granules at the plasma membrane, they exhibit a significant *increase* in the stimulus-invoked secretion of these granules [93].

In the second report on Slp2a, Menasche *et al.* [94] identified a novel splice variant of the protein (Slp2a-hem) using the pull-down of TAP-tagged Rab27a expressed in an NK cell line. Moreover, they provided evidence that (i) Slp2a-hem is the only splice version of Slp2a expressed in T cells and NK cells, (ii) that the protein is stabilized by the presence of Rab27a, and (iii) that Slp2a-hem-Rab27a interaction can be detected by immunoprecipitation prior to T cell engagement (arguing that they are associated constitutively). Consistent with this later observation, as well as with the results of Menager *et al.* [90], GFP-tagged Slp2a-hem was found to co localize extensively and in an SHD-dependent manner with Rab27a-positive, perforin-negative vesicles in unengaged T cells. Moreover, this vesicular targeting of Slp2a was abolished in T cells from Griscellis Syndrome Type 2 (Rab27a null) T cells. Upon target cell engagement, these Rab27a- and Slp2a-hem-positive vesicles were then seen to move to the IS along with perforin-positive, Rab27a-negative secretory lysosomes. Importantly, a subset of the Rab27a- and Slp2a-hem-positive structures that were in the immediate vicinity of the plasma membrane at the IS were found to stain for perforin. This observation suggests that this subset of Rab27a- and Slp2a-hem-positive structures had fused with the Munc13-4-positive recycling compartment to form exocytic vesicles (although Munc13-4 and Rab11 were not imaged), and then had fused with perforin-containing secretory lysosomes to form mature lytic granules. Over expression of the C2 domain of Slp2a-hem appeared to block the close apposition of the Rab27a-positive vesicles with the IS, arguing that Slp2a-hem serves to dock these vesicles *via* its ability to bridge them to the plasma membrane. A more convincing experiment, however, would have been to over express full length Slp2a-hem containing function-blocking point mutations in its C2 domains. The authors also performed essentially the same dominant negative experiment done by Holt *et al.* [92] (over expression of Slp2a-hem's SHD domain), and got the same result (partial inhibition of degranulation). This experiment is subject to the same criticisms mentioned above. Finally, the authors stated that the knockdown of Slp2a-hem did not impair degranulation, suggesting the presence of a redundant molecule in T cells (almost certainly Slp1). In summary, it seems likely that Slp2a-hem and Slp1 facilitate in an overlapping, Rab27a-dependent, and C2 domain-dependent manner the docking of Rab27a-positive vesicles at the IS, thereby supporting the subsequent fusion of these vesicles with secretory lysosomes to create mature lytic granules. That said, a "smoking gun" experiment (*e.g.*, the characterization of T cells from Slp2a/Slp1 double knockout mice) has yet to be done.

2.5. Syntaxin 11 and Syntaxin Binding Protein 2/Munc18-2 may Act in Concert with Rab27a in T Cells

Recent studies have identified two additional molecules involved in lytic granule formation/secretion that should be mentioned, as they probably act either in concert with or down steam from Rab27a (although not through direct interaction with the Rab). The first is the SNARE syntaxin 11 (Stx11), which is mutated in another form of FHL, FHL4 [95]. NK cells from FHL4 patients fail to degranulate upon target cell engagement, although oddly enough this granule release defect is partially suppressed by IL-2 stimulation

[95]. Similarly, wild type NKs in which Stx11 had been knocked down by shRNA exhibit decreased cytotoxicity against a variety of NK-sensitive target cell types. While T cells also express syntaxin 11, T cells from FHL4 patients do not exhibit a defect in lytic granule exocytosis [95-97]. At present, the membrane compartment within NKs/CTLs where syntaxin 11 resides and functions in membrane fusion (*e.g.,* vesicle compartments involved in the formation of lytic granules, the plasma membrane at the IS) is unclear. That said, the fact that secretion induced by phorbol esters and ionomycin is also reduced in NK cells in which Stx11 has been knocked down argues that this SNARE protein functions at a distal step in lytic granule release. The second molecule is syntaxin binding protein 2 (STXBP2; also known as Munc18-2), which is mutated in, and the causative agent of, yet another form of FHL, FHL5 [96]. NK cells from FHL5 patients exhibit impaired lytic granule secretion and target cell killing, and these defects can be rescued by ectopic expression of STXBP2. Interestingly, the stability of syntaxin-11 depends on the presence of STXBP2/Munc18-2, arguing that their interaction is physiologically important. This result, and the fact that FHL5 NK cells polarize their lytic granules properly, argues that STXBP2 is required for a late step in lytic granule secretion such as docking, priming and/or fusion. This step would presumably involve the interaction of STXBP2/Munc18-2 with syntaxin-11 [96]. For example, STXBP2/Munc18-2 in lymphocytes may serve the same function as Munc18-1 in neurons (locking the T SNARE in its closed conformation, and, later, facilitating the fusion reaction by clasping the 4-helix trans SNARE complex), just as Munc13-4 in lymphocytes may serve the same function as Munc13-1 in neurons (releasing the T SNARE from its closed conformation in the priming step) [74, 76]. In other words, the STXBP2/STX11 complex may serve the same functions in the docking, priming and fusion of lytic granules at the IS of cytotoxic cells that the STXBP1/STX1 complex serves for synaptic vesicles at the neuronal synapse [74]. That said, just as T cells from FHL4 (syntaxin-11 mutated) patients degranulate normally, T cells from FHL5 patients release their lytic granules normally [97]. As pointed out in [97], *in vitro* cytotoxic assays may not always reflect the *in vivo* functions of cytotoxic cells. Nevertheless, taken together with the previous studies regarding Rab27a and Munc13-4 in T cells, a fairly comprehensive understanding of the regulation of lytic granule docking, priming and fusion/exocytosis in NK and T cells appears in sight. Indeed, the numerous parallels that can already be drawn between the machineries regulating lytic granule secretion and synaptic vesicle exocytosis were the focus of a recent and comprehensive review [74].

3. CONCLUSIONS

Review of the history of Rab27a studies in melanocytes and T cells makes it extremely clear that the positional cloning of mouse coat color mutants and of genes mutated in patients with FHL have been enormously valuable in defining the mechanisms of action of Rab27a in these two cells types. These studies have provided several "firsts". For example, the studies of Rab27a function in melanocytes resulted in the first definition at the molecular level of an organelle receptor for an actin-based motor protein. Recent studies of Rab27a function in T cells have gotten us much closer to a mechanistic description of how Rab27a drives the docking and secretion of lytic granules. That said, much remains to be determined as regards Rab27a function in these two cells types. In the case of melanocytes, efforts to further define the role of Rab27a effectors other than melanophilin (*e.g.,* Slp2a) are needed. Moreover, a greater understanding of the regulation of Rab27a function in pigment cells could provide an important portal for developing skin treatments to prevent melanoma. In the case of T cells, while recent work on Rab27a and lytic granule secretion has moved the story significantly closer to an understanding that rivals our understanding of secretion in the nervous system, many "holes" in the T cell secretion story still exist. Moreover, the current, rather complex model of lytic granule maturation, which requires a high degree of temporal and spatial coordination in the movement and fusion of multiple membrane compartments, requires further examination.

CONFLICT OF INTEREST

There is no conflict of interest from any of the authors.

REFERENCES

[1] Yamaguchi Y, Hearing VJ. Physiological factors that regulate skin pigmentation. Biofactors 2009 Mar-Apr;35(2):193-9.

[2] Marks MS, Seabra MC. The melanosome: membrane dynamics in black and white. Nat Rev Mol Cell Biol 2001 Oct;2(10):738-48.

[3] Van Den Bossche K, Naeyaert JM, Lambert J. The quest for the mechanism of melanin transfer. Traffic 2006 Jul;7(7):769-78.

[4] Wu X, Bowers B, Rao K, Wei Q, Hammer JA, 3rd. Visualization of melanosome dynamics within wild-type and dilute melanocytes suggests a paradigm for myosin V function *In vivo*. J Cell Biol 1998 Dec 28;143(7):1899-918.

[5] Fukuda M. Versatile role of Rab27 in membrane trafficking: focus on the Rab27 effector families. J Biochem 2005 Jan;137(1):9-16.

[6] Seabra MC, Coudrier E. Rab GTPases and myosin motors in organelle motility. Traffic 2004 Jun;5(6):393-9.

[7] Seabra MC, Mules EH, Hume AN. Rab GTPases, intracellular traffic and disease. Trends Mol Med 2002 Jan;8(1):23-30.

[8] Nascimento AA, Roland JT, Gelfand VI. Pigment cells: a model for the study of organelle transport. Annu Rev Cell Dev Biol 2003;19:469-91.

[9] Hammer JA, 3rd, Wu XS. Rabs grab motors: defining the connections between Rab GTPases and motor proteins. Curr Opin Cell Biol 2002 Feb;14(1):69-75.

[10] Barral DC, Seabra MC. The melanosome as a model to study organelle motility in mammals. Pigment Cell Res 2004 Apr;17(2):111-8.

[11] Izumi T, Gomi H, Kasai K, Mizutani S, Torii S. The roles of Rab27 and its effectors in the regulated secretory pathways. Cell Struct Funct 2003 Oct;28(5):465-74.

[12] Wu XS, Rao K, Zhang H, *et al*. Identification of an organelle receptor for myosin-Va. Nat Cell Biol 2002 Apr;4(4):271-8.

[13] Provance DW, James TL, Mercer JA. Melanophilin, the product of the leaden locus, is required for targeting of myosin-Va to melanosomes. Traffic 2002 Feb;3(2):124-32.

[14] Nagashima K, Torii S, Yi Z, *et al*. Melanophilin directly links Rab27a and myosin Va through its distinct coiled-coil regions. FEBS Lett 2002 Apr 24;517(1-3):233-8.

[15] Strom M, Hume AN, Tarafder AK, Barkagianni E, Seabra MC. A family of Rab27-binding proteins. Melanophilin links Rab27a and myosin Va function in melanosome transport. J Biol Chem 2002 Jul 12;277(28):25423-30.

[16] Fukuda M, Kuroda TS, Mikoshiba K. Slac2-a/melanophilin, the missing link between Rab27 and myosin Va: implications of a tripartite protein complex for melanosome transport. J Biol Chem 2002 Apr 5;277(14):12432-6.

[17] Mercer JA, Seperack PK, Strobel MC, Copeland NG, Jenkins NA. Novel myosin heavy chain encoded by murine dilute coat colour locus. Nature 1991 Feb 21;349(6311):709-13.

[18] Wilson SM, Yip R, Swing DA, *et al*. A mutation in Rab27a causes the vesicle transport defects observed in ashen mice. Proc Natl Acad Sci U S A 2000 Jul 5;97(14):7933-8.

[19] Matesic LE, Yip R, Reuss AE, *et al*. Mutations in Mlph, encoding a member of the Rab effector family, cause the melanosome transport defects observed in leaden mice. Proc Natl Acad Sci U S A 2001 Aug 28;98(18):10238-43.

[20] Akhmanova A, Hammer JA, 3rd. Linking molecular motors to membrane cargo. Curr Opin Cell Biol 2010 Aug;22(4):479-87.

[21] Wu X, Sakamoto T, Zhang F, Sellers JR, Hammer JA, 3rd. *In vitro* reconstitution of a transport complex containing Rab27a, melanophilin and myosin Va. FEBS Lett 2006 Oct 30;580(25):5863-8.

[22] Chabrillat ML, Wilhelm C, Wasmeier C, Sviderskaya EV, Louvard D, Coudrier E. Rab8 regulates the actin-based movement of melanosomes. Mol Biol Cell 2005 Apr;16(4):1640-50.

[23] Sellers JR, Veigel C. Walking with myosin V. Curr Opin Cell Biol 2006 Feb;18(1):68-73.

[24] Scott G, Leopardi S, Printup S, Madden BC. Filopodia are conduits for melanosome transfer to keratinocytes. J Cell Sci 2002 Apr 1;115(Pt 7):1441-51.

[25] Rogers SL, Gelfand VI. Myosin cooperates with microtubule motors during organelle transport in melanophores. Curr Biol 1998 Jan 29;8(3):161-4.

[26] Rodionov VI, Hope AJ, Svitkina TM, Borisy GG. Functional coordination of microtubule-based and actin-based motility in melanophores. Curr Biol 1998 Jan 29;8(3):165-8.

[27] Rogers SL, Karcher RL, Roland JT, Minin AA, Steffen W, Gelfand VI. Regulation of melanosome movement in the cell cycle by reversible association with myosin V. J Cell Biol 1999 Sep 20;146(6):1265-76.

[28]   Kural C, Serpinskaya AS, Chou YH, Goldman RD, Gelfand VI, Selvin PR. Tracking melanosomes inside a cell to study molecular motors and their interaction. Proc Natl Acad Sci U S A 2007 Mar 27;104(13):5378-82.

[29]   Hume AN, Tarafder AK, Ramalho JS, Sviderskaya EV, Seabra MC. A coiled-coil domain of melanophilin is essential for Myosin Va recruitment and melanosome transport in melanocytes. Mol Biol Cell 2006 Nov;17(11):4720-35.

[30]   Fukuda M, Kuroda TS. Slac2-c (synaptotagmin-like protein homologue lacking C2 domains-c), a novel linker protein that interacts with Rab27, myosin Va/VIIa, and actin. J Biol Chem 2002 Nov 8;277(45):43096-103.

[31]   Fukuda M. Synaptotagmin-like protein (Slp) homology domain 1 of Slac2-a/melanophilin is a critical determinant of GTP-dependent specific binding to Rab27A. J Biol Chem 2002 Oct 18;277(42):40118-24.

[32]   Fukuda M, Kuroda TS. Missense mutations in the globular tail of myosin-Va in dilute mice partially impair binding of Slac2-a/melanophilin. J Cell Sci 2004 Feb 1;117(Pt 4):583-91.

[33]   Kuroda TS, Ariga H, Fukuda M. The actin-binding domain of Slac2-a/melanophilin is required for melanosome distribution in melanocytes. Mol Cell Biol 2003 Aug;23(15):5245-55.

[34]   Geething NC, Spudich JA. Identification of a minimal myosin Va binding site within an intrinsically unstructured domain of melanophilin. J Biol Chem 2007 Jul 20;282(29):21518-28.

[35]   Kukimoto-Niino M, Sakamoto A, Kanno E, *et al.* Structural basis for the exclusive specificity of Slac2-a/melanophilin for the Rab27 GTPases. Structure 2008 Oct 8;16(10):1478-90.

[36]   Passeron T, Bahadoran P, Bertolotto C, *et al.* Cyclic AMP promotes a peripheral distribution of melanosomes and stimulates melanophilin/Slac2-a and actin association. FASEB J 2004 Jun;18(9):989-91.

[37]   Kardon JR, Reck-Peterson SL, Vale RD. Regulation of the processivity and intracellular localization of Saccharomyces cerevisiae dynein by dynactin. Proc Natl Acad Sci U S A 2009 Apr 7;106(14):5669-74.

[38]   Wu XS, Tsan GL, Hammer JA, 3rd. Melanophilin and myosin Va track the microtubule plus end on EB1. J Cell Biol 2005 Oct 24;171(2):201-7.

[39]   Hume AN, Ushakov DS, Tarafder AK, Ferenczi MA, Seabra MC. Rab27a and MyoVa are the primary Mlph interactors regulating melanosome transport in melanocytes. J Cell Sci 2007 Sep 1;120(Pt 17): 3111-22.

[40]   Wu X, Xiang X, Hammer JA, 3rd. Motor proteins at the microtubule plus-end. Trends Cell Biol 2006 Mar;16(3):135-43.

[41]   Jiang K, Akhmanova A. Microtubule tip-interacting proteins: a view from both ends. Curr Opin Cell Biol 2011 Feb;23(1):94-101.

[42]   Beach DL, Thibodeaux J, Maddox P, Yeh E, Bloom K. The role of the proteins Kar9 and Myo2 in orienting the mitotic spindle of budding yeast. Curr Biol 2000 Nov 30;10(23):1497-506.

[43]   Yin H, Pruyne D, Huffaker TC, Bretscher A. Myosin V orientates the mitotic spindle in yeast. Nature 2000 Aug 31;406(6799):1013-5.

[44]   Figueiredo AC, Wasmeier C, Tarafder AK, Ramalho JS, Baron RA, Seabra MC. Rab3GEP is the non-redundant guanine nucleotide exchange factor for Rab27a in melanocytes. J Biol Chem 2008 Aug 22;283(34):23209-16.

[45]   Itoh T, Fukuda M. Identification of EPI64 as a GTPase-activating protein specific for Rab27A. J Biol Chem 2006 Oct 20;281(42):31823-31.

[46]   Karcher RL, Roland JT, Zappacosta F, *et al.* Cell cycle regulation of myosin-V by calcium/calmodulin-dependent protein kinase II. Science 2001 Aug 17;293(5533):1317-20.

[47]   Krementsov DN, Krementsova EB, Trybus KM. Myosin V: regulation by calcium, calmodulin, and the tail domain. J Cell Biol 2004 Mar 15;164(6):877-86.

[48]   Sellers JR, Thirumurugan K, Sakamoto T, Hammer JA, 3rd, Knight PJ. Calcium and cargoes as regulators of myosin 5a activity. Biochem Biophys Res Commun 2008 Apr 25;369(1):176-81.

[49]   Li XD, Ikebe R, Ikebe M. Activation of myosin Va function by melanophilin, a specific docking partner of myosin Va. J Biol Chem 2005 May 6;280(18):17815-22.

[50]   Kuroda TS, Fukuda M. Rab27A-binding protein Slp2-a is required for peripheral melanosome distribution and elongated cell shape in melanocytes. Nat Cell Biol 2004 Dec;6(12):1195-203.

[51]   Wu X, Bowers B, Wei Q, Kocher B, Hammer JA, 3rd. Myosin V associates with melanosomes in mouse melanocytes: evidence that myosin V is an organelle motor. J Cell Sci 1997 Apr;110 (Pt 7):847-59.

[52]   Saegusa C, Tanaka T, Tani S, Itohara S, Mikoshiba K, Fukuda M. Decreased basal mucus secretion by Slp2-a-deficient gastric surface mucous cells. Genes Cells 2006 Jun;11(6):623-31.

[53]   El-Amraoui A, Schonn JS, Kussel-Andermann P, *et al.* MyRIP, a novel Rab effector, enables myosin VIIa recruitment to retinal melanosomes. EMBO Rep 2002 May;3(5):463-70.

[54]   Futter CE, Ramalho JS, Jaissle GB, Seeliger MW, Seabra MC. The role of Rab27a in the regulation of melanosome distribution within retinal pigment epithelial cells. Mol Biol Cell 2004 May;15(5):2264-75.

[55]   Gibbs D, Azarian SM, Lillo C, *et al.* Role of myosin VIIa and Rab27a in the motility and localization of RPE melanosomes. J Cell Sci 2004 Dec 15;117(Pt 26):6473-83.

[56]   Kuroda TS, Fukuda M. Functional analysis of Slac2-c/MyRIP as a linker protein between melanosomes and myosin VIIa. J Biol Chem 2005 Jul 29;280(30):28015-22.

[57]   Lopes VS, Ramalho JS, Owen DM, *et al.* The ternary Rab27a-Myrip-Myosin VIIa complex regulates melanosome motility in the retinal pigment epithelium. Traffic 2007 May;8(5):486-99.

[58]   Klomp AE, Teofilo K, Legacki E, Williams DS. Analysis of the linkage of MYRIP and MYO7A to melanosomes by RAB27A in retinal pigment epithelial cells. Cell Motil Cytoskeleton 2007 Jun;64(6):474-87.

[59]   Ramalho JS, Lopes VS, Tarafder AK, Seabra MC, Hume AN. Myrip uses distinct domains in the cellular activation of myosin VA and myosin VIIA in melanosome transport. Pigment Cell Melanoma Res 2009 Aug;22(4):461-73.

[60]   Jordens I, Westbroek W, Marsman M, *et al.* Rab7 and Rab27a control two motor protein activities involved in melanosomal transport. Pigment Cell Res 2006 Oct;19(5):412-23.

[61]   Johansson M, Rocha N, Zwart W, *et al.* Activation of endosomal dynein motors by stepwise assembly of Rab7-RILP-p150Glued, ORP1L, and the receptor betaIII spectrin. J Cell Biol 2007 Feb 12;176(4):459-71.

[62]   Fooksman DR, Vardhana S, Vasiliver-Shamis G, *et al.* Functional anatomy of T cell activation and synapse formation. Annu Rev Immunol 2010 Mar;28:79-105.

[63]   Huse M, Quann EJ, Davis MM. Shouts, whispers and the kiss of death: directional secretion in T cells. Nat Immunol 2008 Oct;9(10):1105-11.

[64]   Stinchcombe JC, Griffiths GM. Secretory mechanisms in cell-mediated cytotoxicity. Annu Rev Cell Dev Biol 2007;23:495-517.

[65]   Martina JA, Wu XS, Catalfarmo M, Sakamoto T, Yi C, Hammer JA. Imaging of lytic granule exocytosis in CD8+ cytotoxic T lymphocytes revels a modified form of full fusion. Cell Immunol 2011;271(2):267-79.

[66]   Dustin ML, Long EO. Cytotoxic immunological synapses. Immunol Rev 2010 May;235(1):24-34.

[67]   Kuhn JR, Poenie M. Dynamic polarization of the microtubule cytoskeleton during CTL-mediated killing. Immunity 2002 Jan;16(1):111-21.

[68]   Stinchcombe JC, Majorovits E, Bossi G, Fuller S, Griffiths GM. Centrosome polarization delivers secretory granules to the immunological synapse. Nature 2006 Sep 28;443(7110):462-5.

[69]   Combs J, Kim SJ, Tan S, *et al.* Recruitment of dynein to the Jurkat immunological synapse. Proc Natl Acad Sci U S A 2006 Oct 3;103(40):14883-8.

[70]   Quann EJ, Merino E, Furuta T, Huse M. Localized diacylglycerol drives the polarization of the microtubule-organizing center in T cells. Nat Immunol 2009 Jun;10(6):627-35.

[71]   Martin-Cofreces NB, Robles-Valero J, Cabrero JR, *et al.* MTOC translocation modulates IS formation and controls sustained T cell signaling. J Cell Biol 2008 Sep 8;182(5):951-62.

[72]   Jenkins MR, Tsun A, Stinchcombe JC, Griffiths GM. The strength of T cell receptor signal controls the polarization of cytotoxic machinery to the immunological synapse. Immunity 2009 Oct 16;31(4):621-31.

[73]   Mentlik AN, Sanborn KB, Holzbaur EL, Orange JS. Rapid lytic granule convergence to the MTOC in natural killer cells is dependent on dynein but not cytolytic commitment. Mol Biol Cell 2010 Jul;21(13):2241-56.

[74]   de Saint Basile G, Menasche G, Fischer A. Molecular mechanisms of biogenesis and exocytosis of cytotoxic granules. Nat Rev Immunol 2010 Aug;10(8):568-79.

[75]   Kurschus FC, Jenne DE. Delivery and therapeutic potential of human granzyme B. Immunol Rev 2010 May;235(1):159-71.

[76]   Sudhof TC, Rothman JE. Membrane fusion: grappling with SNARE and SM proteins. Science 2009 Jan 23;323(5913):474-7.

[77]   Sanborn KB, Rak GD, Mentlik AN, Banerjee PP, Orange JS. Analysis of the NK cell immunological synapse. Methods Mol Biol 2010;612:127-48.

[78]   Stinchcombe JC, Barral DC, Mules EH, *et al.* Rab27a is required for regulated secretion in cytotoxic T lymphocytes. J Cell Biol 2001 Feb 19;152(4):825-34.

[79]   Haddad EK, Wu X, Hammer JA, 3rd, Henkart PA. Defective granule exocytosis in Rab27a-deficient lymphocytes from Ashen mice. J Cell Biol 2001 Feb 19;152(4):835-42.

[80]   Menasche G, Pastural E, Feldmann J, *et al.* Mutations in RAB27A cause Griscelli syndrome associated with haemophagocytic syndrome. Nat Genet 2000 Jun;25(2):173-6.

[81] Bossi G, Booth S, Clark R, *et al.* Normal lytic granule secretion by cytotoxic T lymphocytes deficient in BLOC-1, -2 and -3 and myosins Va, VIIa and XV. Traffic 2005 Mar;6(3):243-51.

[82] Pastural E, Barrat FJ, Dufourcq-Lagelouse R, *et al.* Griscelli disease maps to chromosome 15q21 and is associated with mutations in the myosin-Va gene. Nat Genet 1997 Jul;16(3):289-92.

[83] Menasche G, Ho CH, Sanal O, *et al.* Griscelli syndrome restricted to hypopigmentation results from a melanophilin defect (GS3) or a MYO5A F-exon deletion (GS1). J Clin Invest 2003 Aug;112(3):450-6.

[84] Feldmann J, Callebaut I, Raposo G, *et al.* Munc13-4 is essential for cytolytic granules fusion and is mutated in a form of familial hemophagocytic lymphohistiocytosis (FHL3). Cell 2003 Nov 14;115(4):461-73.

[85] Pachlopnik Schmid J, Cote M, Ménager MM, *et al.* Inherited defects in lymphocyte cytotoxic activity. Immunol Rev 2010 May;235(1):10-23.

[86] Hong W. Cytotoxic T lymphocyte exocytosis: bring on the SNAREs! Trends Cell Biol 2005 Dec;15(12):644-50.

[87] Shirakawa R, Higashi T, Tabuchi A, *et al.* Munc13-4 is a GTP-Rab27-binding protein regulating dense core granule secretion in platelets. J Biol Chem 2004 Mar 12;279(11):10730-7.

[88] Neeft M, Wieffer M, de Jong AS, *et al.* Munc13-4 is an effector of rab27a and controls secretion of lysosomes in hematopoietic cells. Mol Biol Cell 2005 Feb;16(2):731-41.

[89] Tolmachova T, Anders R, Stinchcombe J, *et al.* A general role for Rab27a in secretory cells. Mol Biol Cell 2004 Jan;15(1):332-44.

[90] Menager MM, Menasche G, Romao M, *et al.* Secretory cytotoxic granule maturation and exocytosis require the effector protein hMunc13-4. Nat Immunol 2007 Mar;8(3):257-67.

[91] Wood SM, Meeths M, Chiang SC, *et al.* Different NK cell-activating receptors preferentially recruit Rab27a or Munc13-4 to perforin-containing granules for cytotoxicity. Blood 2009 Nov 5;114(19):4117-27.

[92] Holt O, Kanno E, Bossi G, *et al.* Slp1 and Slp2-a localize to the plasma membrane of CTL and contribute to secretion from the immunological synapse. Traffic 2008 Apr;9(4):446-57.

[93] Gomi H, Mizutani S, Kasai K, Itohara S, Izumi T. Granuphilin molecularly docks insulin granules to the fusion machinery. J Cell Biol 2005 Oct 10;171(1):99-109.

[94] Menasche G, Menager MM, Lefebvre JM, *et al.* A newly identified isoform of Slp2a associates with Rab27a in cytotoxic T cells and participates to cytotoxic granule secretion. Blood 2008 Dec 15;112(13):5052-62.

[95] Bryceson YT, Rudd E, Zheng C, *et al.* Defective cytotoxic lymphocyte degranulation in syntaxin-11 deficient familial hemophagocytic lymphohistiocytosis 4 (FHL4) patients. Blood 2007 Sep 15;110(6):1906-15.

[96] Arneson LN, Brickshawana A, Segovis CM, Schoon RA, Dick CJ, Leibson PJ. Cutting edge: syntaxin 11 regulates lymphocyte-mediated secretion and cytotoxicity. J Immunol 2007 Sep 15;179(6):3397-401.

[97] Cote M, Menager MM, Burgess A, *et al.* Munc18-2 deficiency causes familial hemophagocytic lymphohistiocytosis type 5 and impairs cytotoxic granule exocytosis in patient NK cells. J Clin Invest 2009 Dec;119(12):3765-73.

CHAPTER 7

Early Endocytosis: Rab5, Rab21, and Rab22

Guangpu Li[*]

University of Oklahoma Health Sciences Center, USA

Abstract: Early steps of endocytosis involve budding and formation of endocytic vesicles from the plasma membrane, movement of the vesicles along cytoskeleton, and fusion with early endosomes. The Rab5 subfamily of GTPases (Rab5, Rab21, and Rab22) are localized to early endosomes and plasma membrane and play an important role in regulation of early endocytosis and signal transduction, *via* interactions with multiple effector proteins. Rab5, Rab21, and Rab22 share some common effectors, but differentially interact with other effectors and regulators, and exhibit distinct functions in early endocytosis. The interplay of the early endocytic Rabs is illustrated by the Rab22-Rab5 cascade in which Rab22-GTP recruits Rabex-5 (a Rab5 GEF) to early endosomes for activation of Rab5 and stimulation of early endosome fusion. This and other functions of Rab5, Rab21, and Rab22 in the formation and uncoating of endocytic vesicles, the vesicle movement on cytoskeleton, and the early endosome fusion are discussed in this review.

Keywords: Rab5, Rab21, Rab22, Endosome, Endocytosis.

1. INTRODUCTION

Endocytosis is an essential function of all eukaryotic cells for the uptake of extracellular nutrients, regulation of cell surface receptors and signal transduction, and maintenance of cell membrane homeostasis. Depending on how endocytic vesicles are formed, there are clathrin-mediated endocytosis [1], caveolin-mediated endocytosis [2], and clathrin/caveolin-independent endocytosis [3], with the latter being less well understood. Phagocytosis is a specialized form of endocytosis employed by professional immune cells such as macrophages to internalize and kill invading microorganisms and to clear apoptotic cells and cell debris in the organism [4]. Endocytosed materials are initially delivered to early endosomes *via* fusion between endocytic vesicles and early endosomes, and are then sorted for recycling or degradation in late endosomes and lysosomes.

Rab5 is localized to the cytoplasmic side of plasma membrane, endocytic vesicles, and early endosomes [5], and regulate the formation and uncoating of endocytic vesicles [6, 7] vesicle movement [8], and early endosome fusion [9-11]. Rab21 and Rab22 are also localized to early endosomes and regulate early endocytic trafficking and recycling [12-16]. These three Rabs show close phylogenetic relationship and belong to the Rab5 subfamily, which also includes the *trans*-Golgi network (TGN)-localized Rab31 (a.k.a. Rab22b) [17, 18]. Rab5 is the prototype of this Rab subfamily and appears first on the phylogenetic tree. Rab5 orthologs exist in all eukaryotes from yeast to man, likely through gene duplication. Rab21 begins to emerge in Amoebozoa while Rab22 appears in Metazoa and higher eukaryotes (Chapter 12), reflecting increasing complexity of regulation in the early endosomes.

Rab5 and Rab22 exhibit a similar active GTP-bound conformation in the switch regions that interact with effectors [19]. Indeed, Rab5, Rab21, and Rab22 share some of the regulators and effectors. However, they are not functionally redundant because each Rab also interacts with a set of specific effectors. The crosstalk of the three Rabs is mediated by Rabex-5, which is an effector of Rab22 but also serves as a guanine nucleotide exchange factor (GEF) for Rab5 and Rab21. Rab22-GTP recruits Rabex-5 to the early endosomes for activation of Rab5 and Rab21, establishing an early endocytic Rab cascade [20].

Rab5 is the best characterized among the early endosomal Rabs, partly due to the fact that Rab5 was discovered first. In addition, Rab5 is one of the most ancient Rabs in evolution and controls a housekeeping

---

*Address correspondence to Guangpu Li: Department of Biochemistry and Molecular Biology, University of Oklahoma Health Sciences Center, Oklahoma City, OK 73104, USA; Tel: 405-271-2227; Fax: 405-271-3910; E-mail: guangpu-li@ouhsc.edu

Guangpu Li and Nava Segev (Eds)

function (endocytosis) in all eukaryotes, which likely contributes to its rediscovery and investigation in various organisms. Indeed, there have been over 900 papers on Rab5 since its discovery and cloning in 1989 [21], in contrast to less than 40 papers on Rab21 and Rab22 combined (Chapter 12). In this review, I will discuss our current understanding of Rab5, Rab21 and Rab22, including their genes, structures, regulators such as GEFs and GTPase-activating proteins (GAPs), and effectors that mediate their functions in membrane budding, vesicle movement along the cytoskeleton, and membrane fusion.

2. RAB5

2.1. Rab5 Isoforms

Rab5 has three isoforms (Rab5a, Rab5b, and Rab5c) encoded by three genes [22]. They share over 80% amino acid sequence homology, with differences mainly in the N-terminal region and the C-terminal hypervariable region. All three isoforms are ubiquitously expressed, although their relative abundance in some tissues may vary [23].They appear functionally redundant in stimulation of endocytosis and early endosome fusion [22]. However, the activities of the isoforms may be regulated differently, as they show differential interactions with kinases (ERK1 and cdc2) [24], and a GEF (RIN1) [25]. In the mouse genome, Rab5a and Rab5b genes are mapped to chromosomes 17 and 2, respectively, while the Rab5c gene is suggested to be on chromosome 11 [26].

In yeast, there are also three Rab5 isoforms, namely Ypt51, Ypt52 and Ypt53 [27]. The amino acid sequence homology between the mammalian and the yeast Rab5 isoforms ranges from 41.3% to 48.9%, with the highest homology between Rab5b and Ypt51. However, it's unclear if one or more of the yeast isoforms may represent the ortholog(s) for Rab5a, Rab5b and Rab5c in higher eukaryotes. In the following sections, Rab5 refers to Rab5a unless indicated otherwise.

2.2. Rab5 Structure

Rab5 is among the earliest Rabs discovered and cloned, through screening cDNA libraries for genes containing the Ras-like GTP/GDP-binding domain [21]. Rab5 contains 215 amino acid residues, with a conserved GTP/GDP-binding domain and a C-terminal cysteine motif (-CCSN) [5] where the cysteines are isoprenylated for membrane attachment [28]. Mutational studies have identified critical residues involved in GTP binding and hydrolysis. The GTP/GDP-binding domain consists of four conserved motifs: GESAVGKSS^{34} (for interaction with the β- and γ- phosphates of GTP), WDTAGQ^{79}E (for interaction with the γ phosphate), N^{133}KAD (for interaction with the guanine base), and ETSA (for interaction with the guanine base). Rab5 mutants with Ser^{34} to Asn [29] and Asn^{133} to Ile [9, 30] changes show reduced affinity for GTP and GDP [10] but increased affinity for Rab5 GEFs [20, 31]. As a result, the S34N and N133I mutants are well known dominant negative Rab5 mutants that may sequester the Rab5 GEFs and block the activation and function of endogenous Rab5 in the cell [11, 30]. In contrast, a Gln^{79} to Leu mutation in the second motif [29, 30] and mutations at Ser^{29} in the first motif [33] can dramatically reduce the rate of GTP hydrolysis [10, 33]. The resulting Q79L and Ser^{29} mutants are locked in the active GTP-bound conformation and exhibit a dominant positive phenotype in stimulation of endocytosis and endosome fusion [29, 32, 34].

Crystal structures of Rab5 complexed with GDP, GTP analog and effectors have provided insights into the mechanisms of GTP/GDP binding, GTP hydrolysis, and GTP hydrolysis-induced conformational changes in Rab5. Crystal structures of Rab5 complexed with GppNHp (a nonhydrolyzable analog of GTP) [35-38] GDP [37, 38] and Rabaptin-5 (an effector) [38], have been determined. The structure of a transition state mimic of a Rab5 mutant (A30P) complexed with GDP and AlF$_3$ has also been determined [37]. Like other Ras-like small GTPases, the canonical structure of GTP/GDP-binding domain of Rab5 consists of a six-stranded β-sheet surrounded by five α helices, with interspersed loops. The three-dimensional structure of Rab5 defines the boundaries of the α helices and β strands as follows: α1, residues 33-42; α2, 79-90; α3, 105-121; α4, 145-154; α5, 170-179; β1, 17-26; β2, 55-64; β3, 67-76; β4, 95-100; β5, 127-133; β6, 158-161 [36]. In addition, a comparison of the GppNHp- and GDP-bound Rab5 structures reveals the boundaries of switch I and switch II, which undergo conformational changes during the GTP hydrolysis cycle, as residues 44-64 and 76-94 respectively [38]. Another form of GDP-bound structure appears a GTP hydrolysis intermediate that exhibits conformational changes in switch I but not in switch II [38].

The crystal structures of Rab5 explain the defective GTP binding or hydrolysis of the dominant negative and positive mutants (S34N, N133I, and Q79L). Ser[34] in the first conserved motif is located in the phosphate-binding loop, and is critical for coordination of a Mg^{2+} that stabilizes the β and γ phosphates of bound GTP [35, 37]. Asn[133] in the third conserved motif forms hydrogen bonds to the guanine base of GTP/GDP and is critical for GTP/GDP binding [35, 37]. Gln[79] in the second conserved motif is important for GTP hydrolysis because its side chain is involved in positioning a water molecule for in-line nucleophilic attack of the γ phosphate of GTP [35, 37]. In addition, the specificity for GTP as opposed to ATP and other nucleoside triphosphates is determined by the Asp[136] residue in the third conserved motif. The side chain of Asp[136] forms two hydrogen bonds with N1/N2 on the guanine group of GTP [35-37]. Indeed, the D136N mutant in which the Asp[136] residue is changed to Asn shows switched specificity for XTP (xanthosine triphosphates) [39].

2.3. Rab5 Regulators

2.3.1. GEFs

The Rab5 GTPase cycle is facilitated by GEFs and GAPs in the cell, to accelerate the slow intrinsic rates of GDP dissociation and GTP hydrolysis, respectively (Table 1). Rab5 GEFs all contain a so-called Vps9 domain, which was originally identified in yeast and shown to have GEF activity for Vps21/Ypt51 [40, 41] a Rab5 homolog in yeast. Vps9 domain-containing proteins have been identified in higher eukaryotes including *Caenorhabditis elegans* (*e.g.*, RME-6) [42], *Drosophila melanogaster* (*e.g.*, Sprint) [43] and mammals (*e.g.*, Rabex-5, RIN1, RIN2, RIN3, Alsin, ALS2CL, RME-6/GAPex-5) (Table 1). The Rab5 GEFs are large multi-domain proteins and interact with various upstream regulators. For example, RIN1 contains RA and SH2 domains for binding to Ras-GTP and activated/phosphorylated EGFR (epidermal growth factor receptor), which contributes to the recruitment of RIN1 to the plasma membrane and early endosomes for activation of Rab5 during EGF signal transduction [44, 45]. The large number of Rab5 GEFs might reflect various signal transduction pathways that regulate Rab5 activity through recruiting one of the GEFs. Rabex-5 is of particular interest, because it is the only Rab5 GEF that forms a complex with a Rab5 effector (Rabaptin-5) and can target directly to the Rab5 membrane domain *via* Rabaptin-5 binding to Rab5-GTP, leading to positive feedback activation of Rab5 on the membrane [46, 48].

Rabex-5 was first isolated as part of a complex with Rabaptin-5 from bovine brain extracts, and the complex exhibited GEF activity for Rab5 [46]. It is a multi-domain protein and biochemical and structural studies have indicated that its GEF activity resides in the middle region encompassing the helical bundle (HB) and Vps9 domains [49]. Rabex-5 contains 862 amino acid residues including N-terminal A20-like Zn^{2+} finger domain, UIM (ubiquitin interacting motif) [50], EET (early endosomal targeting) domain [31], HB, Vps9 domain, and CC (coiled-coil) domain near the C-terminus for binding Rabaptin-5 [31, 51]. The catalytic core consists of the tandem HB and Vps9 domains (residues 132-391) and its crystal structure indicates that the HB domain interacts with and stabilizes the Vps9 domain that contains six α-helical structures (αV1-6) [49]. A conserved surface including an invariant acidic Asp[313] residue near a hydrophobic groove between αV4 and αV6, opposite to the HB-interacting surface, is critical for the GEF activity towards Rab5. It interacts specifically with Rab5 as well as Rab21, another member of the Rab5 subfamily, in the switch I, switch II, and inter-switch regions [51]. Mutations at the conserved Phe[57] residue in switch I reduce the GEF activity of Rebex-5 for Rab5, consistent with functional studies of the dominant negative Rab5 mutants (Rab5:S34N and Rab5:N133I), which showed that secondary mutations at Phe[57] can abolish the inhibitory effect of Rab5:S34N and Rab5:N133I on endocytosis and early endosome fusion [11], presumably by relieving the sequestration of Rabex-5. A small side chain residue (Ala or Ser) N-terminal to Phe[57] in the switch I of Rab5 and Rab21 is critical for the specific recognition by Rabex-5 in a small hydrophobic pocket [49]. Other Rabs usually have an acidic residue at the corresponding position and thus are incompatible for interaction with Rabex-5. The mechanism of Rabex-5-facilitated GDP dissociation from Rab5 and Rab21 appears to be stabilization of the nucleotide-free form *via* disruption of the Mg^{2+}-binding site and electrostatic repulsion of the β-phosphate of GDP by Asp[313] of Rabex-5 [49, 51].

The Rabex-5 GEF activity is regulated by Rabaptin-5. The catalytic core (*i.e.* the tandem HB and Vps9 domains) is highly active in promoting GDP dissociation from Rab5 *in vitro*, with a k_{cat}/K_m of 2.3 x 10^4 $M^{-1}s^{-1}$ [49]. The full-length Rabex-5 shows little GEF activity *in vitro*, due to a block by the downstream CC domain, but Rabaptin-5 binding to the CC domain effectively relieves the block [31, 51]. In the cell, however, Rabex-5 can be active by directly targeting to early endosomes *via* binding to Rab22-GTP *via* the EET domain, leading to activation of Rab5 and enhanced fusion and enlargement of early endosomes in a Rab22-Rab5 cascade [20]. Elevated levels of Rab5-GTP in turn recruits Rabex-5/Rabaptin-5 complexes to the membrane, *via* Rabaptin-5 binding to Rab5-GTP, to convert more Rab5-GDP to Rab5-GTP in a positive feedback loop for establishment of functional Rab5 domains and stimulation of early endosome fusion during endocytosis [20]. Rabex-5 is essential for steady state levels of Rab5-GTP in the cell, and Rabex-5 knockout mice die early and develop severe skin inflammation [52], suggesting a non-redundant function *in vivo*. Rabex-5-deficient mast cells show enhanced IgE receptor-mediated degranulation and cytokine release due to the loss of Rabex-5 GEF activity for Rab5 [52].

The RIN proteins (RIN1, RIN2, and RIN3) comprise another family of Vps9 domain-containing Rab5 GEFs, which exhibit different tissue distributions [53, 55]. Their activity depends on stimulation by growth factors or hormones that bind to cell surface receptors, and regulates the endocytosis/degradation and/or intracellular signaling of specific ligand-receptor complexes [45]. RIN1 is highly expressed in the brain and is the best characterized of the three RIN proteins. RIN1 was originally discovered as a Ras effector that interferes with Ras function in cell proliferation and transformation [53]. It contains a N-terminal SH-2 domain that can bind to activated epidermal growth factor receptor (EGFR) [44], a Vps9 domain that can activate Rab5 [45], and a C-terminal RA domain that interacts with activated Ras-GTP [53]. During EGF-mediated signal transduction, the interactions of RIN1 with the activated EGFR and Ras may recruit RIN1 to the membrane for activation of Rab5, which in turn promotes endosome fusion and endocytic trafficking of the signaling complexes to late endosomes and lysosomes for degradation and down-regulation [45, 56-58].

RME-6 is a *Caenorhabditis elegans* protein containing a RasGAP domain at N-terminus and a Vps9 domain at the C-terminus [42]. It preferentially binds to GDP-bound Rab5 mutant in yeast two-hybrid and GST pull-down assays and is required for endocytosis in multiple tissues of *C. elegans* [42]. In contrast to Rabex-5, RME-6 does not co-localize with Rab5 in the early endosomes but is mainly localized in clathrin-coated pits *via* interaction with AP2. Indeed, its mammalian ortholog hRME-6 is shown to interact with AP2 and the interaction is important for its Rab5 GEF activity as well as for uncoating of clathrin-coated vesicles [7]. In addition, the mammalian ortholog (also called GAPex-5) may regulate the trafficking of EGFR [59] and Glut4 [60] in the cell.

Alsin is the 180 kDa protein product of ALS2 gene whose mutated forms are responsible for the juvenile recessive form of amyotrophic lateral sclerosis (ALS) [61, 62], which is a neurodegenerative disease losing the upper motor neuron functions [63]. Alsin contains an RCC1-like domain (RLD), a tandem DH and PH domain, a MORN (membrane occupation and recognition nexus) domain, and a Vps9 domain at the C-terminus. The Vps9 domain is shown to contain Rab5 GEF activity [64, 65] but requires the upstream MORN domain for targeting to early endosomes in the cell [65]. Importantly, ALS-associated mutations in Alsin abolish the Rab5 GEF activity, suggesting that Alsin-mediated Rab5 activation and enhanced endosome fusion/endocytosis is critical for normal neuron function. In support of this idea, mouse embryo fibroblasts from ALS2 knockout mice show impaired endocytosis of EGF and BDNF [66, 67]. However, a recent report shows that ALS2 knockout neurons actually exhibit increased Rab5-mediated endosome fusion and endocytosis/degradation of glutamate receptors [68].

2.3.2. GAPs

The catalytic activity of Rab GAPs resides in the so-called TBC (Tre-2, Bud2, Cdc16) domain, which utilizes "a dual finger" mechanism to accelerate GTP hydrolysis by Rab proteins [69].Two conserved motifs IxxDxxR and YxQ provide an Arg finger and a Gln finger, respectively, to interact with the β and γ phosphates of GTP to stabilize the transition state and accelerate the rate of GTP hydrolysis. The Gln finger also positions the catalytic water for in-line attack of the γ phosphate to facilitate GTP hydrolysis, substituting for the function of the conserved Gln residue in the second GTP/GDP-binding motif (WDxxGQE) of Rab proteins [69].

RabGAP5 is a well-documented Rab5 GAP, which contains a N-terminal TBC domain, a SH3 domain, and a C-terminal RUN domain [70]. It was discovered by yeast two-hybrid screening of TBC domain-containing proteins for interactions with a family of Rabs [70]. RabGAP5 specifically interacts with and accelerates GTP hydrolysis of Rab5. Overexpression of RabGAP5 in Hela cells reduces EEA1 localization on early endosomes and blocks the endocytic traffic of EGF to lysosomes, consistent with its function in lowering the Rab5-GTP level and decreasing Rab5 activity [70, 71].

RN-Tre is another TBC domain-containing protein shown to contain Rab5 GAP activity [72]. However, recent studies reveal that it is a more potent GAP for a different Rab, Rab43, in the Golgi complex, and indeed suppresses Rab43 function in intracellular trafficking of Shiga toxin to the Golgi complex in Hela cells [70, 71].

2.4. Rab5 Effectors and Biological Functions

Rab5 effectors by definition are proteins that interact with the active, GTP-bound form of Rab5, and determine the functions of Rab5. There are multiple Rab5 effectors (Table **1**), suggesting multiple Rab5 functions in regulation of early endocytosis and signal transduction. In a classical pull-down assay by GST-Rab5-GTP analog, 22 potential Rab5 effectors were identified in bovine brain cytosol [73] which suggested new effectors and confirmed previously isolated ones. Several Rab5 effectors have been well characterized, and their functions are described below.

Table 1: Regulators and effectors of Rab5, Rab21 and Rab22*.

	Rab5	Rab21	Rab22	References
Rabex-5	GEF	GEF	Effector	[20, 46, 49]
RIN1	GEF			[45]
RIN2	GEF			[98]
RIN3	GEF			[54]
Alsin/ALS2	GEF			[65]
RME-6/GAPex-5	GEF			[42]
RabGAP5	GAP			[70]
RN-Tre	GAP			[72]
Rabaptin-5	Effector	Effector		[46, 74]
APPL1	Effector	Effector		[86, 87]
Integrin	Effector	Effector		[93]
EEA1	Effector		Effector	[81, 99]
Rabenosyn-5	Effector		Effector	[82, 99]
PI 3-Kinases	Effector			[78]
Rabankyrin-5	Effector			[80]
KIF-16B	Effector			[79]
PI 4-phosphatases	Effector			[77]
PI 5-phosphatases	Effector			[77]

*The protein names and their functions as GEFs, GAPs, or effectors for the three early endosomal Rabs are indicated. Blank, not determined or no interaction.

2.4.1. Rabaptin-5 and Recruitment of Rabex-5 to Early Endosomes for Rab5 Activation

Rabaptin-5 was the first Rab5 effector identified, through yeast two-hybrid screening of a Hela cell cDNA library for proteins interacting with the GTP hydrolysis defective Rab5:Q79L mutant [74]. Although Rabaptin-5 was initially thought to be an effector for endosome fusion, based on an observation that overexpression of Rabaptin-5 can enlarge early endosomes like Rab5:Q79L [74], overexpression of Rabaptin-5 alone generally does not enlarge early endosomes in the cell and the function of Rabaptin-5 appears a binding partner of the Rab5 GEF, Rabex-5, for recruiting Rabex-5 to the membrane for Rab5 activation [46, 47].

Rabaptin-5 is a long helical molecule with 862 amino acid residues, which forms stable homodimers. The Rab5-binding domain (R5BD) is at the C-terminus encompassing residues 789-862 [38, 74]. Crystal structure of R5BD and Rab5-GTP analog complex reveals that the R5BD dimer adopts a coiled-coil conformation interacting with two Rab5 molecules in the switch regions as well as the inter-switch region, suggesting a mode of recruiting Rabaptin-5 to the early endosomes by Rab5-GTP [38]. Rabaptin-5 usually exists in the cytoplasm as a complex with Rabex-5 [46], through the interaction between residues 551-661 of Rabaptin-5 and the coiled-coil domain (residues 401-480) of Rabex-5 [31, 50, 51]. The Rabaptin-5/Rabex-5 complex can enhance early endosome fusion, because the complex targets to early endosomes *via* Rabaptin-5 binding to Rab5-GTP so that Rabex-5 can further activate Rab5 in a positive feedback loop [47, 48]. Rabaptin-5 contains multiple isoforms. In addition to Rabaptin-5α and Rabaptin-5β [75], there are a number of splicing isoforms with different tissue distribution but all contain the C-terminal R5BD domain [76].

2.4.2. Rab5 Function in Formation and Uncoating of Endocytic Vesicles

Rab5 has been shown to be important for the formation of clathrin-coated vesicles [6] as well as the subsequent uncoating process [7]. A cytosolic factor necessary for the assembly of clathrin-coated pits on the plasma membrane is identified to be the Rab5-GDI complex in an *in vitro* budding assay [6]. It does not affect the number of newly formed clathrin-coated pits but may facilitate sequestration of cargoes such as transferrin receptor into coated pits [6]. The Rab5-GDI complex is expected to deliver Rab5 to the plasma membrane in order to exert such a function. Once the clathrin-coated vesicles are formed, Rab5 activity is important for subsequent AP2 uncoating from the coated vesicles. In this case, the Rab5 GEF, hRME-6, binds to the α-adaptin ear to displace the m2 kinase AAK1, and activates Rab5 on the vesicles [7]. As a result, the m2 subunit of the AP2 complex is dephosphorylated and there is decrease in the level of PI [4, 5].P$_2$ on the membrane, leading to AP2 uncoating (7). PI 4-phosphatases and PI 5-phosphatases are known Rab5 effectors that may be recruited by Rab5-GTP to the vesicles to reduce PI [4, 5] P$_2$ [77].

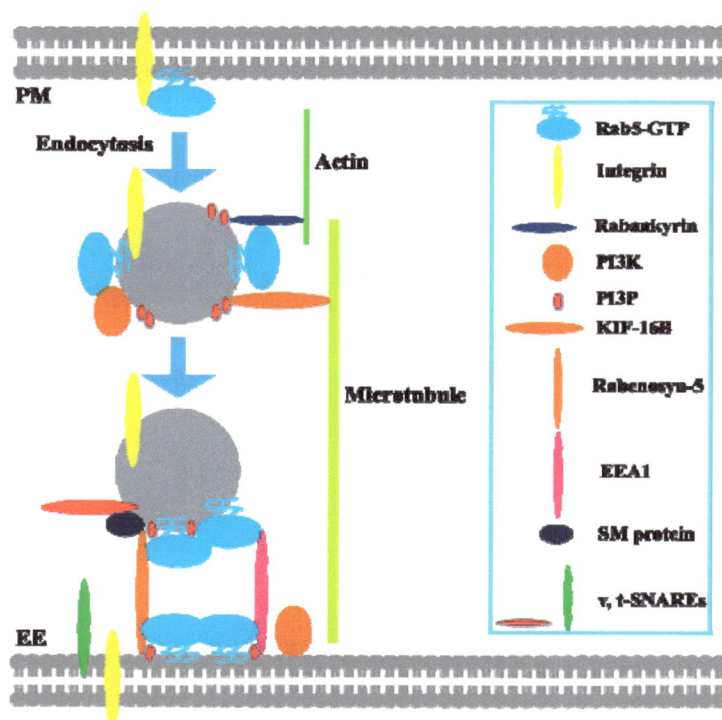

Figure 1: Rab5 functional domain and regulation of vesicle movement and early endosome fusion. Rab5-GTP molecules on endocytic vesicles and early endosomes temporally recruit a number of effectors to establish Rab5 functional domains on the membrane to facilitate vesicle movement along the cytoskeleton and tethering to early endosomes (EE). Integrin is a cargo protein that directly interacts with Rab5 and is likely endocytosed into Rab5-containing vesicles. One of the Rab5 effectors is hVps34, which is a PI 3-kinase (PI3K) and produces PI3P in the Rab5

membrane domain. The kinesin motor KIF-16B is recruited to the Rab5 membrane domain *via* binding to PI3P to facilitate vesicle movement along the microtubule, while Rabankyrin-5 is recruited by binding to both PI3P and Rab5-GTP to facilitate trafficking along the actin cytoskeleton. EEA1 and Rabenosyn-5 are two Rab5 effectors responsible for tethering the endocytic vesicles to early endosomes for membrane fusion. Rabenosyn-5 forms a complex with hVps45, a SM protein, to activate the SNAREs to promote the complex formation between v- and t- SNAREs, leading to membrane fusion.

2.4.3. Phosphotidylinositol-3-OH Kinases (PI 3-K) and Establishment of Functional Rab5 Domain

Both type I (p85α-p110β) and type III (hVps34) PI 3-Ks are Rab5 effectors [78]. Active, GTP-bound Rab5 recruits the two PI 3-Ks to early endosomes *via* binding to p110β and p150 (a binding partner of hVps34), respectively. Only hVps34 is essential for Rab5 function in endosome fusion, as specific antibodies against hVps34 but not p110β can inhibit endosome fusion *in vitro* [78]. The hVps34 produces PI(3)P on the membrane, which together with Rab5-GTP recruits additional Rab5 effectors to the membrane and establishes a functional Rab5 domain enriched with Rab5-GTP, PI3P, and Rab5 effectors for endosomal movement and fusion [78] (Fig. **1**).

2.4.4. Rab5 Function in Early Endosomal Movement and Macropinocytosis: Interactions with Cytoskeleton

The Rab5 domain on early endosomes can recruit effectors that interact with cytoskeleton and facilitate movement of early endosomes in the cell (Fig. **1**). One such effector is KIF16B, which is a plus end kinesin motor traveling on microtubules [79]. KIF16B contains a PX domain near the C-terminus that binds to PI(3)P on early endosomes. Furthermore, it is critical for the peripheral distribution and recycling of early endosomes in the cell, since knockdown of KIF16B by RNAi accumulated endocytosed transferrin in early endosomes and delayed its recycling back to the cell surface [79]. Another effector is Rabankyrin-5, which binds to PI(3)P on early endosomes *via* a FYVE domain and also binds to active, GTP-bound Rab5 [80]. An important characteristic of Rabankyrin-5 distinct from EEA1 and Rabenosyn-5 is the multiple ankyrin repeats in the protein, suggesting interactions with membrane proteins and linkage to actin cytoskeleton. Indeed, Rabankyrin-5 is also localized to macropinosomes and promotes macropinocytosis and it is shown to regulate apical fluid-phase endocytosis in epithelial cells [80]. Both processes are dependent on actin cytoskeleton.

2.4.5. Rab5 Function in Early Endosome Fusion

EEA1 [81] and Rabenosyn-5 [82] are two Rab5 effectors that contain both R5BD for binding to Rab5-GTP and FYVE domain for binding to PI(3)P and both interactions are important for efficient targeting to early endosomes (Fig. **1**). EEA1 is a well documented tethering factor that docks endocytic vesicles to early endosomes as well as clusters early endosomes together [73, 83]. EEA1 contains two R5BD domains: one at the N-terminus and one near the C-terminus upstream of the FYVE domain [81].While the C-terminal R5BD and FYVE domains are responsible for binding to early endosomes, the N-terminal R5BD is expected to seek out and bind to Rab5-GTP on fusion partners, including endocytic vesicles and early endosomes.

Rabenosyn-5 associates with hVps45, which is a Sec-1-like protein that is believed to facilitate membrane fusion by interacting with and activating the SNARE complex [82] (Fig. **1**). Recently, Rab5-regulated membrane fusion has been reconstituted in a liposome system, providing insight into the mechanism of early endosome fusion [84]. These experiments directly demonstrate the functional roles of Rabenosyn-5-hVps45 and EEA1 in promoting efficient and specific SNARE-mediated membrane fusion. Early endosomal SNAREs include three Q-SNAREs (syntaxin 13, VTI1a, and syntaxin 6) and one R-SNARE (VAMP4) [85]. Although proteoliposomes containing only the SNAREs can fuse, the efficiency is much lower. Inclusion of Rab5 and its effectors in the liposome system, at physiological concentrations, can achieve fusion efficiency similar to that of early endosome fusion [84]. Four Rab5 effectors are necessary for fusion. In addition to EEA1 and Rabenosyn-5-hVps45, the Rabaptin-5-Rabex-5 complex and the hVps34-p150 complex are important for production of Rab5-GTP and PI(3)P, respectively, which are essential for the recruitment of EEA1 and Rabenosyn-5-hVps45 to the membrane [84].

2.4.6. Rab5 Function in Signal Transduction

Endocytosis and receptor-mediated signal transduction are intertwined. Some signaling processes occur on early endosomes and Rab5 plays an important role in recruiting and activating the signaling molecules such as APPL proteins (*Adaptor* Protein containing *PH* domain, *PTB* domain, and *Leucine* zipper motif) [86]. APPL1 and APPL2 contain a BAR domain N-terminal to the PH domain and the BAR-PH tandem forms a crescent-shaped, symmetrical dimer for interaction with Rab5-GTP [87]. Growth factors such as EGF can stimulate Rab5-mediated recruitment of APPL to early endosomes, which promotes APPL activation and its subsequent translocation to the nucleus for chromosome remodeling and regulation of development [86]. APPL may compete with EEA1 for binding to Rab5 on the early endosomes, since Rab5- and APPL-positive endosomes lack EEA1 and represent a population of specialized endosomes for signal transduction [86]. The endosomal localization is critical for APPL to function as a signaling molecule in cell survival and development in zebrafish, *via* recruiting Akt to the endosomes and directs its specificity towards phosphorylation of GSK-3β rather than TSC2 [88].

Rab5 activation also induces changes in cell morphology and actin reorganization [34, 89-91]. In epithelial cells, Rab5 is important for phorbol ester-induced reassembly of actin stress fibers and focal adhesions [89]. In fibroblasts, activated Rab5 promotes actin reorganization and cell surface membrane ruffles [90, 91]. The Rab5 signaling pathway leading to actin cytoskeleton remodeling and circular ruffles for macropinocytosis appears to involve RN-Tre as an effector, which interacts with F-actin and actinin-4 to crosslink actin fibers at the plasma membrane [90].

3. RAB21

Rab21 is related to Rab5 with nearly 50% amino acid sequence identity, mostly in the nucleotide-binding domain, and belongs to the Rab5 subfamily [17]. Rab21 co-localizes with Rab5 in early endosomes and parallels Rab5 function in early endocytosis [13]. The phenotypes of dominant positive and negative mutants of Rab21 are similar to those of corresponding Rab5 mutants in the cell. While expression of Rab21:Q78L (the GTP hydrolysis defective, dominant positive mutant) in Hela cells leads to enlarged early endosomes, expression of Rab21:T33N (the GTP-binding defective, dominant negative mutant) inhibits the endocytosis of transferrin as well as EGF [13]. The data suggest a role for Rab21 in early endosome fusion and endocytosis, like Rab5. However, the data with the dominant negative mutant Rab21:T33N need to be interpreted with caution, since Rab21 and Rab5 are activated by the same GEF (Rabex-5) and the expected Rab21:T33N-mediated sequestration of Rabex-5 may block the functions of both Rab21 and Rab5. In addition to Rabex-5, another Vps9 domain-containing protein Varp is suggested to be a GEF for Rab21 as well as Rab5, based on the observation that Varp preferentially binds to the dominant negative mutants of Rab21 and Rab5 and promotes enlargement of endosomes [92]. Although the N-terminal region of Varp appears to stimulate GTP loading on Rab21 and to a lesser extent Rab5 *in vitro*, the nucleotide exchange activity is much lower than Rabex-5 [92], which itself exhibits two orders of magnitude lower activity than the core catalytic domain (*i.e.* HB-Vps9 domain) due to auto-inhibition by a downstream coiled-coil domain [31, 51] suggesting that the *in vitro* GEF activity of Varp towards Rab21 and Rab5, if any, is low.

In addition to the GEFs, many effector proteins also interact with both Rab21 and Rab5, such as Rabaptin-5, EEA1, and APPL1 [87], indicating overlapping functions of Rab21 and Rab5 in regulation of early endocytosis and signal transduction. One exception is Rabenosyn-5 that is specific for Rab5 and Rab22 but not Rab21 [19]. Because of the important role of Rabenosyn-5 in activation of SNAREs and early endosome fusion [84], the Rab21-EEA1 interaction may function only in tethering the endosomes, like Rab5, but may not directly participate in the downstream events, which depend on Rab5-Rabenosyn-5 interaction and lead to efficient and specific SNARE complex formation and membrane fusion.

Recent studies show that the active, GTP-bound Rab21 and Rab5 can interact with β1-integrins by binding to the cytoplasmic tails of a number of α subunits [93] (Fig. **1**). Expression of dominant positive and negative Rab21 mutants in a number of cell types significantly affects the endocytosis and recycling of integrins and consequently impairs integrin-mediated cell adhesion and cell motility [93], which has ramifications in cancer metastasis. Although the studies emphasize Rab21 function in these cellular processes, the dominant negative Rab21 mutant should also block Rab5 function, as discussed above, and thus both Rab21 and Rab5 may play a role in the remodeling of cell surface integrins and in regulation of cell adhesion and motility. A recent screening of compounds that block carcinoma-associated fibroblast-driven cancer invasion has identified Rab21 as a necessary factor in this process, possibly by remodeling cell surface integrins and extracellular matrix [94].

4. RAB22

Rab22 here refers to Rab22a and it is another member of the Rab5 subfamily localized on early endosomes and involved in early endocytosis [14-16, 20, 95]. Rab31 is sometimes called Rab22b because of high sequence homology and close phylogenetic relationship with Rab22 [17]. However, Rab31 is functionally distinct from Rab22. It is localized in TGN and regulates membrane traffic from TGN to the plasma membrane and/or endosomes [96, 97]. Because of the functional difference, Rab31 is a more appropriate name for this Rab than Rab22b to avoid confusion.

In contrast to Rab21, the function of Rab22 is clearly distinct from that of Rab5. First, Rab22 activity appears to promote membrane recycling from early endosomes back to the cell surface but not endocytic traffic to late endosomes and lysosomes. Knockdown of Rab22 by RNAi or dominant negative mutants can abrogate the recycling of endocytosed cargoes such as transferrin receptor (a clathrin-dependent process) [15] as well as MHC-I (a clathrin-independent process) [16]. Enhancement of Rab22 activity by overexpression shows inhibitory rather than stimulatory effect on EGF degradation in the cell [14]. Second, Rab22 appears to interact only with a subset of Rab5 effectors involved in membrane fusion such as EEA1 [14] and Rabenosyn-5 [19] but not other Rab5 effectors tested including Rabaptin-5 (Zhu and Li, unpublished data) and APPL1 [87]. The data suggest that both Rab22 and Rab5 may use EEA1 and Rabenosyn-5 to promote endosome fusion, but Rab22-mediated fusion may lead to membrane recycling while Rab5-mediated fusion may promote endocytic traffic to late endosomes and lysosomes. Finally, Rab22 and Rab5 are regulated differently. The Rab5 regulators such as GEF (Rabex-5) and GAP (RabGAP5) cannot act on Rab22 effectively.

In the case of Rabex-5, its nucleotide exchange activity towards Rab22 is two orders of magnitude lower than towards Rab5 and Rab21 [49]. Indeed, Rabex-5 does not function as a Rab22 GEF in the cell, instead it interacts more efficiently with active, GTP-bound Rab22 and is thus a Rab22 effector [20]. Activated Rab22 recruits Rabex-5 to the endosomal membrane for activation of Rab5, establishing a Rab22-Rab5 cascade to promote Rab5-mediated endosome fusion and endocytic traffic to late endosomes and lysosomes [20]. Rabex-5 may compete with EEA1 and Rabenosyn-5 for interaction with Rab22-GTP and consequently block Rab22-mediated endosome fusion, suggesting that Rabex-5 levels determine whether Rab5 or Rab22 pathway is dominant in different cell types.

5. CROSS-TALK AMONG RAB5, RAB21 AND RAB22

Early endosomes are sorting stations that sort endocytosed ligands and receptors to different intracellular destinations: the degradative late endosomes/lysosomes, various recycling compartments, or signaling endosomes. The relative activity of Rab5, Rab21, and Rab22 is likely to control the strength of each sorting pathway, with Rab5 and Rab21 promoting sorting to late endosomes/lysosomes and Rab22 promoting sorting to recycling and/or signaling endosomes (Fig. **2**). On the one hand, the three Rabs compete for some of the common effectors, *e.g.*, EEA1, to promote their respective trafficking pathways. On the other hand, they cooperate in a Rab22-Rab5/Rab21 cascade mediated by Rabex-5 (Fig. **2**).

Rabex-5 mediates the cross-talk among Rab5, Rab21 and Rab22, since Rabex-5 is not only an effector of Rab22 [20] but also a potent GEF for Rab5 and Rab21 [49].When Rab22 is activated and there are high levels of Rab22-GTP and Rabex-5 in the cell, Rabex-5 can be recruited to the endosomes by binding to Rab22-GTP on the endosomal membrane where Rabex-5 can function as a GEF for activation of Rab5 and Rab21, leading to temporal convergence of the three Rab membrane domains (Fig. **2**). Indeed, Rabex-5 is an essential linker for the co-localization of Rab22 and Rab5 on the same endosomes, since Rabex-5 knockout cells show separation of Rab22 and Rab5 into different populations of endosomes [20]. It remains to be established if there are functional differences between the endosomes containing the three converged Rab domains and the endosomes containing only Rab5 or Rab21. When Rabex-5 level is low and limiting, the Rab22-Rab5/Rab21 cascade cannot be established and Rab22 and Rab5 may separate into different membrane domains and compete for effectors to promote sorting to recycling compartments and late endosomes, respectively. The availability of Rabex-5 is also regulated by Rabaptin-5, which forms a cytosolic complex with Rabex-5 [46]. The higher Rabaptin-5 level is in the cell, the less Rabex-5 is available for the Rab22-Rab5/Rab21 cascade. Since high levels of Rabex-5/Rabaptin-5 complex and other Rab5 GEFs can activate Rab5 independent of the Rab22-Rab5/Rab21 activation cascade, it's possible that the Rab22-Rab5/Rab21 cascade mediated by Rabex-5 may specify a new sorting function for the endosomes, and an attractive possibility is for the biogenesis of signaling endosomes (Fig. **2**) that contain active ligand/receptor complexes and recruit signaling molecules to propagate the signal transduction processes for gene expression and cell growth and differentiation.

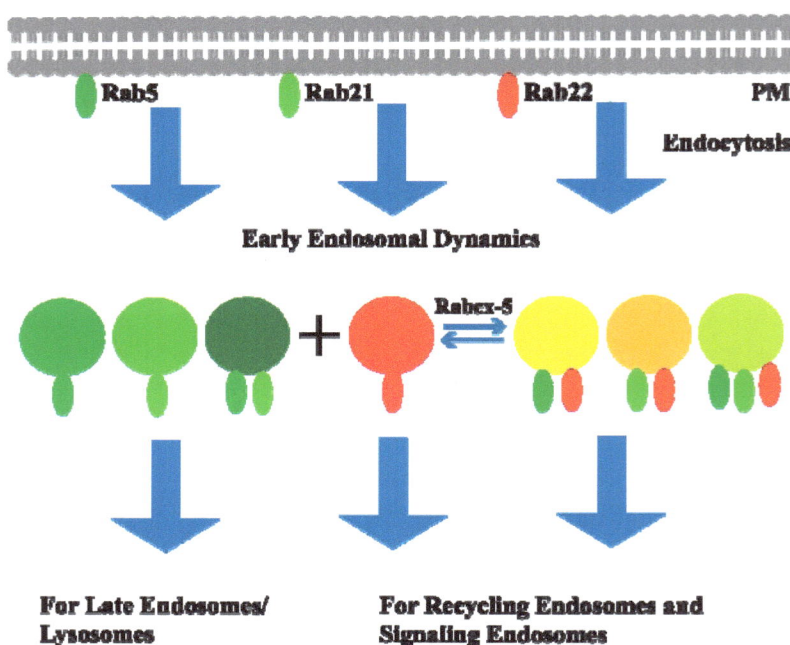

Figure 2: Functions and cross-talk of early endosomal sorting pathways mediated by Rab5, Rab21, and Rab22. Early endosomes consist of heterogeneous membrane compartments marked by Rab5, Rab21 or Rab22 or combinations of these three Rabs. The Rab5- and/or Rab21-positive endosomes promote sorting to late endosomes/lysosomes for degradation, while the Rab22-positive endosomes facilitate sorting to recycling compartments and possibly signaling endosomes. The three Rab membrane domains are likely to converge or segregate *via* fusion and fission. Rabex-5 is necessary for the convergence of the Rab22 domain with those of Rab5 and Rab21, by serving as an effector for Rab22 and an GEF for Rab5 and Rab21.

6. CONCLUDING REMARKS

The Rab5 subfamily members include Rab5, Rab21, Rab22, and Rab31 (aka Rab22b) and are phylogenetically related [17], with over 50% amino acid sequence identity. While Rab31 is localized at TGN [96, 97], Rab5, Rab21 and Rab22 are largely co-localized on early endosomes and regulate early endosomal fusion and sorting, which is the focus of this review.

Rab5 is involved in multiple steps of early endocytosis, including formation and uncoating of clathrin-coated endocytic vesicles [6, 7], early endosome fusion [9-11], endosomal movement on cytoskeleton [79], and signal transduction [86, 88] through interactions with a surprisingly large and diverse group of effectors

[73]. Rab21 appears to perform many of the same functions as Rab5 in endocytosis by interacting with the same effectors and regulators. However, there is one exception that Rab21 cannot interact with Rabenosyn-5 effectively and thus is unlikely to be involved directly in interaction with the SNAREs and subsequent membrane fusion reaction. In addition, there are quantitative differences between Rab5 and Rab21 in interacting with some of the effectors. For example, Rab21 binds more strongly to the integrins than Rab5, which may have functional implications in cell motility and cancer metastasis. Rab22, in contrast, shows distinct function from Rab5 in the sense that it blocks endocytic traffic to late endosomes and lysosomes and promotes instead recycling to the plasma membrane. Rab22 interacts with EEA1 and Rabenosyn-5 but not other Rab5 effectors tested so far. Further investigation into the interactions of Rab22 against a complete set of Rab5 effectors should help explain the functional bifurcation between Rab22 and Rab5 in the regulation of early endosomal sorting.

The functions of Rab5, Rab21, and Rab22 are linked by Rabex-5 [20] which is a well-documented GEF that efficiently activates Rab5 and Rab21 but not Rab22 [46, 49]. Surprisingly Rabex-5 turns out to be a Rab22 effector, which binds to Rab22-GTP and gets recruited by Rab22-GTP to the early endosomal membrane [20]. The cross-talk among the three early endosomal Rabs establishes a Rab22-Rab5/Rab21 cascade in regulation of early endocytosis.

An important open question is if the convergence of the three Rab domains in early endosomes plays a role in the biogenesis of specific functional compartments for signal transduction. Indeed, the nature of signaling endosomes remains unclear, especially their compositional differences from those destined for late endosomes/lysosomes and for recycling compartments. An attractive idea is that the convergence of Rab22, Rab5 and Rab21 domains or combinations of Rab22 with Rab5 or Rab21 may specify signaling endosomes whose biogenesis is regulated by ligands and receptors (Fig. **2**). Along this line, Rab5 and/or Rab21 domain alone without Rab22 would specify conventional early endosomes that transition and mature into late endosomes/lysosomes for degradation. Consistent with this idea, there is no Rab22 in yeast whereas there are also no signaling endosomes. In higher eukaryotes, early endosomes are heterogenous and marked by Rab5, Rab21 or Rab22 or combinations of these three Rabs. Separation and functional characterization of the different populations of early endosomes should clarify the questions and make a great stride at the interface of membrane trafficking and signal transduction. In addition, some of the most challenging questions in the field remain the mechanisms of how Rab-effector interactions lead to specific functions in membrane budding, vesicle movement, and membrane fusion. The recent reconstitution of Rab5-dependent liposome fusion *in vitro* has made significant progress in this direction [84]. Further incorporation of Rab21 and Rab22 into the reconstituted liposome system may yield insights into the synergy and/or competition among Rab5, Rab21 and Rab22 in membrane budding, vesicle movement, and membrane fusion.

ACKNOWLEDGEMENTS

The author's research program is supported by the National Institutes of Health (R01 GM074692).

CONFLICT OF INTEREST

There is no conflict of interest from the author.

REFERENCES

[1]   Doherty GJ, McMahon HT. Mechanisms of endocytosis. Annu Rev Biochem 2009;78:857-902.

[2]   Kirkham M, Parton RG. Clathrin-independent endocytosis: new insights into caveolae and non-caveolar lipid raft carriers. Biochim Biophys Acta 2005 Dec 30;1746(3):349-63.

[3]   Mayor S, Pagano RE. Pathways of clathrin-independent endocytosis. Nat Rev Mol Cell Biol 2007 Aug;8(8):603-12.

[4]   Stuart LM, Ezekowitz RA. Phagocytosis: elegant complexity. Immunity. 2005 May;22(5):539-50.

[5]   Chavrier P, Parton RG, Hauri HP, Simons K, Zerial M. Localization of low molecular weight GTP binding proteins to exocytic and endocytic compartments. Cell 1990 Jul 27;62(2):317-29.

[6]   McLauchlan H, Newell J, Morrice N, Osborne A, West M, Smythe E. A novel role for Rab5-GDI in ligand sequestration into clathrin-coated pits. Curr Biol 1998 Jan 1;8(1):34-45.

[7]   Semerdjieva S, Shortt B, Maxwell E, *et al.* Coordinated regulation of AP2 uncoating from clathrin-coated vesicles by rab5 and hRME-6. J Cell Biol 2008 Nov 3;183(3):499-511.

[8]   Nielsen E, Severin F, Backer JM, Hyman AA, Zerial M. Rab5 regulates motility of early endosomes on microtubules. Nat Cell Biol 1999 Oct;1(6):376-82.

[9]   Gorvel JP, Chavrier P, Zerial M, Gruenberg J. rab5 controls early endosome fusion *in vitro*. Cell 1991 Mar 8;64(5):915-25.

[10]   Hoffenberg S, Sanford JC, Liu S, *et al.* Biochemical and functional characterization of a recombinant GTPase, Rab5, and two of its mutants. J Biol Chem 1995 Mar 10;270(10):5048-56.

[11]   Li G, Barbieri MA, Colombo MI, Stahl PD. Structural features of the GTP-binding defective Rab5 mutants required for their inhibitory activity on endocytosis. J Biol Chem 1994 May 20;269(20):14631-5.

[12]   Khurana T, Brzostowski JA, Kimmel AR. A Rab21/LIM-only/CH-LIM complex regulates phagocytosis *via* both activating and inhibitory mechanisms. EMBO J 2005 Jul 6;24(13):2254-64.

[13]   Simpson JC, Griffiths G, Wessling-Resnick M, Fransen JA, Bennett H, Jones AT. A role for the small GTPase Rab21 in the early endocytic pathway. J Cell Sci 2004 Dec 15;117(Pt 26):6297-311.

[14]   Kauppi M, Simonsen A, Bremnes B, *et al.* The small GTPase Rab22 interacts with EEA1 and controls endosomal membrane trafficking. J Cell Sci 2002 Mar 1;115(Pt 5):899-911.

[15]   Magadan JG, Barbieri MA, Mesa R, Stahl PD, Mayorga LS. Rab22a regulates the sorting of transferrin to recycling endosomes. Mol Cell Biol 2006 Apr;26(7):2595-614.

[16]   Weigert R, Yeung AC, Li J, Donaldson JG. Rab22a regulates the recycling of membrane proteins internalized independently of clathrin. Mol Biol Cell 2004 Aug;15(8):3758-70.

[17]   Pereira-Leal JB, Seabra MC. The mammalian Rab family of small GTPases: definition of family and subfamily sequence motifs suggests a mechanism for functional specificity in the Ras superfamily. J Mol Biol 2000 Aug 25;301(4):1077-87.

[18]   Pereira-Leal JB, Seabra MC. Evolution of the Rab family of small GTP-binding proteins. J Mol Biol 2001 Nov 2;313(4):889-901.

[19]   Eathiraj S, Pan X, Ritacco C, Lambright DG. Structural basis of family-wide Rab GTPase recognition by rabenosyn-5. Nature 2005 Jul 21;436(7049):415-9.

[20]   Zhu H, Liang Z, Li G. Rabex-5 is a Rab22 effector and mediates a Rab22-Rab5 signaling cascade in endocytosis. Mol Biol Cell 2009 Nov;20(22):4720-9.

[21]   Zahraoui A, Touchot N, Chardin P, Tavitian A. The human Rab genes encode a family of GTP-binding proteins related to yeast YPT1 and SEC4 products involved in secretion. J Biol Chem 1989 Jul 25;264(21):12394-401.

[22]   Bucci C, Lutcke A, Steele-Mortimer O, *et al.* Co-operative regulation of endocytosis by three Rab5 isoforms. FEBS Lett 1995 Jun 5;366(1):65-71.

[23]   Gurkan C, Lapp H, Alory C, Su AI, Hogenesch JB, Balch WE. Large-scale profiling of Rab GTPase trafficking networks: the membrome. Mol Biol Cell 2005 Aug;16(8):3847-64.

[24]   Chiariello M, Bruni CB, Bucci C. The small GTPases Rab5a, Rab5b and Rab5c are differentially phosphorylated *in vitro*. FEBS Lett 1999 Jun 18;453(1-2):20-4.

[25]   Chen PI, Kong C, Su X, Stahl PD. Rab5 isoforms differentially regulate the trafficking and degradation of epidermal growth factor receptors. J Biol Chem 2009 Oct 30;284(44):30328-38.

[26]   Barbosa MD, Wakeland EK, Zerial M, Kingsmore SF. Genetic mapping of the Rab5a and Rab5b genes on mouse Chromosomes 17 and 2, respectively. Mamm Genome 1996 Feb;7(2):166-7.

[27]   Singer-Kruger B, Stenmark H, Dusterhoft A, *et al.* Role of three rab5-like GTPases, Ypt51p, Ypt52p, and Ypt53p, in the endocytic and vacuolar protein sorting pathways of yeast. J Cell Biol 1994 Apr;125(2):283-98.

[28]   Li G, Stahl PD. Post-translational processing and membrane association of the two early endosome-associated rab GTP-binding proteins (rab4 and rab5). Arch Biochem Biophys 1993 Aug 1;304(2):471-8.

[29]   Li G, Stahl PD. Structure-function relationship of the small GTPase rab5. J Biol Chem 1993 Nov 15;268(32):24475-80.

[30]   Bucci C, Parton RG, Mather IH, *et al.* The small GTPase rab5 functions as a regulatory factor in the early endocytic pathway. Cell 1992 Sep 4;70(5):715-28.

[31]   Zhu H, Zhu G, Liu J, Liang Z, Zhang XC, Li G. Rabaptin-5-independent membrane targeting and Rab5 activation by Rabex-5 in the cell. Mol Biol Cell 2007 Oct;18(10):4119-28.

[32]   Stenmark H, Parton RG, Steele-Mortimer O, Lutcke A, Gruenberg J, Zerial M. Inhibition of rab5 GTPase activity stimulates membrane fusion in endocytosis. EMBO J 1994 Mar 15;13(6):1287-96.

[33]   Liang Z, Mather T, Li G. GTPase mechanism and function: new insights from systematic mutational analysis of the phosphate-binding loop residue Ala30 of Rab5. Biochem J 2000 Mar 1;346 Pt 2:501-8.

[34]   Li G, Liang Z. Phosphate-binding loop and Rab GTPase function: mutations at Ser29 and Ala30 of Rab5 lead to loss-of-function as well as gain-of-function phenotype. Biochem J 2001 May 1;355(Pt 3):681-9.

[35]   Merithew E, Hatherly S, Dumas JJ, Lawe DC, Heller-Harrison R, Lambright DG. Structural plasticity of an invariant hydrophobic triad in the switch regions of Rab GTPases is a determinant of effector recognition. J Biol Chem 2001 Apr 27;276(17):13982-8.

[36]   Terzyan S, Zhu G, Li G, Zhang XC. Refinement of the structure of human Rab5a GTPase domain at 1.05 A resolution. Acta Crystallogr D Biol Crystallogr. 2004 Jan;60(Pt 1):54-60.

[37]   Zhu G, Liu J, Terzyan S, Zhai P, Li G, Zhang XC. High resolution crystal structures of human Rab5a and five mutants with substitutions in the catalytically important phosphate-binding loop. J Biol Chem 2003 Jan 24;278(4):2452-60.

[38]   Zhu G, Zhai P, Liu J, Terzyan S, Li G, Zhang XC. Structural basis of Rab5-Rabaptin5 interaction in endocytosis. Nat Struct Mol Biol 2004 Oct;11(10):975-83.

[39]   Rybin V, Ullrich O, Rubino M, *et al*. GTPase activity of Rab5 acts as a timer for endocytic membrane fusion. Nature. 1996 Sep 19;383(6597):266-9.

[40]   Burd CG, Mustol PA, Schu PV, Emr SD. A yeast protein related to a mammalian Ras-binding protein, Vps9p, is required for localization of vacuolar proteins. Mol Cell Biol 1996 May;16(5):2369-77.

[41]   Hama H, Tall GG, Horazdovsky BF. Vps9p is a guanine nucleotide exchange factor involved in vesicle-mediated vacuolar protein transport. J Biol Chem 1999 May 21;274(21):15284-91.

[42]   Sato M, Sato K, Fonarev P, Huang CJ, Liou W, Grant BD. Caenorhabditis elegans RME-6 is a novel regulator of RAB-5 at the clathrin-coated pit. Nat Cell Biol 2005 Jun;7(6):559-69.

[43]   Szabo K, Jekely G, Rorth P. Cloning and expression of sprint, a Drosophila homologue of RIN1. Mech Dev 2001 Mar;101(1-2):259-62.

[44]   Barbieri MA, Kong C, Chen PI, Horazdovsky BF, Stahl PD. The SRC homology 2 domain of Rin1 mediates its binding to the epidermal growth factor receptor and regulates receptor endocytosis. J Biol Chem 2003 Aug 22;278(34):32027-36.

[45]   Tall GG, Barbieri MA, Stahl PD, Horazdovsky BF. Ras-activated endocytosis is mediated by the Rab5 guanine nucleotide exchange activity of RIN1. Dev Cell 2001 Jul;1(1):73-82.

[46]   Horiuchi H, Lippe R, McBride HM, *et al*. A novel Rab5 GDP/GTP exchange factor complexed to Rabaptin-5 links nucleotide exchange to effector recruitment and function. Cell 1997 Sep 19;90(6):1149-59.

[47]   Lippe R, Miaczynska M, Rybin V, Runge A, Zerial M. Functional synergy between Rab5 effector Rabaptin-5 and exchange factor Rabex-5 when physically associated in a complex. Mol Biol Cell 2001 Jul;12(7):2219-28.

[48]   Zhu H, Qian H, Li G. Delayed onset of positive feedback activation of rab5 by rabex-5 and rabaptin-5 in endocytosis. PLoS One 2010;5(2):e9226.

[49]   Delprato A, Merithew E, Lambright DG. Structure, exchange determinants, and family-wide rab specificity of the tandem helical bundle and Vps9 domains of Rabex-5. Cell 2004 Sep 3;118(5):607-17.

[50]   Mattera R, Tsai YC, Weissman AM, Bonifacino JS. The Rab5 guanine nucleotide exchange factor Rabex-5 binds ubiquitin (Ub) and functions as a Ub ligase through an atypical Ub-interacting motif and a zinc finger domain. J Biol Chem 2006 Mar 10;281(10):6874-83.

[51]   Delprato A, Lambright DG. Structural basis for Rab GTPase activation by VPS9 domain exchange factors. Nat Struct Mol Biol 2007 May;14(5):406-12.

[52]   Kalesnikoff J, Rios EJ, Chen CC, *et al*. Roles of RabGEF1/Rabex-5 domains in regulating Fc epsilon RI surface expression and Fc epsilon RI-dependent responses in mast cells. Blood 2007 Jun 15;109(12):5308-17.

[53]   Han L, Colicelli J. A human protein selected for interference with Ras function interacts directly with Ras and competes with Raf1. Mol Cell Biol 1995 Mar;15(3):1318-23.

[54]   Kajiho H, Saito K, Tsujita K, *et al*. RIN3: a novel Rab5 GEF interacting with amphiphysin II involved in the early endocytic pathway. J Cell Sci 2003 Oct 15;116(Pt 20):4159-68.

[55]   Saito K, Murai J, Kajiho H, Kontani K, Kurosu H, Katada T. A novel binding protein composed of homophilic tetramer exhibits unique properties for the small GTPase Rab5. J Biol Chem 2002 Feb 1;277(5):3412-8.

[56]   Li G, D'Souza-Schorey C, Barbieri MA, Cooper JA, Stahl PD. Uncoupling of membrane ruffling and pinocytosis during Ras signal transduction. J Biol Chem 1997 Apr 18;272(16):10337-40.

[57] Barbieri MA, Roberts RL, Gumusboga A, *et al.* Epidermal growth factor and membrane trafficking. EGF receptor activation of endocytosis requires Rab5a. J Cell Biol 2000 Oct 30;151(3):539-50.

[58] Bar-Sagi D, Feramisco JR. Induction of membrane ruffling and fluid-phase pinocytosis in quiescent fibroblasts by ras proteins. Science 1986 Sep 5;233(4768):1061-8.

[59] Su X, Kong C, Stahl PD. GAPex-5 mediates ubiquitination, trafficking, and degradation of epidermal growth factor receptor. J Biol Chem 2007 Jul 20;282(29):21278-84.

[60] Lodhi IJ, Chiang SH, Chang L, *et al.* Gapex-5, a Rab31 guanine nucleotide exchange factor that regulates Glut4 trafficking in adipocytes. Cell Metab. 2007 Jan;5(1):59-72.

[61] Yang Y, Hentati A, Deng HX, *et al.* The gene encoding alsin, a protein with three guanine-nucleotide exchange factor domains, is mutated in a form of recessive amyotrophic lateral sclerosis. Nat Genet 2001 Oct;29(2):160-5.

[62] Hadano S, Hand CK, Osuga H, *et al.* A gene encoding a putative GTPase regulator is mutated in familial amyotrophic lateral sclerosis 2. Nat Genet 2001 Oct;29(2):166-73.

[63] Ben Hamida M, Hentati F, Ben Hamida C. Hereditary motor system diseases (chronic juvenile amyotrophic lateral sclerosis). Conditions combining a bilateral pyramidal syndrome with limb and bulbar amyotrophy. Brain 1990 Apr;113 (Pt 2):347-63.

[64] Topp JD, Carney DS, Horazdovsky BF. Biochemical characterization of Alsin, a Rab5 and Rac1 guanine nucleotide exchange factor. Methods Enzymol 2005;403:261-76.

[65] Otomo A, Hadano S, Okada T, *et al.* ALS2, a novel guanine nucleotide exchange factor for the small GTPase Rab5, is implicated in endosomal dynamics. Hum Mol Genet 2003 Jul 15;12(14):1671-87.

[66] Devon RS, Orban PC, Gerrow K, *et al.* Als2-deficient mice exhibit disturbances in endosome trafficking associated with motor behavioral abnormalities. Proc Natl Acad Sci U S A 2006 Jun 20;103(25):9595-600.

[67] Hadano S, Benn SC, Kakuta S, *et al.* Mice deficient in the Rab5 guanine nucleotide exchange factor ALS2/alsin exhibit age-dependent neurological deficits and altered endosome trafficking. Hum Mol Genet 2006 Jan 15;15(2):233-50.

[68] Lai C, Xie C, Shim H, Chandran J, Howell BW, Cai H. Regulation of endosomal motility and degradation by amyotrophic lateral sclerosis 2/alsin. Mol Brain 2009;2:23.

[69] Pan X, Eathiraj S, Munson M, Lambright DG. TBC-domain GAPs for Rab GTPases accelerate GTP hydrolysis by a dual-finger mechanism. Nature 2006 Jul 20;442(7100):303-6.

[70] Haas AK, Fuchs E, Kopajtich R, Barr FA. A GTPase-activating protein controls Rab5 function in endocytic trafficking. Nat Cell Biol 2005 Sep;7(9):887-93.

[71] Fuchs E, Haas AK, Spooner RA, Yoshimura S, Lord JM, Barr FA. Specific Rab GTPase-activating proteins define the Shiga toxin and epidermal growth factor uptake pathways. J Cell Biol 2007 Jun 18;177(6):1133-43.

[72] Lanzetti L, Rybin V, Malabarba MG, *et al.* The Eps8 protein coordinates EGF receptor signalling through Rac and trafficking through Rab5. Nature 2000 Nov 16;408(6810):374-7.

[73] Christoforidis S, McBride HM, Burgoyne RD, Zerial M. The Rab5 effector EEA1 is a core component of endosome docking. Nature 1999 Feb 18;397(6720):621-5.

[74] Stenmark H, Vitale G, Ullrich O, Zerial M. Rabaptin-5 is a direct effector of the small GTPase Rab5 in endocytic membrane fusion. Cell 1995 Nov 3;83(3):423-32.

[75] Gournier H, Stenmark H, Rybin V, Lippe R, Zerial M. Two distinct effectors of the small GTPase Rab5 cooperate in endocytic membrane fusion. EMBO J 1998 Apr 1;17(7):1930-40.

[76] Korobko EV, Kiselev SL, Korobko IV. Multiple Rabaptin-5-like transcripts. Gene 2002 Jun 12;292(1-2):191-7.

[77] Shin HW, Hayashi M, Christoforidis S, *et al.* An enzymatic cascade of Rab5 effectors regulates phosphoinositide turnover in the endocytic pathway. J Cell Biol 2005 Aug 15;170(4):607-18.

[78] Christoforidis S, Miaczynska M, Ashman K, *et al.* Phosphatidylinositol-3-OH kinases are Rab5 effectors. Nat Cell Biol 1999 Aug;1(4):249-52.

[79] Hoepfner S, Severin F, Cabezas A, *et al.* Modulation of receptor recycling and degradation by the endosomal kinesin KIF16B. Cell 2005 May 6;121(3):437-50.

[80] Schnatwinkel C, Christoforidis S, Lindsay MR, *et al.* The Rab5 effector Rabankyrin-5 regulates and coordinates different endocytic mechanisms. PLoS Biol 2004 Sep;2(9):E261.

[81] Simonsen A, Lippe R, Christoforidis S, *et al.* EEA1 links PI(3)K function to Rab5 regulation of endosome fusion. Nature 1998 Jul 30;394(6692):494-8.

[82] Nielsen E, Christoforidis S, Uttenweiler-Joseph S, *et al.* Rabenosyn-5, a novel Rab5 effector, is complexed with hVPS45 and recruited to endosomes through a FYVE finger domain. J Cell Biol 2000 Oct 30;151(3):601-12.

[83]   Rubino M, Miaczynska M, Lippe R, Zerial M. Selective membrane recruitment of EEA1 suggests a role in directional transport of clathrin-coated vesicles to early endosomes. J Biol Chem 2000 Feb 11;275(6):3745-8.

[84]   Ohya T, Miaczynska M, Coskun U, *et al.* Reconstitution of Rab- and SNARE-dependent membrane fusion by synthetic endosomes. Nature 2009 Jun 25;459(7250):1091-7.

[85]   Brandhorst D, Zwilling D, Rizzoli SO, Lippert U, Lang T, Jahn R. Homotypic fusion of early endosomes: SNAREs do not determine fusion specificity. Proc Natl Acad Sci U S A 2006 Feb 21;103(8):2701-6.

[86]   Miaczynska M, Christoforidis S, Giner A, *et al.* APPL proteins link Rab5 to nuclear signal transduction *via* an endosomal compartment. Cell 2004 Feb 6;116(3):445-56.

[87]   Zhu G, Chen J, Liu J, *et al.* Structure of the APPL1 BAR-PH domain and characterization of its interaction with Rab5. EMBO J 2007 Jul 25;26(14):3484-93.

[88]   Schenck A, Goto-Silva L, Collinet C, *et al.* The endosomal protein Appl1 mediates Akt substrate specificity and cell survival in vertebrate development. Cell 2008 May 2;133(3):486-97.

[89]   Imamura H, Takaishi K, Nakano K, *et al.* Rho and Rab small G proteins coordinately reorganize stress fibers and focal adhesions in MDCK cells. Mol Biol Cell 1998 Sep;9(9):2561-75.

[90]   Lanzetti L, Palamidessi A, Areces L, Scita G, Di Fiore PP. Rab5 is a signalling GTPase involved in actin remodelling by receptor tyrosine kinases. Nature 2004 May 20;429(6989):309-14.

[91]   Spaargaren M, Bos JL. Rab5 induces Rac-independent lamellipodia formation and cell migration. Mol Biol Cell 1999 Oct;10(10):3239-50.

[92]   Zhang X, He X, Fu XY, Chang Z. Varp is a Rab21 guanine nucleotide exchange factor and regulates endosome dynamics. J Cell Sci 2006 Mar 15;119(Pt 6):1053-62.

[93]   Pellinen T, Arjonen A, Vuoriluoto K, Kallio K, Fransen JA, Ivaska J. Small GTPase Rab21 regulates cell adhesion and controls endosomal traffic of beta1-integrins. J Cell Biol 2006 Jun 5;173(5):767-80.

[94]   Hooper S, Gaggioli C, Sahai E. A chemical biology screen reveals a role for Rab21-mediated control of actomyosin contractility in fibroblast-driven cancer invasion. Br J Cancer 2010 Jan 19;102(2):392-402.

[95]   Mesa R, Salomon C, Roggero M, Stahl PD, Mayorga LS. Rab22a affects the morphology and function of the endocytic pathway. J Cell Sci 2001 Nov;114(Pt 22):4041-9.

[96]   Ng EL, Wang Y, Tang BL. Rab22B's role in trans-Golgi network membrane dynamics. Biochem Biophys Res Commun 2007 Sep 28;361(3):751-7.

[97]   Rodriguez-Gabin AG, Cammer M, Almazan G, Charron M, Larocca JN. Role of rRAB22b, an oligodendrocyte protein, in regulation of transport of vesicles from trans Golgi to endocytic compartments. J Neurosci Res 2001 Dec 15;66(6):1149-60.

[98]   Kimura T, Sakisaka T, Baba T, Yamada T, Takai Y. Involvement of the Ras-Ras-activated Rab5 guanine nucleotide exchange factor RIN2-Rab5 pathway in the hepatocyte growth factor-induced endocytosis of E-cadherin. J Biol Chem 2006 Apr 14;281(15):10598-609.

[99]   Mishra A, Eathiraj S, Corvera S, Lambright DG. Structural basis for Rab GTPase recognition and endosome tethering by the C2H2 zinc finger of Early Endosomal Autoantigen 1 (EEA1). Proc Natl Acad Sci U S A 2010 Jun 15;107(24):10866-71.

CHAPTER 8

Role of Rab4 in Transport through the Endosomal System

Emma Martinez Sanchez, Ioana Popa, Magda Deneka and Peter van der Sluijs[*]

University Medical Center Utrecht, The Netherlands

Abstract: The endosomal system is a mosaic of highly dynamic interconnected organelles that serve many physiological functions including nutrient uptake, regulation of cell polarity, migration and plasma membrane remodeling. These widely divergent processes rely on the flawless transfer of cargo molecules between the distinct endosomal subcompartments and is regulated by members of the rab family of small GTPases and their effector networks. Here we provide an overview of the function of rab4, and the machinery it deploys to regulating exocytic membrane trafficking from early endosomes.

Keywords: Rab4a, Rab4b, Endosomes, Recycling, Domain.

1. INTRODUCTION

Endosomes form a highly dynamic and pleiopmorphic membrane system where internalized molecules are sorted for recycling and degradation [1-3]. Endosomal subcompartments include early and recycling endosomes, whose functions are typically associated with sorting and returning cargo back to the plasma membrane, and a branch that is primarily designed for degradation. This latter is comprised of multivesicular endosomes, late endosomes, and lysosomes. The distinct clathrin dependent- and independent internalization routes converge at the level of early endosomes which provide the common entry point into the endosomal system [4]. Endosomes also connect the compartments of endocytic and biosynthetic membrane systems for exchange and transfer of content and which represents a popular route amongst certain toxins [5]. While the endocytic pathway was traditionally perceived as a conduit for eating and drinking, it is now clear that endosomal membranes serve vastly broader functions in such diverse processes as memory and learning [6], immune defense [7], cell polarity [8] and cytokinesis [9]. Particularly the analysis of intact organisms has started to reveal a complex array of cellular processes that are regulated by endosomes during development and morphogenesis [10, 11].

Great progress has been made in understanding the cellular machinery that is required for the orderly transfer of cargo between the distinct endosomal subcompartments. Members of such highly conserved protein families as rab [12] and ARF [13] small GTPases, SNAREs, and SNARE regulators [14] have all been found to control various aspects of the endosomal trafficking pathways. In this review we focus on one of these regulators namely rab4. We describe its role in the endosomal recycling pathway and the effectors it utilizes to do this job. We will also discuss the current understanding how rab4 activity might be controlled and provide examples of processes that are dependent on rab4 function. Since surface expression of many proteins is regulated by endocytic recycling, an appreciable number of papers describing a role for rab4 in receptor trafficking pathways and signaling has appeared. We therefore limit ourselves in this review to a few of the more recent ones.

2. RAB4 ISOFORMS

A number of rabs including rab4, have two or more isoforms that are encoded by the same or different genes. Humans have two forms of rab4 that are encoded by the *RAB4A* and *RAB4B* genes, located on chromosome 1 (1q42) and chromosome 19 (19q13), respectively. The genome of Danio rerio also contains two *RAB4* genes, while Drosophila melanogaster has one, and Caenorhabditis elegans and Saccharomyces

*Address correspondence to Peter van der Sluijs:** Department of Cell Biology, University Medical Center Utrecht, 3584 CX Utrecht, The Netherlands; Tel: +31-88-755-7578; Fax: +31-30-254-1797; E-mail: p.vandersluijs@umcutrecht.nl

cerevisiae do not have rab4, perhaps bearing testimony to the less specialized organization of endocytic pathways in these species. Early knockout experiments on the isoforms of vps21 (rab5) in yeast [15] and rab3 [16] in mice led to the concept that isoforms likely perform redundant functions. Recent data on rab27 however show that although rab27a and rab27b are expressed in the same cells, they are nevertheless involved in different trafficking events [17, 18]. Both rab27a and rab27b are important for multivesicular endosome (MVE) docking at the plasma membrane [17]. However, whereas rab27a silencing strongly increases the size of MVEs, rab27b knock down redistributes MVEs towards the perinuclear region showing that the two rab27 isoforms have different roles in the exosomal pathway [17]. Human rab4a and rab4b have 93% homology and are ubiquitously expressed. Nevertheless there are tissues with high expression of either one. For instance rab4a is highly expressed in brain [19], while rab4b is the predominant form in B cells [20]. Since rab4a was discovered first, most of what we know about rab4 function derives from transfection studies in which mutants of this isoform are employed. The possibility to selectively knock-down rab4 isoforms with siRNA, greatly enhanced the opportunities to probe for rab4b function. Exciting new evidence begins to reveal specific functions for rab4b in antigen presentation in immune cells and glucose uptake in fat cells, as discussed below.

3. LOCALIZATION OF RAB4A

Free flow electrophoresis of membrane fractions enriched in endosomes and lysosomes showed that rab4 co-purified with endocytosed ^{125}I-Transferrin (Tf), and partially with the trans Golgi marker galactosyltransferase [21]. This suggested that rab4 was associated with the early endosomal system and likely with a subcompartment involved in recycling to the plasma membrane. Expression experiments in HeLa cells confirmed and extended this notion because rab4 was predominantly localized in a perinuclear structure, together with the transferrin receptor (TfR), a marker for recycling endosomes. A more precise localization was gained from pulse chase experiments with fluorescently labeled Tf. These showed that internalized Tf first reaches endosomes enriched in rab4 and subsequently moved to an organelle in the perinuclear area that was decorated with antibodies against the v-SNARE cellubrevin and relatively depleted in rab4 [22]. Thus rab4 is functionally organized proximal to the recycling compartment characterized by cellubrevin. Polarized epithelia endocytose both at the apical and basolateral plasma membranes, and early and recycling endosomes are found beneath these plasma membrane domains [23]. Analysis of the distribution of rab4 over the endosomal system in polarized filter-grown MDCK epithelial cells, showed that rab4 containing endosomes are mainly localized in the sub-apical region [24]. A quantitative immuno EM study of endogenous rab4 in PC12 cells showed that it was localized on the limiting membrane and tubulovesicular structures of endosomes with little or no internal vesicles in the vacuolar part [25, 26]. These are typical features of early endosomes and recycling tubules. Interestingly, a small amount (~ 5%) of endogenous rab4 is associated with the Golgi complex [26], perhaps *via* association with one or more of the Golgins that interact with rab proteins *via* extended coiled-coil regions [27, 28]. This observation might suggest that a limited pool of rab4 is involved in transport between endosomes and the Golgi complex.

4. RAB4 FUNCTION IN RECEPTOR RECYCLING FROM ENDOSOMES

Overexpresssion studies revealed that rab4 regulates endosomal recycling of endocytosed Tf and TfR from endosomes back to the plasma membrane [29]. Neither endocytic internalization nor transport to late endocytic structures are affected by the inhibitory rab4-N121I mutant that does not bind guanine nucleotide [30]. In polarized MDCK cells, recycling of TfR occurs mainly to the basolateral surface, while ~ 20% is recycled to the apical plasma membrane [31, 32]. Expression of rab4 or the hydrolysis-deficient rab4-Q67L mutant causes mistargeting of basolaterally endocytosed Tf to the apical cell surface, while treatment of non-transfected MDCK cells with brefeldin A phenocopies the effect on Tf recycling [24]. Brefeldin A is a small molecule inhibitor of ARF guanine nucleotide exchange factors [33]. Because this small GTPase is required for AP-1 recruitment to post Golgi membranes [34-37], it is likely that ARF-1 and rab4 act in a pathway that controls the AP-1 dependent formation of endocytic vesicles. Brefeldin A treatment of MDCK cells expressing rab4-Q67L does not increase the extent of TfR mistargeting to the apical plasma membrane showing that rab4 function and the target of BFA act in the same pathway [24]. A membrane-cytoplasm cycle is required for the function of rab4 in exocytic release of Tf to the apical plasma membrane as was found in expression studies with NHrab4cbvn in which the carboxy-terminal prenylation motif was replaced by the transmembrane domain of cellubrevin. The chimeric protein is permanently attached to membranes, properly targeted to early endosomes, and bound guanine nucleotide to the same extent as wild type rab4 [38]. Nevertheless the rab4 chimera does not effcienty support this pathway showing that a endosome-cytosol cycle is required for normal rab4 function [38].

5. CELL TYPE SPECIFIC FUNCTION OF RAB4B IN FAT CELLS

Recently the first papers reporting on the functions of rab4b have appeared. Cormont's group found that adipocytes express both rab4a and rab4b [39]. More interestingly, they observed a 2-fold decrease in rab4b expression during obese diabetic states in adipose tissue. The correlation between rab4b expression and the disease state suggests an isoform specific pathway controlling transcription and or degradation regulatory mechanisms. In the 3T3-L1 model system for the differentiation of fibroblasts into fat cells, ectopically expressed rab4a and rab4b co-localize in the perinuclear region of 3T3-L1 fibroblasts. Upon differentiation to adipocytes rab4a and rab4b distribute to distinct structures dispersed throughout the cytoplasm, suggesting that rab4b is partly segregated from rab4a labeled early endosomes. While rab4b localizes with the insulin regulated glucose transporter GLUT4 and the v-SNARE VAMP2, rab4a is mainly found with TfR in housekeeping endosomes. Silencing rab4b expression two-fold with specific siRNAs, produced a small increase in basal deoxyglucose uptake and a more sustained increase in the presence of insulin. This phenotype correlates with a relocation of GLUT4 to the cell surface of basal and insulin-stimulated adipocytes. Cell surface expression of TfR however was not affected by the rab4b knock down . These experiments suggest that rab4b is involved in targeting GLUT4 towards its non-endosomal sequestration-storage compartment, from where GLUT4 is translocated to the plasma membrane. Rab4a on the other hand rather regulates the house-keeping recycling pathway, that in non-differentiated 3T3-L1 cells might coincide with the rab4b pathway [39].

6. CELL TYPE SPECIFIC FUNCTION OF RAB4B IN ANTIGEN PRESENTING CELLS

A second paper addresses the function of rab4b in the immune system and also revealed a role in specialized cell types. Krawczyk *et al.* screened for proteins involved in antigen presentation [20, 40]. The approach they took was to search for targets of MHC class II transcriptional coactivator (CIITA). CIITA is a master regulator of genes involved in antigen presentation, such as MHC class II, Ii, HLA–DO, HLA–DM and MHC-I. They combined *in silico* searches for genes that contain the characteristic MHC-II-like S-Y enhancers in their upstream regions with ChIP-on-chip screens to identify direct targets of CIITA in dendritic cells and B cells. These approaches converged on *RAB4B*, documenting that its transcription is controlled by the same regulatory machinery that is critical for the expression of MHC-II genes, and enhances the expression of MHC-I genes. The molecular link between the transcriptional activation of *RAB4B* and genes involved in antigen presentation suggests that dendritic cells and B cells boost their antigen presentation capacity by increasing the efficiency of endocytic recycling. Importantly, rab4 has been implicated directly in MHC class II restricted antigen presentation. Expression of dominant negative rab4-N121I in the mouse A20 B cell line blocks presentation of antigens internalized *via* the B cell receptor or ectopically expressed FcRIIγ-B2 [41]. Since rab4a does not seem to be expressed in A20 B cells [20], it is a distinct possibility that the effects of rab4a-N121I are caused by interaction of the mutant with proteins that normally bind to rab4b. Often times rab isoforms bind to the same effectors, although the binding characteristics might be somewhat different. For instance both rab4a and rab4b [42], as well as rab27a and rab27b [43] each interact with different affinities to rabaptin-5α, and munc13-4, respectively.

7. OTHER RABS REGULATING ENDOSOME RECYCLING

Multiple sequence alignment and bioinformatic analysis of all rab sequences allows the clustering of rabs within subfamilies whose members might share related properties and functions [44, 45]. Rab4 is in a subfamily together with rab2, rab11 and rab25, and rab14. All of these are localized to early endosomal compartments, however only the comparative localization of rab4 and rab11 has been assessed [19, 46, 47].

It is not known for instance whether these rabs have a preference for similar or different domains on the endosomal membrane. The problem here and with most rabs is that the available antibodies often are not suited to detect the endogenous proteins by morphological methods in double label experiments. As a consequence, most of the distributions of rabs have been addressed using ectopic expression. Because these can of course recruit or stabilize additional effector molecules at their steady state localization, such a strategy might alter the properties of endosomal membrane domains. For example, expression of wild type rab4 and rab5 affect the size of early endosomes and their morphology [26]. When the cellular level of rab4 is increased, endosomes become bigger and generate more, and longer associated tubules [48]. With the availability of RNAi-mediated knock down, this matter can now be resolved using knock-in approaches with genetically encoded fluorescent proteins using controlled expression systems. Studies with the individual members of the rab2 subgroup show that they are all involved in exocytic events from endosomes. Although rab35 does not belong to this subfamily, it is also localized on endosomes and important for endosome recycling [49], cytokinesis [50] and exosome secretion [50]. Recently additional endosomal recycling routes have been discovered that do not appear to return the archetype TfR back to the cell surface. One of these involves ARF-6 dependent recycling of MHC class I and interleukin 2 receptor alpha subunit [3]. It is possible that these novel non-conventional pathways rely on other specific rab proteins [3]. Finally, it is also possible that they are a reflection of cell type or tissue specialization, in combination with transport of specific cargo molecules.

8. RAB4 FUNCTION AND THE FAST DIRECT *VS.* SLOW INDIRECT RECYCLING PATHWAY

The classical recycling route delivers TfR cargo *via* two pathways from endosomes to the cell surface [51]. A short circuit pathway is thought to return receptors directly from the early endosome to the plasma membrane, although it is possible that it is a composite route that also includes recycling of vesicles that have not yet reached early endosomes [51]. The indirect recycling route involves passage through a intermediate organelle known as the recycling compartment, before recepors are re-inserted in the plasma membrane. Because endocytosed transferrin was reported to transit through rab4 endosomes before it reaches rab11 endosomes [46], it was thought that rab4 regulates the direct pathway, while rab11 is important for the slow recycling route. Although intuitively appealing, this idea likely is not correct. First, recent high speed double label live cell imaging shows that recycling vesicles containing both rab4 and rab11 leave from the very same early endosome [47]. After some time, rab4 is lost from this vesicle, while the remnant containing rab11 fuses with the plasma membrane [47]. Secondly a population of rab4 molecules is found in recycling endosomes by biochemical and microscopy methods both in fibroblasts and in polarized epithelial cells, which is not compatible with a role in the direct pathway [52]. Third, several groups found that knock down of rab4 enhances the rate of transferrin recycling [53, 54], which does not square with a function as positive regulator of the fast recycling pathway. Fourth, in cells expressing rab4 and rabaptin-5α, the effector colocalizes strongly with cellubrevin, a marker of recycling endosomes [42]. A fifth argument builds on the concept that the flux of cargo through the direct and indirect pathways is determined by the amount of cargo molecules present in the early endosome. Any experimental manipulation such as addition of drugs or expression of inhibitory mutants that affect one exit pathway will lead to accumulation of cargo in the donor compartment and enhance a compensatory flux through the other pathway. In other words, measurements on the amount of recycled molecules will invariably represent secondary effects. An alternative and more likely scenario therefore posits that rab4 regulates cargo transfer from early endosomes to recycling endosomes.

9. RAB4 AND BUDDING ENDOCYTIC VESICLES FROM ENDOSOMES

The machinery involved in the formation of endosomal carriers and the molecules regulating endosomal recycling pathways are not well understood. Elegant whole amount EM experiments showed that AP-1 is localized on early endosomal buds and associated with clathrin coated carriers having a diameter of 100 nm diameter [55]. Supposedly these could serve as endocytic vesicles returning proteins and membrane back to the plasma membrane. The involvement of clathrin and the AP-1 adaptor protein complex in endosome recycling has been controversial [51], but it is clear now that both clathrin-dependent and independent recycling pathways exist [3, 56]. The lab of Spiess developed an *in vitro* reconstitution assay to analyze the requirements for the formation of endocytic vesicles from endosomes [57]. Permeablized MDCK cells expressing the asialoglycoprotein receptor (serving as recycling cargo), generated vesicles containing the receptor, in a cytosol and ATP dependent manner. Performing the budding assay with cytosol that was depleted of candidate proteins showed that the formation of these recycling vesicles requires rab4, AP-1, and clathrin, but not AP-2, AP-3 and rab5 [57]. Earlier results from *in vitro* reconstitution assays [58] with membranes of the neuroendocrine PC12 cell line are in accord with this function of rab4 and showed in addition that the active form of the GTPase is needed for endosomal budding of synaptic like microvesicles and HRP-Tf containing endocytic vesicles [26]. Thus various lines of evidence suggest that rab4 acts in the same pathway as ARF-1 namely the formation of clathrin/AP-1 dependent vesicles from endosomes that contain receptors for recycling to the plasma membrane.

10. BIVALENT EFFECTORS LINKING RAB5 AND RAB4 DOMAINS

To understand how the rab4 GTPase switch contributes to the dynamic properties of a specific domain involved in recycling, requires identification and understanding of its effector network. Rab4 might serve as a platform for recruitment of cytoplasmic effectors that cooperatively create a discrete endosomal microdomain that is involved in cargo selection, coat recruitment, local membrane remodeling, membrane deformation and budding.

The use of two hybrid screens and affinity-based protein purification methods yielded a increasing collection of rab4-interacting proteins. Members of the rabaptin-5 family were the first cytoplasmic rab4 effectors to be identified [42, 59]. Rabaptin-5, rabaptin-5β (representing a different gene product) and rabaptin-5α also bind rab5 [59, 60], which qualifies the rabaptins therefore as the first bivalent rab effectors interacting with rab4 and rab5. Rabaptin-5 is an essential protein in early endosome fusion. It is present in a cytoplasmic complex with the exchange factor rabex-5 and initially recruited to early endosomes by rab5 [61]. Once at this location, it is thought to associate with rab4, and become engaged in rab4-dependent recycling. Additional rabaptin-5 variants with N-terminal deletions have been described, one of which lost the ability to interact with rab4 [62, 63]. The presence of separate rab4 and rab5 binding domains allows rabaptins to form a tethering link between the two rabs and integrate the activities of the rab5 domains in regulating delivery of incoming membrane, and the rab4 domain for removal of membrane and proteins by budding of endocytic vesicles. Since effector binding reduces hydrolysis rate of bound GTP [42, 64], the attractiveness of this model is further enhanced by the possibility of positive cooperativity. Rab5 also binds vps34/vps15 (PI3-kinase) which generates a local pool of PI-3P that is a critical factor in subsequent recruitment of FYVE and PX domain proteins [65-67]. These include EEA1 and two additional bivalent effectors of rab5 and rab4; rabenosyn-5 [68] and rabip4/rabip4' [69, 70], strengthening the coupling of the two domains.

The precise molecular details are not known, and many questions still need to be answered concerning this mechanism. For instance whether a single rabaptin-5 molecule binds simultaneously to a active rab4 and rab5 molecule to bring about rab domain coupling. Biochemical *in vivo* studies have shown that rabaptin-5 can oligomerize [59], while crystallographic structures of the rab5-rabaptin-5 complex reveal a rabaptin-5 dimer of which each rabaptin-5 has a molecule of rab5 bound [71]. Conceptually dimerization of the two rabaptin-5 monomers might also form the physical link between the endosomal rab4 and rab5 domains. Another question that has not been rigorously addressed is whether rabaptin-5 function in the rab4 domain is dependent on previous recruitment by rab5. The rab5 binding domain on rabaptin-5 has been mapped to a relatively short sequence in the C-terminus [60, 71]. It should be straightforward to make point mutants in the rab5 binding domain that lost the ability to bind rab5 [71], and test whether these rescue rab4 dependent recycling in a pan rabaptin-5 knock-down model. Interestingly, recent experiments in mast cells show that rabaptin-5 function is not needed for rab5-dependent internalization of the high affinity IgE receptor (FcεRI) and β1-integrin, but markedly reduced the amount of these proteins on the plasma membrane [72]. Supposedly rab4-rabaptin-5 dependent recycling of these signaling receptors is impaired in the absence of rabaptin-5, suggesting that rab4-rabaptin-5 tether in this signaling pathway functions autonomously, independent of rab5.

11. UNCOUPLING RAB5 AND RAB4 DOMAINS

Rabaptin-5 function as bivalent effector within the endosomal system can be switched-off irreversibly by cleavage of caspase sites within the hinge between the coiled-coil regions [73-75]. The conditions and molecular mechanisms that lead to cleavage are diverse and two will be discussed here. *L. pneumophila*, the causative agent of Legionnaire's disease hijacks rabaptin-5 to avoid degradation [76]. Infection of macrophages or peripheral blood monocyte cells with *L. pneumophila* causes the activation of host cell executioner caspase-3 through the Dot/Icm effectors IcmS, IcmR and IcmQ during the early stages of infection [77]. The Dot/Icm type IV secretion system is essential for the evasion of fusion between nascent *L. pneumophila* phagosomes and host cell endosomes [78]. This causes their subversion at a very early stage of the endocytic pathway, escape from transport into degradative lysosomes and allows to create a niche for replication. A second process that causes a specific cleavage of rabaptin-5 by capase-3 is programmed cell death. Apoptosis of a cell leads to the rapid phagocytosis and destruction by its neighboring cells, which is important for multicellular organisms because it avoids inflammatory responses. Cells that go into apoptosis display a series of cytoarchitectural changes reflecting the loss of organized endomembrane structures such as the Golgi complex [79, 80]. Rabaptin-5 cleavage occurs solely through caspase-3 which together with caspase -7 is a typical executioner caspase. This suggests that the cleavage of rabaptin-5 is a late event when the cell is already programmed to die, instead of an early signal that triggers initiator activation and subsequent apoptosis from within.

12. ORGANIZATION OF 'EXOCYTIC' ENDOSOMAL RAB4 DOMAIN

In preceding paragraphs we discussed that the endosomal rab4 localization is important for budding of clathrin and AP-1 coated endocytic vesicles. The molecular link between rab4 and the machinery that is forming these vesicles has not yet been completely worked out, but a conceptual framework is emerging in which rab4, the AP-1 adaptor complex, and a kinesin motor cooperate. Some time ago we and others discovered that rabaptin-5 interacts with the ear domain of γ-adaptin, a large subunit of AP-1 [53, 81, 82]. The binding domain for γ-adaptin on rabaptin-5 is in the hinge between the N and C-terminal coiled coil regions, and is positioned immediately C-terminal of the rab4 binding site. AP-1 and rabaptin-5 colocalize on endosomal membranes but not on the trans Golgi network [53]. Since rabaptin-5 is associated with endosomes, it is likely that rabaptin-5 in cooperation with other factors, recruits or stablizes AP-1 to a membrane but not vice versa. The association of the complex with endosomes requires the γ-adaptin binding site in the context of the rab4 binding domain. A rabaptin-5 construct without the rab4 binding domain revealed that the rab5 binding domain cannot replace the rab4 binding domain since it failed to recruit AP-1 to endosomes. The endosomal domain that AP-1 is recruited to is defined by rab4 because co-transfection of rab4 but not rab5 recruits rabaptin-5 and AP-1 [53]. In a biochemical correlate, rab4 and rab5 both interact with rabaptin-5, but only rab4 binds both rabaptin-5 and AP-1.

Recently, we uncovered a second link between the rab4 domain and AP-1. In two hybrid interaction assays we found that rab4 in the GDP form bound directly and independently of rabaptin-5 to AP-1 (Fig. **1A**). Other endosomal rabs that we tested did not bind to AP-1 (Fig. **1B**). The interaction occurred through the γ-adaptin subunit, and was not seen with the large subunits of AP-2 and AP-3 (Fig. **1C**). GST pull down experiments confirmed the association of AP-1 with rab4 (D), and also showed that GGA1 and GGA2 did not interact (Fig. **1E**). The weak signal for GGA3-S likely represents non-specific binding since rab4 was reported not to associate with GGA3 [83]. What might be the function of this interaction between rab4GDP and AP-1? It is possible that AP-1 acts as an accessory factor assisting in delivery of cytoplasmic rab4GDP to the endosomal membrane. Upon guanine nucleotide exchange, rab4 recruit rabaptin-5, thereby creating favorable conditions for a rabaptin-5 interaction with AP-1 at that site. The two distinct modes for coupling rab4 to AP-1 could provide an additional layer of specificity to control recruitment of machinery for formation of this class of endosomal vesicles on the rab4 domain. These could be different from vesicles with a GGA coat, since GGAs like AP-1 bind to rabaptin-5 [82, 84] but not to rab4. Moreover, rab4 could recruit additional effectors, amongst which are those that regulate the level of endosomal PI4P, a phosphatidylinositol phosphate that emerged as an important membrane component in regulating AP-1 coat assembly *in vivo* and *in vitro* [85, 86].

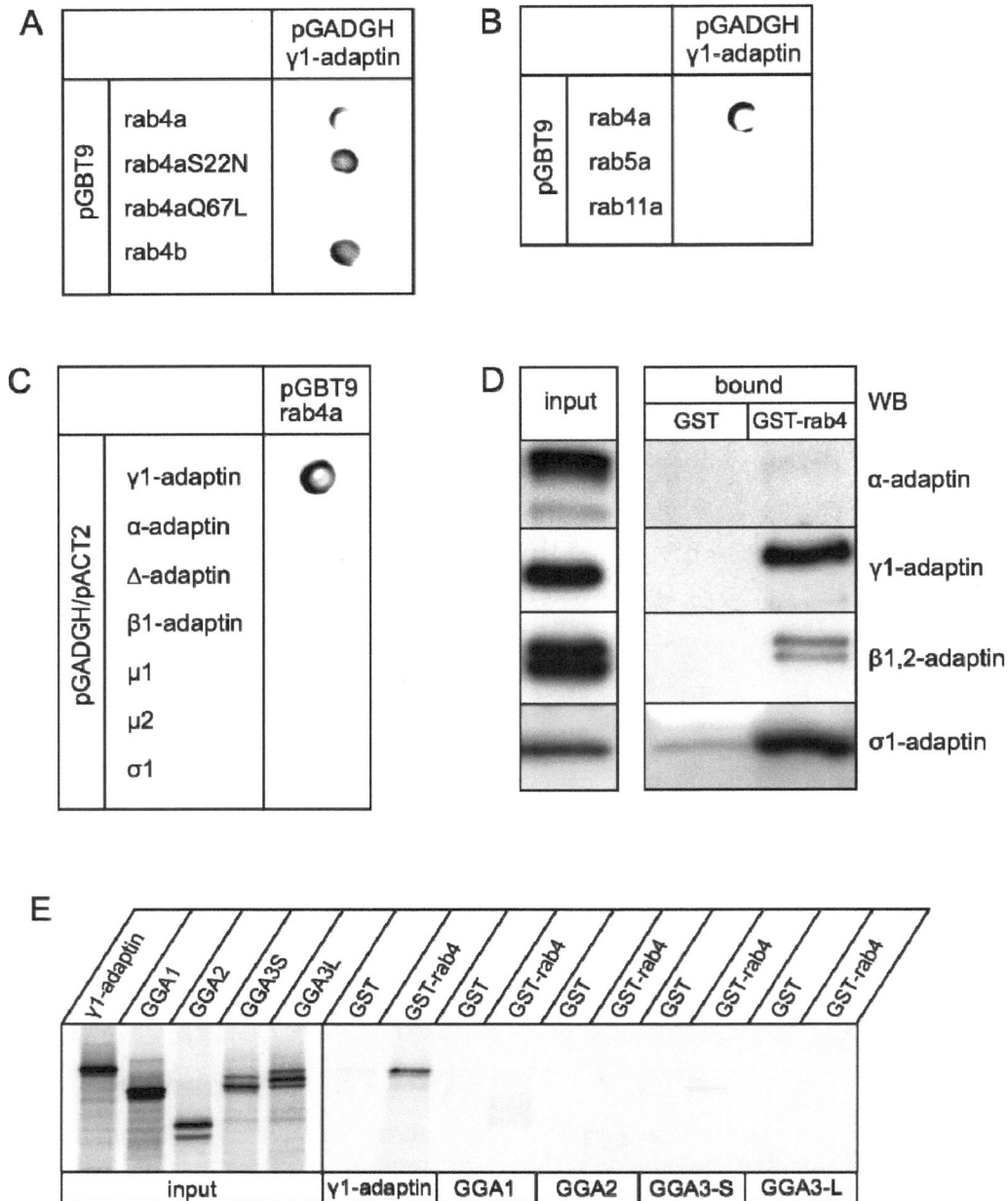

Figure 1: Novel interaction between Rab4 and AP-1. Two hybrid assays between indicated rab4 constructs and γ1-adaptin documenting binding between wild type rab4a/rab4b, and the GDP mutant to γ1-adaptin (**A**). Two hybrid assay showing that rab4, but not the other endosomal rabs, interacts with γ1-adaptin (**B**). Two hybrid assay showing that, with the exception of γ1-adaptin, other large, medium and small subunits do not bind rab4 (**C**). Pull-down binding assay with GST-rab4 and brain cytosol. Bound adaptin subunits were detected by Western blot (**D**). Pull-down binding assay with GST-rab4 and various ^{35}S-labeled adaptor proteins produced in a coupled *in vitro* transcription translation reaction. Bound protein was assayed by phsophorimaging (**E**).

13. RAB4 IN POSITIONING AND MOTILITY OF ENDOCYTIC VESICLES

How do the protein-protein interactions of the rab4 network contribute to the spatial organization of the recycling pathway? A distinct possibility might be *via* involvement of the microtubule cytoskeleton. Such interactions could remodel the rab4 domain by providing support for extension of membrane tubules that either could pinch-off or generate vesicles. So it could be that AP-1 assists in endosome tubulation *via* microtubule binding proteins. Alternatively, AP-1 might link the endosomal vesicles *via* motor proteins to the microtubule network to assist in their movement. An increasing number of studies feature a role for the AP-1 complex in conjunction with kinesin motors in motility and positioning of various classes of endosomal vesicles. For instance KIF13A associates with AP-1 for chronic positioning of recycling endosomes towards the cell periphery and might pull tubules from vacuolar endosomal domains towards melanosomes [87]. KIF13A also associates with β_1-adaptin to re-distribute vesicles containing mannose 6-phosphate receptor from the trans Golgi network to the plasma membrane [88]. Wolkoff *et al.* developed reconstitution assays for endocytic vesicle motility and fission on microtubules *in vitro*. These vesicle preparations contain the interacting kinesins KIF2 and KIF5, and rab4, and their dynamic properties are regulated by the guanine nucleotide status of rab4 [89-91]. KIF5 was recently shown to associate with Gadkin/γBAR, a protein that binds to the ear domain of γ-adaptin [54]. Transfection of Gadkin/γBAR just like rabaptin-5, prevents BFA-induced dissociation of AP-1 from endosomal membranes [53, 92]. Functional experiments suggest that Gadkin acts as a regulator of endosomal membrane traffic. It does so by participating in a ternary complex with AP-1 and KIF5 which provides a direct molecular link between a class of endosomal vesicles and the microtubule-based cytoskeleton [54]. From the above examples and many other studies on AP-1 an AP-2 adaptors, a generic model emerges in which the ear domain of the large adaptin subunits acts as a platform for docking of accessory proteins that assist in adaptor protein complex function [93, 94]. A subgroup of these interacting proteins such as Gadkin/γBAR and rabaptin-5 provides the physical bridge to link the adaptor and a rab that is regulating transport in this pathway. The specificity in this model is conferred by the particular rab/effector combination, since AP-1, cargo molecules it can bind to as well as the motor proteins usually are found at multiple locations in a cell. Additional scenarios for the crosstalk of rab4 with motor proteins can also be envisaged. For instance in fat cells a class of endosome-derived rab4-containing vesicles traffics GLUT4 in a plus end directed manner to the cell surface. This insulin induced GLUT4 exocytosis can be selectively reduced by microinjection of antibody against KIF3, implicating the minus end directed microtubule motor in this process. Co-immunoprecipitation shows that rab4 physically associates with KIF3, suggesting a mechanism for GTP-hydrolysis regulated association of the organelle with KIF3 [95]. Finally rab4 has also been implicated to play a role in minus end directed motility of endosomes given its interaction with cytoplasmic dynein [96].

14. COUPLING OF ENDOSOMAL RAB4 AND RAB11 DOMAINS

As mentioned, rab5, rab4, and rab11 nucleate sequentially organized domains in the endosomal system through which internalized Tf transits. The rab11 and rab4 domain partially overlap, but little if any direct cross-talk occurs between the rab5 and rab11 domains [46, 97]. So far few proteins, including rab coupling protein (RCP) [98] and dual specificity kinase anchoring protein 2 (D-AKAP2) were reported to interact directly with rab4 or with rab11 [99]. More precise evaluations suggest that RCP is a rab11, rather than a rab4 binding protein [100, 101], while a role of D-AKAP2 in integrating rab4 and rab11 activities on endosomes needs additional attention. It remains to be established however whether rab4 and rab11 domains are coupled according to similar principles as the upstream rab5 and rab11 domains. Conceptually this may reflect the observation that at least part of rab11-dependent trafficking occurs *via* endocytic vesicles that also contain rab4 [47].

Recently the neuron-enriched protein GRASP-1 [102] was discovered as a new effector for rab4. GRASP-1 is important in recycling of AMPA receptors to the plasma membrane of spines in postsynaptic neurons [19]. Although it does not bind directly to rab11, it clearly is a regulator of rab4 and rab11 domains. The GRASP-1 expression level directly correlates with the extent of overlapping rab4 and rab11 domains [19]. Increased expression of GRASP-1 enhances rab4 and rab11 colocalization, while rab5 - rab4 coupling is decreased. Knock down of GRASP-1 in neurons produces the opposite effects; increased rab5-rab4 domain coupling, and decreased rab4-rab11 coupling. The link between rab11 domain and GRASP-1 is provided by syntaxin13, that is enriched in the rab11 domain [103] and binds to C-terminus of GRASP-1. GRASP-1 is expressed at high levels in neurons but cannot be detected in most tissue culture cells [19, 102]. We anticipate therefore that additional more ubiquitously expressed proteins are employed to link the rab4 and rab11 domains in less specialized cells.

15. REGULATION OF RAB4

Like other small GTPases, the activity of rab4 is controlled by GTP hydrolysis and GDP exchange. A candidate GTP hydrolysis activating protein (GAP) for rab4 was discovered when Barr *et al.* re-evaluated the specificity of the TBC domain protein GAPCenA towards rab6 [104, 105]. Experiments with the full length protein revealed that GTPase stimulating activity towards rab4 was the highest of all rab proteins tested [105]. A role of GAPCenA in endosomal recycling has not been established, but if silencing of GAPCenA produces a phenotype in the pathway, this would be very strong evidence for a specific function as a GAP for rab4. Cell-type or tissue-specific functions of rab4 can be achieved through the recruitment of cell type specific effectors as we have seen with GRASP-1 in neurons. It is equally possible that tissue or developmentally regulated expression of a GAP can exert specific control over rab4 activity in a particular cell type. For instance an insulin-inducible rab4 GAP activity was found in a plasma membrane fraction isolated from adipocytes [106]. Because the insulin receptor pathways are limited to few cell types, the fat cell rab4 GAP is likely to be different from GAPCenA. Its molecular identity however remains to be determined. A guanine nucleotide exchange factor (GEF) for rab4 is not known. A conserved SEC7 domain as in the ARF GEFs [107] or CDC25 domain in ras GEFs [108] has not been identified in rab GEFs. Instead it is more likely that related rabs may employ GEFs with conserved domains. The GEFs for rab8 [109], rab3 [110], and sec4 [111] contain a SEC2 domain, while *e.g.,* rabex-5 which is a GEF for rab5 and, VARP for rab21 contain a vps9 domain [112-114].

Rab4 protein might be regulated in several other manners. We already mentioned that its expression is downregulated in adipocytes, and can be controlled by the CIITA genetic module in antigen presenting cells. Two other examples are discussed below where rab4 is regulated either at the level of expression, or through hijacking by a micro-organism. Human immunodeficiency virus type 1 (HIV-1) tat gene stimulates transcription of *RAB4A via* trans-activation of its promotor [115]. Transfection of HIV-1 tat in Hela cells or infection of Jurkat cells by HIV-1 increased rab4a expression and inhibits surface expression of CD4, probably through mistargeting to lysosomes [116]. A variation on this theme is provided by Chlamydiae-infected cells. Chlamydiae are obligate intracellular bacteria that replicate in nonlysosomal vacuoles, named reticulate bodies. Chlamydia actively remodels its vacuole to avoid fusion with lysosomes [117]. The interplay between host and chlamydial proteins that are involved in the biogenesis of the inclusion are partially understood, and several rabs including rab4a, and rab11 associate with inclusions shortly after infection. Rab4 is recruited to the inclusion membrane by the chlamydial inclusion protein CT229 which is secreted during the initial stages of infection [118]. Rab4a and rab11a silencing does not affect formation of inclusions, suggesting that this process is independent of rab4A or rab11A [119]. Although the role of rab4 has not been clarified, rab11 is needed for Chlamydia development and Golgin-84-dependent Golgi fragmentation [119, 120]. Thus specific chlamydial proteins recruit key regulators of membrane trafficking to the inclusion, which may function to regulate the maturation and dynamic properties of the inclusion [121, 122].

16. CONCLUDING REMARKS AND PERSPECTIVES

Some 20 years ago we already had a fairly accurate morphological description of endocytc compartments and the pathways connecting them. A major obstacle for further progress was the near complete oblivion about the molecular principles responsible for executing the distinct stages in trafficking beween endosomal compartments. With the discovery of rab GTPases and the recognition of their critical role as coordinators of membrane transport, this situation has changed dramatically. Structures have been determined for several rabs in complex with effectors and regulatory molecules, giving molecular insights how the rab GTPase switch can regulate membrane-associated processes. In a tour de force rab5 dependent fusion was recently reconstituted with completely synthetic components. The identification of multiple effectors for single rabs posed new questions how a rab can manage to coordinate the activity of these effectors in time. Over the years, the role of rab4 as regulator of membrane and protein recycling from early endosomes has been firmly established. This knowledge then contributed to the appreciation of the significance of endosomal recycling pathways for functions in cells and multicellular organisms. Important questions that lie ahead of us concern identification of cascades used to transduce extracellular signals to the activation/termination of rab4 function in endosome recycling. Another challenge is the significance of mitotic phosphorylation of rab4 and several other rabs of biosynthetic and endocytic pathways [30, 123, 124]. Dramatic cyto-architectural changes occur during this stage of the cell cycle [125] where many steps in membrane transport are coordinately inhibited in complex eukaryotes [126]. In other words how does the phosphorylation of rab4 impinge on delivery of membrane from the endosomal system to the cell surface during cytokinesis. Once rab4 has been converted to the GTP-bound form, how does effector recruitment cooperate with rab4 binding an increasing number of membrane proteins like CFTR [127], P-glycoprotein [128], ENaC [129], CD4, and TCRzeta [116]. Supposedly such interactions contribute to the incompletely understood role of rab4-dependent sorting and packaging into carriers for recycling. Further investigations of rab4 in multicellular model organisms will likely lead to a grasp on how rab4a and rab4b isoforms contribute to development and tissue homeostasis.

ACKNOWLEDGEMENTS

This work was supported through grants of the Netherlands Organization for Scientific Research (NWO-CW, NWO-ALW, ZON-MW to PvdS). We would like to thank Marc Peski for help with preparing the figure.

CONFLICT OF INTEREST

There is no conflict of interest from any of the authors.

REFERENCES

[1]   Gruenberg J. The endocytic pathway: a mosaic of domains. Nat Rev Mol Cell Biol 2001;2:721-30.

[2]   Gould GW, Lippincott-Schwartz JL. New roles for endosomes: from vesicular carriers to multipurpose platforms. Nat Rev Mol Cell Biol 2009;10:287-92.

[3]   Grant BD, Donaldson JG. Pathways and mechanisms of endocytic recycling. Nat Rev Mol Cell Biol 2009;10:597-608.

[4]   Mayor S, Pagano S. Pathways of clathrin-independent endocytosis. Nat Rev Mol Cell Biol 2007;8:603-11.

[5]   Johannes L, Popoff V. Tracing the retrograde route in protein trafficking. Cell 2008;135:1175-87.

[6]   Newpher TM, Ehlers MD. Glutamate receptor dynamics in dendritic microdomains. Neuron 2008;58:472-97.

[7]   Trombetta ES, Mellman I. Cell biology of antigen processing *in vitro* and *in vivo*. Annu Rev Immunol 2005;23:975-1028.

[8]   Mellman I, Nelson WJ. Coordinated protein sorting, targeting and distribution in polarized cells. Nat Rev Mol Cell Biol 2008;9:833-45.

[9]   Simon GC, Prekeris R. Mechanisms regulating targeting of recycling endosomes to the cleavage furrow during cytokinesis. Biochem Soc Trans 2008;36:391-4.

[10]   Fürthauer M, Gonzalez-Gaitan M. Endocytosis, asymmetric cell division, stem cells and cancer: Unus pro omnibus, omnes pro uno. Mol Oncol 2009;3:339-53.

[11]   Fürthauer M, Gonzalez-Gaitan M. Endocytic regulation of Notch signaling during development. Traffic 2009;10:792-802.

[12]   Stenmark H. Rab GTPases as coordinators of vesicle traffic. Nat Rev Mol Cell Biol 2009;10:513-25.

[13]   D'Souza-Schorey C, Chavrier P. ARF proteins: roles in membrane traffic and beyond. Nat Rev Mol Cell Biol 2006;7:347-58.

[14]   Südhof TC, Rothman JE. Membrane fusion: grappling with SNARE and SM proteins. Science 2009;233:474-7.

[15]   Singer-Krüger SB, Stenmark H, Dusterhoft A, *et al.* Role of three rab5-like GTPases, Ypt51p, Ypt52p, and Ypt53p, in the endocytic and vacuolar protein sorting pathways of yeast. J Cell Biol 1994;125:283-98.

[16]   Schluter OM, Schmitz F, HJahn R, Rosenmund C, Südhof TC. A complete genetical analysis of neuronal rab3 function. J Neurosci. 2004;24:6629-37.

[17]   Ostrowski M, Carmo NB, Krumeich S, *et al.* Rab27a and rab27b control different steps of the exosome secretion pathway. Nature Cell Biol 2010;12:1-13.

[18]   Johnson JL, Brzezinska AA, Tolmachova T, *et al.* Rab27a and Rab27b regulate neutrophil azurophilic granule exocytosis and NADPH oxidase activity by independent mechanisms. Traffic 2010;11:533-47.

[19]   Hoogenraad CC, Popa I, Futai K, *et al.* Neuron specific rab4 effector GRASP-1 coordinates membrane specialization and maturation of recycling endosomes. PLoS Biol 2010;8:e1000283.

[20]   Krawczyk M, Leimgruber E, Seguín-Estévez Q, Dunand-Sauthier I, Barras E, Reith W. Expression of RAB4B, a protein governing endocytic recycling, is co-regulated with MHC class II genes. Nucl Acid Res 2007;35:595-605.

[21]   van der Sluijs P, Hull M, Zahraoui A, Tavitian A, Goud B, Mellman I. The small GTP binding protein rab4 is associated with early endosomes. Proc Natl Acad Sci USA 1991;88:6313-7.

[22]   Daro E, van der Sluijs P, Galli T, Mellman I. Rab4 and cellubrevin define different early endosome populations on the pathway of transferrin receptor recycling. Proc Natl Acad Sci USA 1996;93:9559-64.

[23]   Folsch H, mattila PE, Weisz OA. Taking the scenic route: biosynthetic traffic to the plasma membrane in polarized epithelial cells. Traffic 2009;10:972-81.

[24]   Mohrmann K, Leijendekker R, Gerez L, van der Sluijs P. Rab4 regulates transport to the apical plasma membrane in madin-darby canine kidney cells. J Biol Chem 2002;277:10474-81.

[25]   de Wit H, Lichtenstein Y, Geuze H, Kelly RB, van der Sluijs P, Klumperman J. Synaptic vesicles form by budding from tubular extensions of sorting endosomes in PC12 cells. Mol Biol Cell 1999;10:4163-76.

[26]   de Wit H, Lichtenstein Y, Kelly RB, Geuze HJ, Klumperman J, van der Sluijs P. Rab4 regulates formation of synaptic-like microvesicles from early endosomes in PC12 cells. Mol Biol Cell 2001;12:3703-15.

[27]   Hayes GL, Brown FC, Hass AK, Nottingham RM, Barr FA, Pfeffer SR. Multiple rab GTPase binding sites in GCC185 suggest a model for vesicle tethering at the trans Golgi. Mol Biol Cell 2009;20:209-17.

[28]   Sinka R, Gillingham AK, Kondylis V, Munro S. Golgi coiled coil proteins contain multiple binding sites for rab family G proteins. J Cell Biol 2008;183:607-15.

[29]   van der Sluijs P, Hull M, Webster P, Goud B, Mellman I. The small GTP binding protein rab4 controls an early sorting event on the endocytic pathway. Cell 1992;70:729-40.

[30]   Gerez L, Mohrmann K, van Raak M, *et al.* Accumulation of rab4 GTP in the cytoplasm and association with the peptidyl-prolyl isomerase Pin1 during mitosis. Mol Biol Cell 2000;11:2201-11.

[31]   Odorizzi G, Pearse A, Domingo D, Trowbridge IS, Hopkins CR. Apical and basolateral endosomes of MDCK cells are interconnected and contain a polarized sorting mechanism. J Cell Biol 1996;135:139-52.

[32]   Odorizzi G, Trowbridge IS. Structural requirements for basolateral sorting of the human transferrin receptor in the biosynthetic and endocytc compartments of MDCK cells. J Cell Biol 1997;137:1255-67.

[33]   Peyroche A, Antonny B, Robineau S, Acker J, Cherfils J, Jackson CL. Brefeldin A acts to stabilize an abortive ARF-GDP-Sec7 domain protein complex: involvement of specific residues of the Sec7 domain. Mol Cell 1999;3:275-85.

[34]   Dittie AS, Hajibagheri N, Tooze SA. The AP-1 adaptor complex binds to immature secretory granules from PC12 cells, and is regulated by ADP-ribosylation factor. J Cell Biol 1996;132:523-36.

[35]   Zhu Y, Traub LM, Kornfeld S. ADP-ribosylation factor 1 transiently activates high affinity adaptor protein complex AP-1 binding sites on Golgi membranes. Mol Biol Cell 1998;9:1323-37.

[36]   Zhu Y, Drake MT, Kornfeld S. ADP-ribosylation factor 1 dependent clathrin-coat assembly on synthetic liposomes. Proc Natl Acad Sci USA 1999;96:5013-8.

[37]   Zhu Y, Traub L, Kornfeld S. High-affinity binding of the AP-1 adaptor complex to trans Golgi network membranes devoid of mannose 6-phosphate receptors. Mol Biol Cell 1999;10:537-49.

[38]   Mohrmann K, Gerez L, Oorschot V, Klumperman J, van der Sluijs P. A cytoplasm-membrane cycle is essential for the function of rab4. J Biol Chem 2002;277:32039-5.

[39]   Kaddai V, Gonzalez T, Keslair F, *et al.* Rab4b is a small GTPase involved in the control of the glucose transporter GLUT4 localization in adipocyte. PLoS One 2009;4:e5257.

[40]   Krawczyk M, Seguin-Estevez Q, Leimgruber E, *et al.* Identiifcation of CIITA regulated genetic module dedicated for antigen presentation. PLoS Genet 2008;4:e1000058.

[41]   Lazzarino DA, Blier P, Mellman I. The monomeric guanosine triphosphatase rab4 controls an essential step on the pathway of receptor mediated antigen processing in B cells. J Exp Med 1998;188:1769-74.

[42]   Nagelkerken B, van Anken E, van Raak M, *et al.* Rabaptin4, a novel effector of rab4a, is recruited to perinuclear recycling vesicles. Biochem J 2000;346:593-601.

[43]   Neeft M, Wieffer M, de Jong AS, *et al.* Munc13-4 is an effector of rab27a and controls secretion of lysosomes in haematopoietic cells. Mol Biol Cell 2005;731-41.

[44]   Pereira-Leal JB, Seabra MC. The mammalian rab family of small GTPases: definition of family and subfamily sequence motifs suggests a mechanism for fundamental specificity in the ras superfamily. J Mol Biol 2000;301:1077-87.

[45]   Pereira-Leal JB, Seabra MC. Evolution of the Rab family of small GTP-binding proteins. J Mol Biol 2001;313:889-901.

[46]   Sönnichsen B, De Renzis S, Nielsen E, Rietdorf J, Zerial M. Distinct membrane domains on endosomes in the recycling pathway visualized by multicolor imaging of rab4, rab5, and rab11. J Cell Biol 2000;149:901-13.

[47]   Ward ES, Martinez C, Vaccaro C, Zhou J, Tang Q, Ober RJ. From sorting endosomes to exocytosis: association of Rab4 and Rab11 GTPases with the Fc receptor, FcRn, during recycling. Mol Biol Cell 2005;16:2028-38.

[48]   de Leeuw HP, Fernandez-Borja M, Reits EA, *et al.* Small GTP binding protein ral modulates regulated exocytosis of von Willebrand factor by endothelial cells. Arteriosc Thromb Vasc Biol 2001;21:899-904.

[49]   Allaire PD, Marat AL, Dall'A Armi C, di Paolo G, McPherson PS, Ritt B. The Connecdenn DENN domain: a GEF for rab35 mediating cargo-specific exit from early endosomes. Mol Cell 2010;37:370-82.

[50]   Kouranti I, Sachse M, Arouche N, Goud B, Echard A. Rab35 regulates an endocytic recycling pathway essential for the terminal steps of cytokinesis. Curr Biol 2006;16:1719-25.

[51]   Maxfield FR, McGraw TEM. Endocytic recycling. Nat Rev Mol Cell Biol 2004;5:121-32.

[52]   Sheff DR, Daro EA, Hull M, Mellman I. The receptor recycling pathway contains two distinct populations of early endosomes with different sorting functions. J Cell Biol 1999;145:123-39.

[53]   Deneka M, Neeft M, Popa I, *et al.* rabaptin-5a/rabaptin-4 serves as a linker between rab4 and γ1-adaptin in membrane recycling from endosomes. EMBO J 2003;22:2645-57.

[54]   Schmidt MR, maritzen T, Kukhtina V, *et al.* Regulation of endosomal membrane traffic by a gadkin/AP-1/kinesin KIF5 complex. Proc Natl Acad Sci USA 2009;106:15344-9.

[55]   Stoorvogel W, Oorschot V, Geuze HJ. A novel class of clathrin coated vesicles budding from endosomes. J Cell Biol 1996;132:21-34.

[56]   Balklava Z, Pant S, Fares H, Grant BD. Genome-wide analysis identifies a general requirement for polarity proteins in endocytic traffic. Nat Cell Biol 2007;9:1066-73.

[57]   Pagano A, Crottet P, Prescianotto-Baschong C, Spiess M. *In vitro* formation of recycling vesicles from endosomes requires AP-1/clathrin and is regulated by rab4 and the connector rabaptin-5. Mol Biol Cell 2004;15:4990-5000.

[58]   Desnos C, O'Grady LC, Kelly RB. Biogenesis of synaptic vesicles *in vitro*. J Cell Biol 1995;130:1041-50.

[59]   Vitale G, Rybin V, Christoforidis S, *et al.* Distinct rab-binding domains mediate the interaction of rabaptin-5 with GTP-bound rab4 and rab5. EMBO J 1998;17:1941-51.

[60]   Stenmark H, Vitale G, Ullrich O, Zerial M. Rabaptin-5 is a direct effector of the small GTPase rab5 in endocytic membrane fusion. Cell 1995;83:423-32.

[61]   Horiuchi H, Lippe R, McBride HM, *et al.* A novel rab5 GDP/GTP exchange factor complexed to rabaptin-5 links nucleotide exchange to effector recruitment and function. Cell 1997;90:1149-59.

[62]   Korobko EV, Kiselev SL, Korobko IV. Multiple rabaptin-5-like transcripts. Gene 2002;292:191-7.

[63]   Korobko E, Kiselev S, Olsnes S, Stenmark H, Korobko I. The rab5 effector rabaptin-5 and its isoform rabaptin-5delta differ in their ability to interact with the small GTPase rab4. FEBS J 2005;272:37-46.

[64]   Jordens I, Fernandez-Borja M, Marsman M, *et al.* The rab7 effector protein RILP controls lysosomal transport by inducing the recruitment of dynein-dynactin motors. Curr Biol 2001;11:1680-5.

[65]   Simonsen A, Lippe R, Christoforidis S, *et al.* EEA1 links PI(3)K function to rab5 regulation of endosome fusion. Nature 1998;394:494-8.

[66]   Christoforidis S, Miaczynska M, Ashman K, *et al.* Phosphatidylinositol-3-OH kinases are rab5 effectors. Nat Cell Biol 1999;1:249-52.

[67]   Ellson CD, Gosse SG, Anderson KE, *et al.* PtdIns(3)P regulates the neutrophil oxidase complex by binding to the PX domain of p40phox. Nat Cell Biol 2001;3:679-82.

[68]   Nielsen E, Christoforidis S, Utenwiler-Joseph S, *et al.* Rabenosyn-5, a novel rab5 effector, is complexed with hVPS45 and recruited to endosomes through a FYVE finger domain. J Cell Biol 2000;151:601-12.

[69]   Cormont M, Mari M, Galmiche A, Hofman P, Le Marchand-Brustel Y. A FYVE finger-containing protein, rabip4, is a rab4 effector protein involved in early endosomal traffic. Proc Natl Acad Sci USA 2001;98:1637-42.

[70]   Fouraux M, Deneka M, Ivan V, *et al.* rabip4' is an effector of rab5 and rab4 and regulates transport through early endosomes. Mol Biol Cell 2004;15:611-24.

[71]   Zhu G, Zhai P, Liu J, Terzyan S, Li G, Zhang C. Structural basis of rab5-rabaptin5 interaction in endocytosis. Nat Struct Mol Biol 2004;11:975-83.

[72]   Rios EJ, Piliponsky AM, Ra C, Kalesnikoff J, Galli SJ. Rabaptin-5 regulates receptor expression and functional activation in mast cells. Blood 2009;112:4148-57.

[73]   Cosulich SC, Horiuchi H, Zerial M, Clarke PR, Woodman PG. Cleavage of rabaptin-5 blocks endosome fusion during apoptosis. EMBO J 1997;16:6182-91.

[74]   Swanton E, Bishop N, Woodman P. Human rabaptin-5 is selectively cleaved by caspase-3 during apoptosis. J Biol Chem 1999;274:37583-90.

[75]   Korobko EV, Palgova IV, Kiselev SL, Korobko IV. Apoptotic cleavage of rabaptin-5-like proteins and a model for rabaptin-5 inactivation in apoptosis. Cell Cycle 2006;5:1854-8.

[76]   Molmeret M, Zink SD, Han L, *et al.* Activation of caspase-3 by the Dot/Icm virulence system is essential for arrested biogenesis of the Legionella-contianing phagosome. Cell Microbiol 2004;6:33-48.

[77]   Ensminger AW, Isberg RR. Legionella pneumophila Dot/Icm translocated substrates: a sum of parts. Curr Opin Microbiol 2009;12:67-73.

[78]   Isberg RR, O'Connor TJ, Heidtman M. The Legionella pneumophila replication vacuole: making a cosy niche inside host cells. Nat Rev Microbiol 2009;7:13-24.

[79]   Machamer CE. Golgi disassembly in apoptosis: cause or effect. Trends Cell Biol 2003;13:279-81.

[80]   Hicks SW, Machamer CE. Golgi structure in stress sensing an apoptosis. Biochim Biophys Acta 2005;1744:406-14.

[81]   Shiba Y, Takatsu H, Shin HW, Nakayama K. γ-adaptin interacts directly with rabaptin-5 through its ear domain. J Biochem 2002;131:327-36.

[82]   Mattera R, Arighi CN, Lodge R, Zerial M, Bonifacino JS. Divalent interaction of the GGAs with the rabaptin-5-rabex-5 complex. EMBO J 2003;22:78-88.

[83]   Dell'Angelica EC, Puertollano R, Mullins C, *et al.* GGAs: a family of ADP ribosylation factor-binding proteins related to adaptors and associated with the Golgi complex. J Cell Biol 2000;149:81-93.

[84]   Miller GI, Mattera R, Bonifacino JS, Hurley JH. Recognition of accessory protein motifs by the g-adaptin ear domain of GGA3. Nat Struct Biol 2003;10:599-606.

[85]   Wang YJ, Wang J, Sun HQ, *et al.* Phosphatidylinositol 4 phosphate regulates targeting of clathrin adaptor AP-1 complexes to the Golgi. Cell 2003;114:299-310.

[86]   Baust T, Czupalla C, Krause E, Bonnet LB, Hoflack B. Proteomic analysis of adaptor protein 1A coats selectively assembled on liposomes. Proc Natl Acad Sci USA 2006;103:3159-64.

[87]   Delevoye C, Hurbain I, Tenza D, *et al.* AP-1 and KIF13A coordinate endosomal sorting and positioning during melanosome biogenesis. J Cell Biol 2009;187:247-64.

[88]   Nakagawa T, Setou M, Seog DH, *et al.* A novel motor, KIF13A, transports mannose 6-phosphate receptor to plasma membrane through direct interaction with AP-1 complex. Cell 2000;103:568-81.

[89]   Bananis E, Murray JW, Stockert RJ, Satir P, Wolkoff AW. Microtubule and motor-dependent endocytic vesicle sorting *in vitro*. J Cell Biol 2000;151:179-86.

[90]   Bananis E, Murray JW, Stockert RJ, Satir P, Wolkoff AW. Regulation of early endocytic vesicle motility and fission in a reconstituted system. J Cell Sci 2003;116:2749-61.

[91]   Nath S, Bananis E, Sarkar S, *et al.* Kif5B and Kifc1 interact and are required for motility and fission of early endocytic vesicles in mouse liver. Mol Biol Cell 2007;18:1839-49.

[92]   Neubrand VE, Will RD, Mobius W, *et al.* gamma-BAR, a novel AP-1 interacting protein involved in post-Golgi trafficking. EMBO J 2005;24:1122-33.

[93]   Schmid EM, Ford MGJ, Burtey A, *et al.* Role of the AP2 beta-appendage hub in recruiting partners for clathrin-coated vesicle assembly. PLoS Biol 2006;4(9):e262.

[94]   Schmid EM, McMahon HT. Integrating molecular network biology to decode endocytosis. Nature 2007;448:883-8.

[95]   Imamura T, Huang J, Usui I, Satoh H, Bever J, Olefsky JM. Insulin-induced GLUT4 translocation involves protein kinase C-lambda mediated functional coupling between rab4 and the motor protein kinesin. Mol Cell Biol 2003;23:4892-900.

[96]   Bielli A, Thörnqvist PO, Hendrick AG, Finn R, Fitzgerald K, McCaffrey MW. The small GTPase rab4a interacts with the central region of cytoplasmic dynein light intermediate chain. Biochem Biophys Res Commun 2001;281:1141-53.

[97]   de Renzis S, Sönnichsen B, Zerial M. Divalent rab effectors regulate the sub-compartmental organization and sorting function of early endosomes. Nat Cell Biol 2002;4:124-33.

[98]   Lindsay AJ, Hendrick AG, Cantalupo G, *et al.* Rab coupling protein (RCP), a novel rab4 and rab11 effector protein. J Biol Chem 2002;277:12190-9.

[99]   Eggers CT, Schafer JC, Goldenring JR, Taylor SS. D-AKAP-2 interacts with rab4 and rab11 through its RGS domains and regulates transferrin receptor recycling. J Biol Chem 2009;284:32869-80.

[100] Lindsay AJ, McCaffrey MW. Characterization of the rab binding properties of rab coupling protein (RCP) by site-directed mutagenesis. FEBS Lett 2004;571:86-92.

[101] Peden AA, Schonteich E, Chun J, Junutala JR, Scheller RH, Prekeris R. The RCP-rab11 complex regulates endocytic protein sorting. Mol Biol Cell 2004;15:3530-41.

[102] Ye B, Liao D, Zhang X, Zhang P, Dong H, Huganir R. GRASP-1: a neuronal rasGEF associated with the AMPA receptor/GRIP complex. Neuron 2000;26:603-17.

[103] Trischler M, Stoorvogel W, Ullrich O. Biochemical analysis of distinct rab5- and rab11 positive endosomes along the transferrin pathway. J Cell Sci 1999;112:4773-83.

[104] Cuif MH, Possmayer F, Zander H, et al. Characterization of GAPC en A, a GTPase activating protein for rab6, part of which associates with the centrosome. EMBO J 1999;18:1772-82.

[105] Fuchs E, Haas AK, Spooner RA, Yoshimura SI, Lord JM, Barr FA. Specific rab GTPase activating proteins define the Shiga toxin and eGF uptaske pathways. J Cell Biol 2007;177:1133-43.

[106] Bortoluzzi MN, Cormont M, Gautier N, Van Obberghen E, Le Marchand Brustel Y. GTPase activating protein activity for rab4 is enriched in the plasma membrane of 3T3-L1 adipocytes. Possible involvement in the regulation of rab4 subcellular distribution. Diabetologia 1996;39:899-906.

[107] Jackson CL, Casanova JE. Turning on ARF: the sec7 family of guanine nucleotide exchange factors. Trends Cell Biol 2000;10:60-7.

[108] Bos JL, Rehmann H, Wittinghofer A. GEFs and GAPs: critical elements in control of small G proteins. Cell 2007;129:865-77.

[109] Hattula K, Furuhjem J, Arffman A, Peränen J. A rab8-specific GDP/GTP exchange factor is involved in actin remodeling and polarized membrane transport. Mol Biol Cell 2002;13:3268-80.

[110] Brondyk WH, McKiernan CJ, Fortner KA, Stabila P, Holz RW, Macara IG. Interaction cloning of rabin3, a novel protein that associates with the ras-like GTPase rab3A. Mol Cell Biol 1995;15:1137-43.

[111] Walch-Solimena C, Collins RN, Novick PJ. Sec2p mediates nucleotide exchange on sec4p and is involved in polarized delivery of post Golgi vesicles. J Cell Biol 1997;137:1495-509.

[112] Hama H, Tall GR, Horazdovsky B. Vps9p is a guanine nucleotide exchange factor involved in vesicle mediated vacuolar protein transport. J Biol Chem 1999;274:15284-91.

[113] Delprato A, Merithew E, Lambright DG. Structure, exchange determinants, and family wide rab specificity of the tandem helical bundle and vps9 domains of rabex-5. Cell 2004;118:607-17.

[114] Zhang X, He X, Fu XY, Chang Z. Varp is a Rab21 guanine nucleotide exchange factor and regulates endosome dynamics. J Cell Sci 2006;119:1053-62.

[115] Nagy G, Ward J, Mosser DD, et al. Regulation of CD4 expression *via* recycling by HRES-1/RAB4 controls susceptibility to HIV infection. J Biol Chem 2006;281:34574-91.

[116] Fernandez DR, Telarico T, Bonilla E, et al. Activation of mammalian target of rapamycin controls the loss of TCRzeta in lupus T cells through HRES-1/Rab4-regulated lysosomal degradation. J Immunol 2009;182:2063-72.

[117] Fields KA, Hackstadt T. The chlamydial inclusion: escape from the endocytic pathway. Annu Rev Cell and Dev Biol 2002;18:221-45.

[118] Rzomp KA, Moorhead AR, Scidmore MA. The GTPase rab4 interacts with chlamydia trachomatis inclusion membrane protein CT229. Infect Immun 2006;74:5362-73.

[119] Rejman Lipinski A, Heymann J, Meissner C, et al. rab6 and rab11 regulate chlamydia trachomatis development and Golgin-84 dependent Golgi fragmentation. PLOS Pathog 2009;5:e1000615.

[120] Heuer D, Rejman Lipinski A, Machuy N, et al. Chlamydia causes fragmentation of the Golgi compartment to ensure reproduction. Nature 2009;457:731-5.

[121] Rzomp KA, Scholtes LD, Briggs BJ, Whittaker GR, Scidmore MA. Rab GTPases are recruited to chlamydial inclusions in both a species-dependent and species-independent manner. Infect Immun 2003;71:5855-70.

[122] Brumell JH, Scidmore MA. Manipulation of rabGTPase function by intracellular bacterial pathogens. Microbiol Rev 2007;71:636-52.

[123] Bailly E, Touchot N, Zahraoui A, Goud B, Bornens M. p34cdc2 protein kinase phosphorylates two small GTP-binding proteins of the rab family. Nature 1991;350:715-8.

[124] van der Sluijs P, Hull M, Huber LA, Male P, Goud B, Mellman I. Reversible phosphorylation-dephosphorylation determines the localization of rab4 during the cell cycle. EMBO J 1992;11:4379-89.

[125] Boucrot E, Kirchhausen T. Endosomal recycling controls plasma area during mitosis. Proc Natl Acad Sci USA 2007;104:7939-44.

[126] Warren G. Membrane partitioning during cell division. Annu Rev Biochem 1993;62:323-48.

[127] Saxena SK, Kaur S, George C. Rab4GTPase modulates CFTR function by impairing channel expression at plasma membrane. Biochem Biophys Res Commun 2006;341:184-91.

[128] Ferrándiz-Huertas C, Fernández-Carvajal A, Ferrer-Montiel A. Rab4 interacts with the human P-glycoprotein and modulates its surface expression in multidrug resistant K562 cells. Int J Cancer 2010;128:192-05.

[129] Saxena SK, Singh M, Shibata H, Kaur S, George C. Rab4 GTP/GDP modulates amiloride-sensitive sodium channel (ENaC) function in colonic epithelia. Biochem Biophys Res Commun 2006;340:726-33.

CHAPTER 9

Rab11a, Rab8a and Myosin V: Regulators of Recycling and Beyond

James R. Goldenring[*], Joseph T. Roland and Lynne A. Lapierre

Vanderbilt University School of Medicine, Nashville VA Medical Center, USA

Abstract: Recycling of endocytic cargoes utilizes multiple pathways employing Rab small GTPases. Rab11a is associated with slow recycling pathways that are responsible for plasma membrane recycling in non-polarized cells and the apical recycling system in polarized cells. Rab11a assembles multiple protein complexes along the recycling pathway utilizing multiple Rab11-Family Interacting Proteins and Class V myosins. Rab8a also assembles regulatory complexes associated with recycling pathways. While myosin V species are involved in both Rab11a- and Rab8a-dependent pathways, the details of interactions between these trafficking pathways remains unclear.

Keywords: Rab11, Rab8, Myosin V, Rab11-FIP, Vesicle Trafficking.

1. INTRODUCTION

Internalization and recycling of plasma membrane protein components is critical for the maintenance of the surface characteristics of mammalian cells. While these processes are often treated as almost bulk flow mechanisms, the correct assignment of proteins to vesicle trafficking pathways requires multiple decision points to establish categorization for internalization within particular endocytic internalization mechanisms, assignment to and trafficking through sorting endosome compartments, and further assignment into either a degradation pathway to the lysosomes or recycling to the plasma membrane through either rapid or slow recycling pathways. Further complications can also be added in for assignment to secretory lysosome pathways (exosomes) or retrograde pathways to the Golgi apparatus. Finally, it seems increasingly likely that *de novo* secretory pathways can deliver newly synthesized proteins into these endosomal pathways to either reload recycling pathways or utilize their trafficking machinery for delivery to the plasma membranes. Thus, intracellular endosomal systems are anything but linear. This review will focus on emerging concepts of the interconnectivity of trafficking systems and the utilization of slow recycling pathways for coordination of recycling system functions.

2. RAB11A AND SLOW RECYCLING ALONG CLATHRIN-DEPENDENT ENDOCYTOTIC PATHWAYS

The role of Rab11a was first delineated in relation to the recycling of transferrin receptors following clathrin-dependent endocytosis [1]. Expression of a dominant-negative mutant of Rab11a (Rab11aS25N), which maintained a constitutively GDP-bound state, inhibited recycling of transferrin. Since that time, a number of cargoes including receptors and transporters that are internalized by clathrin-dependent mechanisms have been associated with recycling through Rab11a-dependent pathways. These pathways of post-endocytic processing appear to follow an orderly progression in non-polarized cells from Rab5-positive early endosomes to Rab4-positive sorting endosomes to Rab11a-positive recycling endosomes [2]. In polarized epithelial cells, pathways appear to be more specialized. Clathrin-dependent endocytosis occurring from both the apical and basolateral poles is processed into early endosomes and then further processing seems to occur in common sorting endosomes (perhaps the best analog of the sorting endosomes in non-polarized cells) and then cargoes destined for trafficking to the apical membrane are further processed into a discrete apical recycling endosome compartment which contains Rab11a [3, 4]. Cargoes destined for the basolateral membrane do not enter the apical recycling endosome and do not encounter

*Address correspondence to James Goldenring:** Department of Surgery and Cell and Developmental Biology, Epithelial Biology Center, Vanderbilt University School of Medicine, 2213 Garland Avenue, Nashville, TN 37232-2733, USA; Tel: 615-936-3726; Fax: 615-343-1351; E-mail: jim.goldenring@vanderbilt.edu

Guangpu Li and Nava Segev (Eds)

Rab11a [5, 6]. Rab11a appears to mediate a slow apical recycling pathway and basolateral to apical transcytosis (as marked by the trafficking of polymeric IgA receptor in MDCK cells). More recent investigations in MDCK have demonstrated that trafficking along an apical to basolateral transcytotic pathway is mediated by Rab25, rather than Rab11a [7]. Other apical recycling cargoes, such as H/K-ATPase [8-11] or Aquaporin-2 [12], also use the Rab11a-mediated slow recycling pathway following clathrin-dependent endocytosis. However, some apically recycling cargoes (*e.g.,* CFTR) utilize a more rapid recycling pathway mediated by Rab4 [13]. How cargoes select for these pathways is not clear. Only a small number of cargoes have ever been implicated in direct association with Rab11a [14], so the decision in pathway selection is likely made at the interface of multi-protein complexes regulating vesicle trafficking.

3. RAB11-FAMILY INTERACTING PROTEINS (RAB11-FIPS)

Since Rab proteins generally serve as the focus for assembly of multi-protein regulatory complexes, direct isolation and yeast two-hybrid studies led to the identification of a family of Rab11-interacting proteins designated as the Rab11-Family Interacting Proteins (Rab11-FIPs) because of their capacity to interact with all three members of the Rab11-family, Rab11a, Rab11b and Rab25 [15]. The Rab11-FIPs are coded off of five separate genes (Fig. **1**). All of the Rab11-FIP genes code for proteins with carboxyl terminal amphipathic alpha-helical domains responsible for Rab11 binding [15, 16]. The Rab11-FIP1 family has multiple splice variants (Rab11-FIP1A-H) that code for proteins with a range of domains. The other Rab11-FIP proteins assemble a range of functional domains including C2 domains (Rab11-FIP1B and C, Rab11-FIP2 and Rab11-FIP5) or ERM domains (Rab11-FIP3 and Rab11-FIP4).

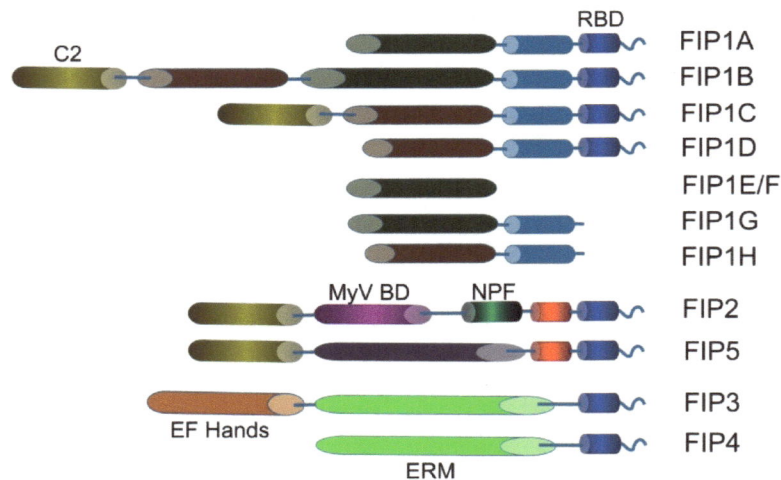

Figure 1: Rab11-Family Interacting Proteins. Schematic of presently identified Rab11-FIP members noting critical regions structural homology: RBD, Rab11 binding domain; C2, C2-domain; ERM, ezrin-radixin-moesin domain. Also noted are the NPF and myosin Vb binding domains in Rab11-FIP2 and the EF-hands in Rab11-FIP5.

Although the absolute amounts of Rab11-FIP proteins seem to vary within individual cells, multiple Rab11-FIPs are usually present within cells. This suggests that these proteins likely define discrete functions within the cell along Rab11-related pathways. In part it appears likely that these functions involve secondary interactions with other proteins. Rab11-FIP1C was originally described as Rab coupling protein (RCP) based on its association with both Rab4 and Rab11 [17]. However, further studies have questioned the functional relevance of Rab4 interactions with Rab11-FIP1C//RCP [18]. More recently, Rab11-FIP1C interactions with Rab14 have also been reported [19], although their exact significance presently remains unclear. The association of Rab11-FIP1C with Golgin-97 has also been associated with the retrograde transport pathway [20]. Rab11-FIP2 uniquely associates with myosin Vb and myosin Va, which also can interact with Rab11a in a functional ternary complex (see below) [15]. In addition, mutants of Rab11-FIP2 have provided some of the most adaptable inhibitors of the plasma membrane recycling system. Rab11-FIP2 (129-512), a truncation lacking its amino-terminal C2 domain, potently inhibits transferrin trafficking and accumulates Rab11a in a collapsed recycling system [21]. Rab11-FIP2(129-512) also potently inhibits basolateral to apical transcytosis [22]. Importantly, Rab11-FIP2(129-512) appears to be extremely specific for Rab11a-dependent trafficking and does not alter Rab8a function or MHC Class I trafficking [23]. Another Rab11-FIP2 mutant, Rab11-FIP2(SARG) also inhibits basolateral to apical transcytosis, but its effects appear to be dependent on the establishment of polarity [22]. Rab11-FIP5, originally described as Rip11 [24], also contains an amino-terminal C2-domain and is associated with recycling systems in both non-polarized and polarized cells. Interestingly, recent investigations indicate that Rab11-FIP5 interacts with kinesin II and regulates early trafficking out of the plasma membrane recycling system [25]. Rab11-FIP3 and Rab11-FIP4 represent a separate sub-class of Rab11-interacting proteins. Rab11-FIP3, originally described as Arfophillin [26], also interacts with Arf6 and has important roles not only in membrane recycling, but also in the delivery of membrane to the cleavage furrow during division [27-30]. Thus, all of these results indicate that the Rab11-FIP proteins are setting up differentiable protein complexes along the Rab11-based recycling system. These complexes may be setting up functional domains within the recycling system. How hand-offs between or coordination of multiple complexes is achieved remains unclear. Nevertheless, it is also important to note that some Rab111-FIP proteins may have functions independent of their interactions with Rab11 family small GTPases. Thus, expression of Rab11-FIP2(S227A), a mutant that lacks the phosphorylation site for Par1b/MARK2, causes a deficit in cell polarization, but has no effect on Rab11a-dependent transcytosis [31]. It seems likely that other Rab11-FIPs will also have Rab11-independent functions. Indeed, three of the Rab11-FIP1 splice variants do not contain the exon with the Rab11-binding domain.

4. RAB8A ASSOCIATION WITH RECYCLING

The investigations of Rab8a-dependent trafficking are far less studied than for Rab11a. In non-polarized cells, Rab8a appears to reside in two separate compartments: a punctate vesicular compartment and an extensive tubular compartment. The long tubular structures marked with Rab8a contain Arf6 and are associated with recycling of the non-clathrin-dependent cargo, MHC Class I [32, 33]. The tubular structures also contain EHD1 and are contiguous and overlapping with tubules containing Rab10 [23, 34]. Rab8a has also been associated with transferrin trafficking [33], but the relationship of Rab8a to Rab11a seems less clear. The two proteins can overlap in their distribution considerably in both the perinuclear region as well as in cell extensions. In live cells, we have observed that Rab11a-containing vesicles moved along Rab8a-containing tubules, but we could not discern any clear point of fusion of vesicles with these tubules [23]. Rab8a distribution is completely unaffected by the expression of Rab11-FIP(129-512), which potently inhibits trafficking through Rab11a-containing recycling systems [23]. Nevertheless, it seems likely that these two systems must share some points of contact, perhaps through shared interactions with myosin V motors.

5. CLASS V MYOSINS AS MULTIFUNCTIONAL REGULATORS OF RECYCLING SYSTEM TRAFFICKING

We originally reported the interaction of myosin Vb with Rab11a in 2000 [6]. This interaction was identified using a yeast two-hybrid screen of a parietal cell library. The interaction of Rab11a with myosin Vb occurs in the cargo-binding domain of the carboxyl globular tail of the myosin Vb molecule. Myosin Vb colocalized with Rab11a in MDCK cells and showed the general dynamics of Rab11a-containing endosomes including dispersal following microtubule disruption with nocodazole and relocation to sub-apical corners with microtubule stabilization with taxol [6]. While these studies were somewhat puzzling in implicating a myosin V motor more as a microtubule-associated protein than as an actin-based motor, continuing results have maintained this association. Myosin Vb carboxyl-terminal tail constructs lacking the neck and motor domains act as powerful dominant negative inhibitors of trafficking through plasma membrane recycling systems and accumulate Rab11a in collapsed membrane cisternae. In non-polarized cells, myosin Vb tail causes accumulation of transferrin receptor and internalized transferrin within these collapsed cisternae and inhibits recycling of transferrin. In contrast, in polarized MDCK cells, transferrin does not accumulate within myosin Vb tail containing cisternae. However, transcytosing polymeric IgA and the polymeric IgA receptor are trapped within myosin Vb tail inhibited membrane cisternae in MDCK cells [6]. These results provide potent data demonstrating the cell specific or epithelial cell specific specialization of Rab11a in regulating trafficking. Since the original studies, the myosin Vb tail has been used in a number of polarized and non-polarized cell systems to demonstrate the association of specific cargoes with membrane recycling systems. In the former case of polarized cells, regulated aquaporin-2 apical recycling is inhibited in renal collecting duct cells by expression of myosin Vb tail [12] and in hepatocytes, expression of myosin Vb tail causes loss of bile canaliculi [35]. Myosin Vb tail also inhibited the assembly and polarized release of Respiratory Synchytial Virus (RSV) [36]. The plasma membrane recycling of a number of cargoes in non-polarized cells including M3-muscarinic receptor [37] and CXCR2 [38] are also inhibited by myosin Vb tail expression. Myosin Va can also interact with Rab11a and expression of myosin Va tail can alter both Rab11a distribution and transferrin trafficking [34]. No interaction of Rab11a can be observed with the third class V myosin, myosin Vc. Interestingly, while both myosin Va and myosin Vb tails can pull in Rab11a, the patterns for these inhibited vesicle systems are strikingly different [34]. Thus, myosin Vb tail causes a dense collapse of the recycling system membranes, while myosin Va tail elicits a more scattered set of membrane puncta throughout the cell. These results suggest that interactions of myosin Va and myosin Vb with Rab11a-containing vesicles may occur at distinct points along the plasma membrane recycling system. Thus, myosin Va and Vb both have potential for regulating Rab11a-mediated recycling system trafficking.

Recently we have also recognized that all three class V myosins can interact with Rab8a, but not Rab8b [23] (Fig. 2). The site of Rab8a interaction with the myosin V species is separable from that for Rab11a. The carboxyl-terminal tails of all three myosin V species inhibit MHC-class I recycling. Interestingly, the patterns for distribution of the myosin V tails are again different. In contrast with the different patterns observed for myosin Va and myosin Vb tails, as described above, myosin Vc tail caused accumulation of Rab8a into large round puncta that were first reported by Cheney and colleagues [39]. Interestingly, movement of myosin Va tail and myosin Vb tail could be observed in low-expressing cells using live cell microscopy [40]. In myosin Vc tail expressing cells, the fluorescent vesicles were generally extremely non-motile, vibrating around a single position. Recently, Cheney and colleagues have described two-different morphologies for myosin Vc-containing elements: a tubular morphology containing Rab8a and also a punctate morphology enriched for secretory vesicle Rabs, such as Rab27b and Rab3D [40]. Klip and colleagues have noted that Rab8a and myosin Vb are both involved in recycling of GLUT4 [41]. Thus, the combination of Rab8a and myosin Vb clearly associate in the recycling of plasma membrane cargoes. It is also important to note that recent studies indicate that Rab10 can also associate with myosin V species [34]. In the case of myosin Va and myosin Vb, Rab10 associates with the alternatively spliced exon D, which is a relatively low abundance isoforms in most epithelial tissues. The role of Rab10 in recycling systems is not clear at this time. However, its association with myosin V species suggests that Rab10 may also impact these pathways through interactions with myosin V motors. All of these findings indicate that myosin V motors are utilized by a number of Rab proteins in trafficking steps along both endocytotic as well as exocytotic pathways (Fig. 2).

It is important to note that both Rab11a and Rab8a interact directly with the myosin Vb tail, as evidenced by positive FRET measurements [23, 34]. While Rab11-FIP2 represents a ternary adapter for Rab11a with myosin Vb, no such adaptor for Rab8a and myosin V has been identified. However, Rab8a does associate with myosin VI through an adaptor protein, optineurin, which directly interacts with myosin VI and is required for localization of myosin VI at the Golgi complex [42]. Rab8a binds to an amino-terminal domain of optineurin (formerly designated FIP-2) [43], while another vesicular protein, Huntingtin, binds to a carboxyl-terminal domain of optineurin. Only the wild-type and the constitutively GTP bound form (Rab8aQ67L) of Rab8 bind to optineurin, while the constitutively GDP bound form (Rab8aT22N) does not bind. Wild-type Rab-8a and Rab8aQ76L co-localized at the Golgi with myosin VI [42]. Interestingly, over-expression of the GTP bound form of Rab8 resulted in long tubular structures that also contained myosin VI, and this recruitment of myosin VI to the long tubular structures requires optineurin. Thus, Rab8a-containing tubules may represent trafficking hubs for a number of sorting events.

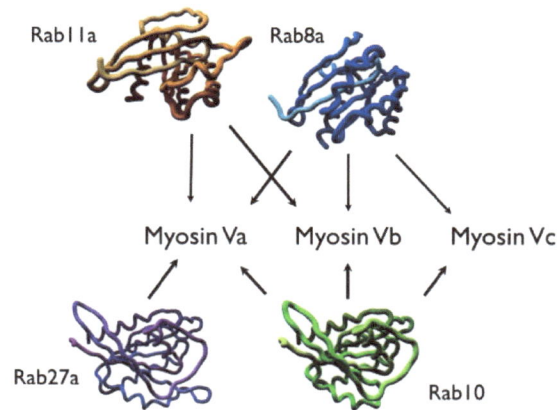

Figure 2: Rab protein binding with myosin motors. Schematic of presently identified interactions among myosin V motors and Rab proteins. Note that the Rab10 interactions are dependent on the presence of the exon D-containing splicing variant of myosin Va and myosin Vb.

6. SWITCHING TRACKS: CAN CERTAIN CARGOES CHANGE POST-ENDOCYTIC TRAFFICKING PATHWAYS?

A critical concept in vesicle trafficking lies in the facilitation of movement through trafficking decision points. For endocytosed cargoes, cells must decide whether to shunt vesicles towards recycling versus degradation. This decision is thought to differentiate movement through a Rab7-containing system versus a Rab4/Rab11a-containing system. Several investigations indicate that this pathway decision may be influenced by both receptor occupancy and by cargo ubiquitination. A particularly good example of the physiology of such a decision point lies in the handling of CXCR2 [38, 44]. Ligand binding for a short amount of time causes internalization of the receptor followed by recycling through a pathway involving Rab11a. However, after prolonged application of the ligand the receptor is redirected to the Rab7 containing late endosomes and the lysosome [38]. Interestingly, expression of myosin Vb tail causes ligand-dependent intracellular accumulation of CXCR2 in Rab11a containing endosomes without shunting to the lysosomes [44]. These results suggest that CXCR2 must go through multiple rounds of internalization and recycling to induce shunting to the lysosome. All of these findings shed light therefore on the dynamic process of decision-making between recycling and degradation. The molecular mechanisms responsible for this decision process remain obscure.

It also appears likely that cargoes can move between the slow and rapid recycling systems. Mercer and colleagues [45] investigated the effects of expression of a myosin Vb mutant that specifically bound a non-hydrolyzable analog of ATP. This chemical mutagenesis was predicted to induce a rigor state.

It had been expected that this mutant would be a potent inhibitor of recycling, similar to the myosin Vb tail. However, instead they observed more transferrin receptor recycling to the surface of the cells. It is interesting to note that knockdown of Rab8a expression also appears to induce shunting of transferrin trafficking into the rapid recycling pathway [33]. These studies indicated that cargoes can be shunted from the slow recycling system to the rapid recycling system.

7. ARE RAB11A AND RAB8A REALLY SPECIFIC TO RECYCLING SYSTEMS?

While the majority of the present literature focuses on the association of Rab11a and Rab8a with endosomal recycling systems, in both cases there is increasing evidence that certain cell systems may utilize these proteins for delivery of *de novo* synthesized proteins to discrete locations in the cell. In the case of Rab11a, this participation in *de novo* trafficking may involve the use of trafficking of newly synthesized proteins from the Golgi apparatus to the recycling system for eventual targeting to the membranes *via* established recycling system pathways. This is especially notable in the case of apical recycling systems in polarized epithelial cells where a portion of newly synthesized apical membrane proteins may find their way to the apical pole of the cell through the recycling system. Alternatively, proteins such as the polymeric IgA receptor make their way to the apical membranes by first trafficking to the basolateral membrane and then relocating through endocytosis and transcytosis from the basal to apical poles through the Rab11a-containing recycling system. In addition, Rab11a is associated with second messenger-regulated recycling systems such as for the H/K-ATPase in gastric parietal cells and aquaporin-2 in kidney collecting duct cells. In these systems, Rab11a is clearly involved in the process of cAMP-dependent regulated exocytosis, although there appears to be a direct connection with the recycling system for forming the regulatory exocytotic apparatus in these systems [46]. Also, at least in the case of parietal cell tubulovesicles, there needs to be a mechanism for refilling the recycling vesicles with newly synthesized intrinsic factor, which is released into the lumen after vesicle fusion. Finally, recent investigations have indicated that Rab11a is involved in trafficking of the exocyst complex in the process of basolateral trafficking in polarized MDCK cells [47]. Thus, Rab11a may be regulating basolateral trafficking of *de novo* synthesized cargoes.

In the case of Rab8a, the question is even more complicated. It is clear that Rab8a participates in recycling of non-clathrin-dependent endocytosed membrane proteins in non-polarized cells, but the functions of Rab8a in polarized cells appear to be distinct and more compatible with *de novo* directed trafficking. Recent investigations have suggested that Rab8a is involved in the directed trafficking of the assembled BBSome, a complex of proteins responsible for mutations seen in Bardet-Biedel Syndrome, to the base of primary cilia in polarized kidney and lung cells [48, 49]. Indeed, recent studies indicate that both Rab11a and Rab8a are involved in the formation and maintenance of the primary cilium [50]. While this pathway may involve piggybacking onto a specialized apical recycling system function, the details of these trafficking steps has not been defined.

Intestinal cells do not form primary cilia. However, Rab8a has been implicated in establishing polarity and the structure of the apical brush border through its association with myosin Vb [51]. Thus, mutations in myosin Vb have been identified as the cause of Microvilllous Inclusion Disease (MVID), a congenital cause of catastrophic neonatal diarrhea, which is caused by an absence of normal brush border and the presence in 10-15% of enterocytes of discernable apical inclusions [52, 53]. While no mutations in Rab8a have been detected in MVID patient, mice with targeted deletion of the Rab8a gene demonstrate some of the morphological and physiological characteristics of MVID, with aberrant apical intestinal membranes [51]. This has led to the concept that Rab8a, perhaps along with Myosin Vb, may be regulating apical trafficking of brush border components. Whether members of the BBSome complex are also involved in this process is not clear, although patients with Bardet-Biedel syndrome do not show defects in intestinal function. Nevertheless, these data all point to the involvement of Rab8a in trafficking pathways that have not traditionally been thought of as recycling.

8. CONCLUSION

Rab11a and Rab8a are multifunctional regulators of vesicle trafficking. While a number of investigations over the past decade have indicated that Rab11a is associated with plasma membrane slow recycling systems, it is clear that Rab11a may also be involved in both the exocytotic and retrograde trafficking pathways. Similarly, while Rab8a is involved in recycling of some cargoes, it is apparent that this small GTPase is also involved in specialized pathways for apical trafficking including the formation of primary cilia. Rab11a and Rab8a both utilize myosin V motors, but how vesicle trafficking pathways interact remains unclear. Future studies will be required to determine whether these Rab proteins can co-regulate individual cargo processing and recycling.

ACKNOWLEDGEMENTS

This work was supported by NIH Grants R01 DK070856 and R01 DK48370 to JRG and and F32 DK072789 to JTR.

CONFLICT OF INTEREST

There is no conflict of interest from any of the authors.

REFERENCES

[1] Ullrich O, Reinsch S, Urbe S, Zerial M, Parton RG. Rab11 regulates recycling through the pericentriolar recycling endosome. J Cell Biol 1996;135:913-24.

[2] Sonnichsen B, De Renzis S, Nielsen E, Reitdorf J, Zerial M. Distinct membrane domains on endosomes in the recycling pahtway visualized by multicolor imaging of Rab4, Rab5 and Rab11. J Cell Biol 2000;149:901-14.

[3] Brown PS, Wang E, Aroeti B, Chapin SJ, Mostov KE, Dunn KW. Definition of distinct compartments in polarized Madin-Darby canine kidney (MDCK) cells for membrane-volume sorting, polarized sorting and apical recycling. Traffic 2000;1:124-40.

[4] Casanova JE, Wang X, Kumar R, *et al.* Rab11a and Rab25 association with the apical recycling system of polarized MDCK cells. Mol Biol Cell 1999;10:47-61.

[5] Wang X, Kumar R, Navarre J, Casanova JE, Goldenring JR. Regulation of vesicle trafficking in Madin-Darby Canine Kidney cells by Rab11a and Rab25. J Biol Chem 2000;275:29138-46.

[6] Lapierre LA, Kumar R, Hales CM, *et al.* Myosin Vb is associated with and regulates plasma membrane recycling systems. Mol Biol Cell 2001;12:1843-57.

[7] Tzaban S, Massol RH, Yen E, *et al.* The recycling and transcytotic pathways for IgG transport by FcRn are distinct and display an inherent polarity. J Cell Biol 2009 May 18;185(4):673-84.

[8] Duman JG, Tyagarajan K, Kolsi MS, Moore HH, Forte JG. Expression of rab11a N124I in gastric parietal cells inhibits stimulatory recruitment of the H+-K+-ATPase. Am J Physiol 1999;277:C361-C72.

[9] Calhoun BC, Goldenring JR. Two Rab proteins, VAMP-2, and SCAMPs are present on immunoisolated gastric tubulovesicles. Biochem J 1997;325:559-64.

[10] Calhoun BC, Lapierre LA, Chew CS, Goldenring JR. Rab11a redistributes to apical secretory canaliculus during stimulation of gastric parietal cells. Am J Physiol 1998;275:C163-C70.

[11] Peng X-P, Yao X, Chow D-C, Forte JG, Bennett MK. Association of syntaxin 3 and vesicle associated membrane protein (VAMP) with H+/K+-ATPase-containing tubulovesicles in gastric parietal cells. Mol Biol Cell 1997;8:399-407.

[12] Nedvetsky PI, Stefan E, Frische S, *et al.* A Role of myosin Vb and Rab11-FIP2 in the aquaporin-2 shuttle. Traffic 2007 Feb;8(2):110-23.

[13] Marino CR, Matovcik LM, Gorelick FS, Cohn JA. Localization of the cystic fibrosis transmembrane conductance regulator in pancreas. J Clin Invest 1991;88:712-6.

[14] van de Graaf SF, Chang Q, Mensenkamp AR, Hoenderop JG, Bindels RJ. Direct interaction with Rab11a targets the epithelial Ca2+ channels TRPV5 and TRPV6 to the plasma membrane. Mol Cell Biol 2006 Jan;26(1):303-12.

[15] Hales CM, Vaerman JP, Goldenring JR. Rab11 family interacting protein 2 associates with Myosin Vb and regulates plasma membrane recycling. J Biol Chem 2002 Dec 27;277(52):50415-21.

[16] Prekeris R, Davies JM, Scheller RH. Identificaition of a novel Rab11/25 binding domain in eferin and Rip proteins. J Biol Chem 2001;276:38966-70.

[17] Lindsay AJ, Hendrick AG, Cantalupo G, *et al.* Rab coupling protein (RCP), a novel Rab4 and Rab11 effector protein. J Biol Chem 2002;277:12190-9.

[18] Peden AA, Schonteich E, Chun J, Junutula JR, Scheller RH, Prekeris R. The RCP-Rab11 complex regulates endocytic protein sorting. Mol Biol Cell 2004 Aug;15(8):3530-41.

[19] Kelly EE, Horgan CP, Adams C, *et al.* Class I Rab11-family interacting proteins are binding targets for the Rab14 GTPase. Biol Cell 2010 Jan;102(1):51-62.

[20] Jing J, Junutula JR, Wu C, *et al.* FIP1/RCP Binding to Golgin-97 Regulates Retrograde Transport from Recycling Endosomes to the Trans-Golgi Network. Mol Biol Cell 2010 Jul 7.

[21] Hales CM, Griner R, Hobdy-Henderson KC, *et al.* Identification and characterization of a family of Rab11-interacting proteins. J Biol Chem 2001 Oct 19;276(42):39067-75.

[22] Ducharme NA, Williams JA, Oztan A, Apodaca G, Lapierre LA, Goldenring JR. Rab11-FIP2 regulates differentiable steps in transcytosis. Am J Physiol Cell Physiol 2007 Sep;293(3):C1059-72.

[23] Roland JT, Kenworthy AK, Peranen J, Caplan S, Goldenring JR. Myosin Vb Interacts with Rab8a on a Tubular Network Containing EHD1 and EHD3. Mol Biol Cell 2007 May 16;18:2828-37.

[24] Prekeris R, Klumperman J, Scheller RH. A Rabll/Rip11 complex regulates apical membrane trafficking *via* recycling endosomes. Mol Cell 2000;6:1437-48.

[25] Schonteich E, Wilson GM, Burden J, *et al.* The Rip11/Rab11-FIP5 and kinesin II complex regulates endocytic protein recycling. J Cell Sci 2008 Nov 15;121(Pt 22):3824-33.

[26] Shin OH, Ross AH, Mihai I, Exton JH. Identification of arfophilin, a target protein for GTP-bound class II ADP-ribosylation factors. J Biol Chem 1999 Dec 17;274(51):36609-15.

[27] Hickson GR, Matheson J, Riggs B, *et al.* Arfophilins are dual Arf/Rab 11 binding proteins that regulate recycling endosome distribution and are related to Drosophila nuclear fallout. Mol Biol Cell 2003 Jul;14(7):2908-20.

[28] Fielding AB, Schonteich E, Matheson J, *et al.* Rab11-FIP3 and FIP4 interact with Arf6 and the exocyst to control membrane traffic in cytokinesis. EMBO J 2005 Oct 5;24(19):3389-99.

[29] Wilson GM, Fielding AB, Simon GC, *et al.* The FIP3-Rab11 protein complex regulates recycling endosome targeting to the cleavage furrow during late cytokinesis. Mol Biol Cell 2005 Feb;16(2):849-60.

[30] Eathiraj S, Mishra A, Prekeris R, Lambright DG. Structural basis for Rab11-mediated recruitment of FIP3 to recycling endosomes. J Mol Biol 2006 Nov 24;364(2):121-35.

[31] Ducharme NA, Hales CM, Lapierre LA, *et al.* MARK2/EMK1/Par-1Balpha phosphorylation of Rab11-family interacting protein 2 is necessary for the timely establishment of polarity in Madin-Darby canine kidney cells. Mol Biol Cell 2006 Aug;17(8):3625-37.

[32] Peranen J, Auvinen P, Virta H, Wepf R, Simons K. Rab8 promotes polarized membrane transport through reorganization of actin and microtubules in fibroblasts. J Cell Biol 1996;135:153-67.

[33] Hattula K, Furuhjelm J, Tikkanen J, Tanhuanpaa K, Laakkonen P, Peranen J. Characterization of the Rab8-specific membrane traffic route linked to protrusion formation. J Cell Sci 2006 Dec 1;119(Pt 23):4866-77.

[34] Roland JT, Lapierre LA, Goldenring JR. Alternative splicing in class V myosins determines association with Rab10. J Biol Chem 2009 Jan 9;284(2):1213-23.

[35] Wakabayashi Y, Dutt P, Lippincott-Schwartz J, Arias IM. Rab11a and myosin Vb are required for bile canalicular formation in WIF-B9 cells. Proc Natl Acad Sci U S A 2005 Oct 18;102(42):15087-92.

[36] Brock SC, Goldenring JR, Crowe JE, Jr. Apical recycling systems regulate directional budding of respiratory syncytial virus from polarized epithelial cells. Proc Natl Acad Sci U S A 2003 Dec 9;100(25):15143-8.

[37] Volpicelli LA, Lah JJ, Fang G, Goldenring JR, Levey AI. Rab11a and myosin Vb regulate recycling of the M4 muscarinic acetylcholine receptor. J Neurosci 2002 Nov 15;22(22):9776-84.

[38] Fan GH, Lapierre LA, Goldenring JR, Richmond A. Differential regulation of CXCR2 trafficking by Rab GTPases. Blood 2003 Mar 15;101(6):2115-24.

[39] Rodriguez OC, Cheney RE. Human myosin-Vc is a novel class V myosin expressed in epithelial cells. J Cell Sci 2002 Mar 1;115(Pt 5):991-1004.

[40] Jacobs DT, Weigert R, Grode KD, Donaldson JG, Cheney RE. Myosin Vc is a molecular motor that functions in secretory granule trafficking. Mol Biol Cell 2009 Nov;20(21):4471-88.

[41] Ishikura S, Klip A. Muscle cells engage Rab8A and myosin Vb in insulin-dependent GLUT4 translocation. Am J Physiol Cell Physiol 2008 Oct;295(4):C1016-25.

[42] Sahlender DA, Roberts RC, Arden SD, *et al.* Optineurin links myosin VI to the Golgi complex and is involved in Golgi organization and exocytosis. J Cell Biol 2005 Apr 25;169(2):285-95.

[43] Hattula K, Peranen J. FIP-2, a coiled-coil protein, links Huntingtin to Rab8 and modulates cellular morphogenesis. Curr Biol 2000 Dec 14-28;10(24):1603-6.

[44] Fan GH, Lapierre LA, Goldenring JR, Sai J, Richmond A. Rab11-family interacting protein 2 and myosin Vb are required for CXCR2 recycling and receptor-mediated chemotaxis. Mol Biol Cell 2004 May;15(5):2456-69.

[45] Provance DW, Jr., Gourley CR, Silan CM, *et al.* Chemical-genetic inhibition of a sensitized mutant myosin Vb demonstrates a role in peripheral-pericentriolar membrane traffic. Proc Natl Acad Sci U S A 2004 Feb 17;101(7):1868-73.

[46] Forte TM, Machen TE, Forte JG. Ultrastructural changes in oxyntic cells associated with secretory function: a membrane recycling hypothesis. Gastroenterology 1977;73:941-55.

[47] Oztan A, Silvis M, Weisz OA, *et al.* Exocyst requirement for endocytic traffic directed toward the apical and basolateral poles of polarized MDCK cells. Mol Biol Cell 2007 Oct;18(10):3978-92.

[48] Nachury MV. Tandem affinity purification of the BBSome, a critical regulator of Rab8 in ciliogenesis. Methods Enzymol 2008;439:501-13.

[49] Nachury MV, Loktev AV, Zhang Q, *et al.* A core complex of BBS proteins cooperates with the GTPase Rab8 to promote ciliary membrane biogenesis. Cell 2007 Jun 15;129(6):1201-13.

[50]   Knodler A, Feng S, Zhang J, *et al.* Coordination of Rab8 and Rab11 in primary ciliogenesis. Proc Natl Acad Sci U S A 2010 Apr 6;107(14):6346-51.

[51]   Sato T, Mushiake S, Kato Y, *et al.* The Rab8 GTPase regulates apical protein localization in intestinal cells. Nature 2007 Jul 19;448(7151):366-9.

[52]   Erickson RP, Larson-Thome K, Valenzuela RK, Whitaker SE, Shub MD. Navajo microvillous inclusion disease is due to a mutation in MYO5B. Am J Med Genet A 2008 Dec 15;146A(24):3117-9.

[53]   Muller T, Hess MW, Schiefermeier N, *et al.* MYO5B mutations cause microvillus inclusion disease and disrupt epithelial cell polarity. Nat Genet 2008 Oct;40(10):1163-5.

CHAPTER 10

Role of Rab7/Ypt7 in Organizing Membrane Trafficking at the Late Endosome

Mirjana Nordmann, Christian Ungermann[*] and Margarita Cabrera

University of Osnabrück, Germany

Abstract: Late endosomal biogenesis depends on the Rab7 GTPase and its interaction with effectors. Within this review, we will focus on the Rab7 activation and inactivation cycle, and identify the critical regulators and interaction partners. We will highlight the role of Rab7 in membrane tethering between late endosome and lysosome, and its function in nutrient sensing, which is coupled to the establishment of membrane contact zones and late endosomal distribution within the cell.

Keywords: Mon1, Ccz1, Rab7, Ypt7, HOPS.

1. INTRODUCTION TO RAB7

The endocytic transport of proteins and lipids is initiated at the plasma membrane. Endocytic vesicles then fuse with the early endosome, which matures into the late endosome prior to its fusion with the lysosome. This maturation is accompanied by a massive remodeling of the endosomal surface, resulting in an organelle that contains intraluminal vesicles and is competent to fuse with the lysosome. Fusion with the lysosome requires the Rab GTPase Rab7, which needs to be loaded onto the maturing endosome prior to fusion. Rab7 is thus part of the fusion machinery that cooperates with tethering complexes and SNAREs to mediate membrane fusion. In addition, Rab7 also coordinates lysosomal biogenesis and its intracellular localization. This review will focus on the interactions and functions of mammalian Rab7 at the late endosome, which are essential for endosomal and lysosomal biogenesis and tightly linked to multiple diseases. We also compare the findings on Rab7 and its homologs in yeast and other organisms.

Rab7 was identified as one of the low molecular weight GTPases and was localized to late endosomes in mammalian cells [1, 2]. Even though several Rab GTPases were found at endocytic structures, including Rab4 and Rab5, Rab7 localization was distinct and overlapped only with the Rab9 GTPase [3, 4]. Further studies then suggested that Rab7 and Rab9 require distinct machineries for their recruitment to late endosomes and may thus fulfill different functions at this organelle [3]. More recent studies then confirmed that GFP-tagged Rab7 colocalizes with the lysosomal markers Lamp-1, Lamp-2 and cathepsin D [5], whereas Rab9 was linked to receptor recycling from the late endosome and lysosome [6]. In yeast, transport to the lysosome equivalent, the vacuole, also depends on the Rab7 homolog Ypt7 [7]. Vacuoles in yeast are large and morphological changes have been taken early as an indicator of deficiencies in vacuole biogenesis [8]. Indeed, loss of Ypt7 resulted in massive vacuole fragmentation and led to multiple sorting defects of vacuolar hydrolases and a deficiency in the degradation of endocytic cargo [7, 9]. Ypt7 was enriched on isolated vacuoles, suggesting that it is directly involved in fusion [10]. In fission yeast, two Rab7 homologs exist (Ypt7 and Ypt71), which seem to conduct antagonistic roles in regulating vacuolar morphology [11, 12]. Similarly, *C. elegans* Rab7 was localized to late endosomes in macrophage-like coelomocytes and its depletion caused defects in transport to the lysosomes [13].

2. RECRUITMENT, ACTIVATION AND TURNOVER OF RAB7

2.1. GDF and GEF

Similar to all known Rab GTPases, Rab7 can exist in two states, the inactive GDP-bound and the active GTP-bound form, in which it binds to its effectors. Within the cytosol, Rab7-GDP is chaperoned by the

---

*Address correspondence to Christian Ungermann: Department of Biology/Chemistry, University of Osnabrück, Barbarastrasse 13, 49076 Osnabrück, Germany; Tel: 49-541-969-2752 Fax: 49-541-969-2884; E-mail: Christian.ungermann@biologie.uni-osnabrueck.de

GDP dissociation inhibitor (GDI). Its recruitment to the late endosome requires GDI-displacement, which may be achieved by a so far unknown GDI displacement factor (GDF). For Rab9, the membrane protein Yip3 functions as a GDF [14]. Recent data on the DrrA protein from *Salmonella* indicate that efficient GEF activity might in some cases be sufficient to displace GDI from the Rab and thus couple GDF and GEF activity [15]. Activation of Rab7 then requires a guanine nucleotide exchange factor (GEF). In yeast, a subunit of the HOPS complex (discussed below), Vps39, has been linked to the Rab7 GEF activity [16]. In agreement with this, knock down of hVps39 results in larger early and late endosomes, and delay in the recruitment of Rab7 onto endosomes [17]. However, other studies rather suggest that Vps39 functions primarily as part of the Rab7 effector, the HOPS complex. Indeed, the expression of a dominant negative Vps39 mutant influenced lysosomal morphology without affecting Rab7-GTP levels, suggesting that Vps39 functions downstream or independent of Rab7 [6, 18]. In agreement with this, Mon1 and Ccz1 have been recently identified as Ypt7 GEF complex instead of Vps39 [19], but a possible role of Vps39 in Ypt7 function was suggested as well [20] (see below). Interestingly, Vps39 also stimulates GTP hydrolysis of the yeast Rag GTPase Gtr1, which is part of the EGO complex linked to the TOR function in microautophagy [21]. How Vps39 performs this postulated dual function is not yet resolved.

2.2. GAP

Inactivation of Rab7 requires a GTPase activating protein (GAP), which is TBC1D15 for Rab7. TBC1D15 is a cytosolic protein with high homology to yeast Gyp7. Even though purified TBC1D15 stimulates the intrinsic GTPase activity of both Rab7 and Rab11 *in vitro* [22], recent *in vivo* experiments point to an exclusive function of TBC1D15 for Rab7. Overexpression of TBC1D15 induces lysosome fragmentation but does not alter internalization and recycling of transferrin, which depend on Rab5 and Rab11, respectively [6]. For yeast Ypt7, Gyp7 was identified as the authentic GAP, which activates the Ypt7 GTPase *in vitro* and on isolated vacuoles, and causes vacuole fragmentation upon overproduction *in vivo* [23-26].

2.3. Nucleotide-Locked Mutants Reveal a Role of Rab7 in the Endocytic Pathway

Rab7 and Ypt7 structures have been resolved. They do not differ much from the known Rab GTPase structures, having only slight changes in the main chain trace [27, 28]. To obtain insight into Rab7 function within the endocytic pathway, GTP or GDP locked mutants have been employed. Two mutants (T22N and N125I) localize poorly to endosomes and cause a dominant negative phenotype. Rab7 T22N displays a reduced affinity for GTP, whereas Rab7 N125I contains a mutation in the guanine nucleotide-binding region (NKXD) and represents the nucleotide-free form. The expression of either mutant results in several defects in the endocytic pathway: accumulation of the vesicular stomatitis virus G glycoprotein in the early endosomes [29], inhibition of low-density lipoprotein (LDL) degradation [5, 30], mislocalization of the cation-independent mannose-6-phosphate receptor (CI-MPR) and its ligand (cathepsin D) to early endosomes [31], delay of cathepsin D processing [31], dispersed lysosomes and impaired lysosome acidification [5], slow epidermal growth factor receptor (EGFR) degradation and EGF-EGFR complex accumulation in late endosomes [32]. A similar phenotype was also obtained, when Rab7 was depleted by RNA interference, indicating that Rab7 mediates fusion of late endosomes with lysosomes [33].

In contrast, the constitutively active mutant Rab7 Q67L exhibits impaired GTP hydrolysis and thus remains in a GTP-bound form. It is basically localized like the wild-type protein, but its expression causes changes in CI-MPR localization [30], increased lysosome fusion [5], and accelerated EGFR degradation [32], in line with the suggested role of Rab7.

3. RAB7 EFFECTORS AT THE LATE ENDOSOME AND LYSOSOME

How does Rab7 then conduct its multiple functions? Rab7 is not only involved in fusion at the late endosome, but also affects the localization of late endosomes and receptor recycling. We will focus here on Rab7 and its known effectors RILP and ORP1L (for late endosomal intracellular positioning and lipid transport), FYCO1 (to redistribute autophagosomes within the cell), Rabring7 (for late endosomal biogenesis), Vps34 (to coordinate early and late endosomal membrane constitution), XAPC7 (a link between endosomes and proteasomal degradation), retromer (to mediate cargo retrieval from the late endosome), Mon1-Ccz1 (to drive endosomal maturation), and Vps41/HOPS (for fusion of late endosomes with lysosomes).

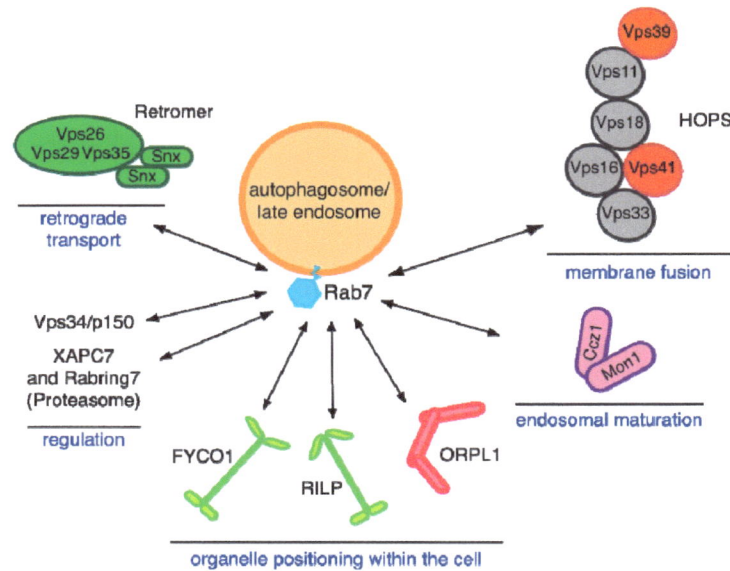

Figure 1: Interactions of Rab7 at the late endosome. For details see text.

3.1. Rab7 Controls Endosome/Autophagosome Positioning

The two proteins RILP and ORP1L both interact with Rab7-GTP (Fig. **1**) and regulate the positioning of late endosomes in response to cellular cholesterol levels. RILP (Rab-interacting lysosomal protein) was identified as a Rab7 effector in a yeast two-hybrid assay [34]. RILP binds specifically to Rab7 and GTP bound Rab7 Q67L and its recruitment on endosomal membranes depends on active Rab7. *In vivo*, RILP colocalizes with endosomal and lysosomal markers. It is required for normal organization of the late endosomes and lysosomes in the perinuclear region and trafficking of LDL and EGF cargo to the lysosomes [34]. Interestingly, RILP can stabilize Rab7 in the GTP-bound state and thus controls the Rab7 cycle [35]. Indeed, the crystal structure of Rab7-GTP in complex with RILP revealed that RILP forms a homodimer that is able to interact simultaneously with two Rab7 molecules [36]. Binding to Rab7 occurs *via* the RILP C-terminal domain, whereas the N-terminal portion interacts with the p150[Glued] subunit of the dynein-dynactin complex responsible for minus-end directed transport and thus connects late endosomes to the cytoskeleton [35, 37]. Rab7-mediated recruitment of RILP occurs in other Rab7-compartments including phagosomes [38], melanosomes [39] and cytosolic granules [40].

RILP is involved in endosomal morphology and multivesicular body biogenesis, potentially by interacting with the ESCRT components Vps22 and Vps36 [41-43]. Another Rab7 effector at the late endosome is ORP1L, a member of the oxysterol-binding protein family, which localizes to late endosomes and lysosomes [44] (Fig. **2**). ORP1L expression affects not only Rab7 nucleotide exchange, but also the distribution of late endosomes and lysosomes. ORP1L is a cholesterol sensor at the late endosome and interacts with the ER protein VAP under conditions of low cholesterol [45]. VAP binding to the Rab7-RILP complex displaces the RILP associated motor and as a consequence, late endosomes move to the plus end of the microtubules and the microtuble-organizing center (MTOC), and establish contact zones with the ER, which presumably facilitate cholesterol exchange. Indeed, cholesterol accumulation on late endosomes stabilizes Rab7 on the organelle [46]. In reverse, if cholesterol levels at the ER are high, the late endosomes accumulate at the minus end due to the RILP-dynein-dynactin interaction. Thus, Rab7-binding to RILP and ORP1L regulates late endosomal mobility and explains the connection between cholesterol sensing, Rab7 and late endosome positioning [46]. Obviously, such a process will be impaired in cells lacking the lysosomal cholesterol transporter NPC1, which leads to Niemann-Pick disease [47]. In fact, the cellular defect of the disease can be compensated partially, if either active Rab7 or Rab9 are overexpressed [48, 49]. Interestingly, the adenoviral protein RIDa mimicks Rab7-GTP and diverts RILP and ORP1L to early endosomes [50]. This interaction allows the selected degradation of cargo like cell surface receptors and may thus facilitate the viral replication inside the cell [50].

It should be noted that a contact zone between ER and the vacuole has also been described in yeast, which also contains oxysterol-binding proteins [51, 52]. This contact zone seems to be independent of Ypt7, and rather linked to microautophagic processes [53].

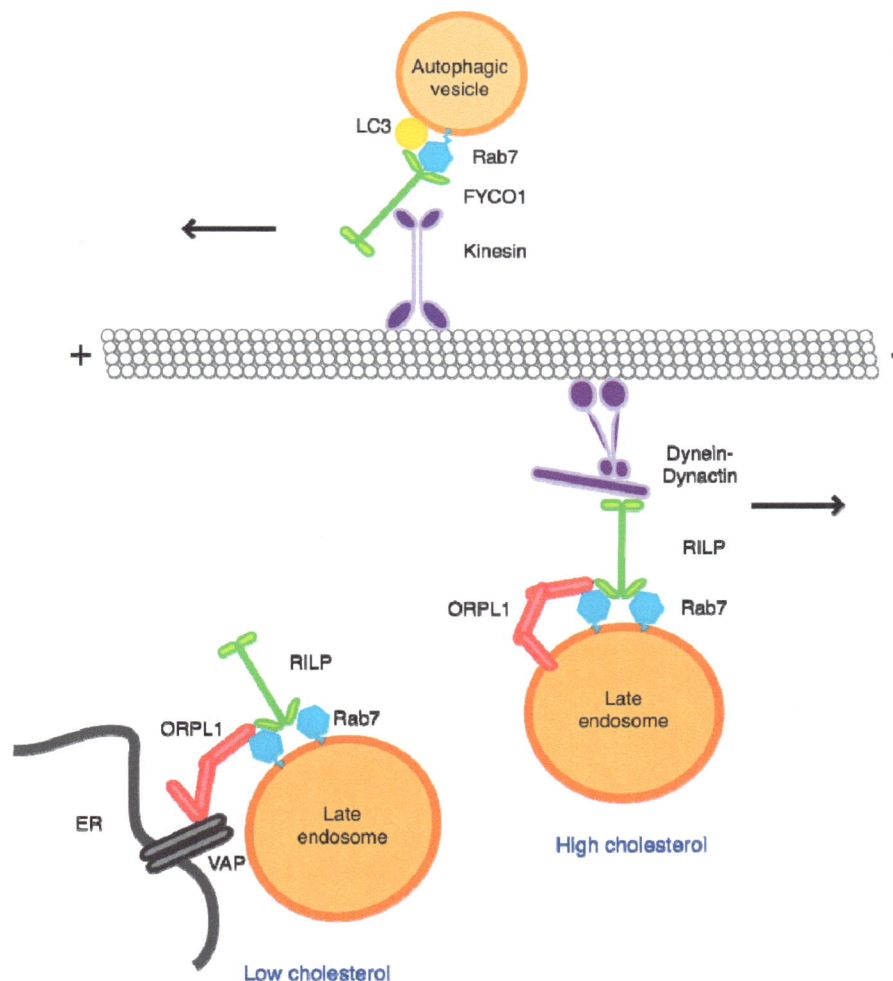

Figure 2: Rab7 controls late endosomes and autophagic vesicle positioning. RILP and FYCO1 effectors connect Rab7 to motor proteins to mediate organelle transport along microtubules [45, 54]. Under low cholesterol conditions, ORPL1 interaction with the ER protein VAP transfers Rab7-RILP complex to the ER and allows ER-late endosomes contacts [45]. For details see text.

Besides RILP, another Rab7 effector was recently identified that combines Rab7-GTP binding with microtubule-dependent localization [54, 55] (Figs. **1** and **2**). FYCO1 (FYVE and coiled-coil domain containing protein 1) is a multidomain protein, which binds the lipidated autophagic marker LC3, Rab7-GTP and phosphoinositide-3-phosphate (PI-3-P). It is very similar in domain composition to EEA1 [56], RILP [34], and RUFY [57] with two predicted Rab binding sites and a long coiled-coil domain. To bind to autophagic vesicles, FYCO1 interacts *via* its LIR (LC-3 interacting region) motif with LC3 and binds PI-3-P with its neighboring FYVE domain. Binding to the Rab7-GTP requires a coiled-coil region, and may occur similarly to the RILP-Rab7-GTP interaction. If FYCO1 is overexpressed, LC3 and Rab7 positive structures redistribute to the cell periphery in a microtubule dependent manner [54]. Moreover, lysosomes and late endosomes expand in diameter, suggesting that FYCO1 also promotes late endosome/autophagosome-lysosome fusion.

FYCO1 seems to be a receptor that is regulated by nutrient supplies. If nutrients are abundant, it resides on perinuclear ER, and does not interact with kinesin. Upon amino acid starvation, FYCO1 interacts to plus-end directed motors and confers the redistribution of preautophagosomal membranes to the sites of autophagosome formation. Upon overexpression, FYCO1 causes distribution of Rab7 and ORPL1 to microtubule plus ends suggesting that FYCO1 could compete with the dynein recruitment complex (containing RILP) for Rab7-GTP and subsequently would permit autophagosomal transport on microtubule tracks [54, 55]. Indeed, depletion of FYCO1 leads to a redistribution of autophagosomes to perinuclear clusters [54]. Thus, RILP and FYCO1 might act as competing motor on microtubule tracks, but would follow similar principles in their membrane association (Fig. **2**).

3.2. Rab7 and Endosomal Biogenesis

Several proteins have been linked to Rab7 during endosomal biogenesis, including the PI-3-kinase Vps34, the proteasomal subunit XAPC7 and an E3 ligase Rabring7 (Fig. **1**), although little is known about their exact targets at the endosome.

Endosomal PI-3-P is generated *via* the PI-3-kinase Vps34 [58]. Mammalian Vps34 binds to the early endosomal Rab5 [59], and to Rab7-GTP *via* its adapter p150 [60]. Whereas human Vps34 localization was Rab7-independent, its activity was dependent on Rab7 nucleotide cycling [60]. This function may be critical for the regulation of early and late endosome biogenesis, which is tightly connected to the levels of phosphoinositides [61, 62].

Proteasome-mediated degradation and endosomal biogenesis seem to be linked *via* Rab7. The proteasomal a-subunit XAPC7 is recruited specifically *via* Rab7-GTP to late endosomes, and impairs late endocytic transport if overproduced [63]. As this effect was compensated by Rab7-GTP, it is likely that XAPC7 negatively regulates endosomal transport. However, Rab7 was not a target of proteasomal transport [63], though a link between the yeast Ypt7 and the proteasome has been suggested [64]. How XAPC7 influences endosomal biogenesis *via* Rab7 is an open issue.

Rabring7 is a Rab7 effector, which associates with late endosomes and lysosomes in a Rab7-dependent manner [65]. Rabring7 binds Rab7-GTP *via* its N-terminal domain, which is also sufficient to induce endosomal clustering. Its C-terminal part contains an H2-type RING finger motif, which has E3 ligase activity and promotes self-ubiquitination [66]. Its direct function is not yet clear as substrates involved in endocytosis remain to be identified. It should be noticed that several proteins of the HOPS and CORVET complex (Vps11, Vps18 and Vps8) also contain C-terminal RING domains of the H2-type [67]. At least for mammalian Vps18, E3-ligase activity has been demonstrated [68]. It is thus possible that Rabring7 and subunits of the HOPS complex perform similar functions, as both bind Rab7-GTP. Of note, the yeast Ypt7 has been identified as an ubiquitinated protein [64].

3.3. Rab7 and Retrograde Transport

Whereas all previously discussed effectors are involved in endosome-lysosome fusion and late endosome mobility, the interaction of Rab7 with retromer extends its function towards recycling processes. Retromer is a membrane-active complex involved in the retrieval of cargo receptors from the late endosome [69, 70]. It consists of two subcomplexes: the Snx dimer (in yeast Vps5 and Vps17; in mammalian cells SNX1, 2, 5, and 6) and a heterotrimer of Vps26, Vps29 and Vps35 [71]. The Snx dimer is recruited to endosomes by binding PI-3-P *via* its PX domain [72]. It has the ability to induce tubule formation by binding highly curved membranes *via* its BAR-domain [69, 70]. The cargo-sorting subcomplex is recruited by Rab7-GTP [73, 74], and the Snx dimer. Cargo retrieval occurs *via* Snx-generated tubules that emerge from late endosomal membranes, where retromer and Rab7 colocalize [74]. Indeed, depletion of Rab7 results in a mislocalization of Vps26 and interferes with the retromer-directed transport of cargo like MPRs to the Golgi. Thus, Rab7-GTP may coordinate receptor recycling and fusion at the late endosome (Fig. **4**).

3.4. The Mon1-Ccz1 Complex and Endosomal Maturation

The maturation of early to late endosomes requires a complex membrane remodeling accompanied by the exchange of Rab5 to Rab7 [17]. This conversion is linked to the function of two Rab7 interacting proteins, Mon1 and Ccz1 [75]. Ccz1 function was initially described in yeast [76] and later in *C.elegans* and metazoans [77]. It forms a complex with Mon1 [78], and has been linked to Rab7/Ypt7 activation [79] and vacuole fusion [80]. Mon1 (also called SAND-1 in worms) and Ccz1 affect endosomal biogenesis, both in yeast and *C.elegans* [13, 81, 82]. Recent data suggest that Mon1 and Ccz1 interact with Rab5-GTP, and are thus recruited to early endosomes [77] (Fig. **3**). Here, Mon1 competes with the GEF of Rab5, Rabex5, and displaces it from the membrane [83]. At the same time, Mon1 together with Ccz1 seems to recruit Ypt7, potentially by displacing it from GDI [77]. Recently, we showed that the Mon1-Ccz1 complex, but neither subunit alone, is the GEF of Ypt7 [19]. Mon1 binds the HOPS subunit Vps39 [19], and additional interactions with HOPS have been reported in other studies [83, 80]. Thus, the Mon1-Ccz1 complex may facilitate endosomal maturation by inhibiting Rab5 re-activation and promoting Rab7 conversion to the active GTP-form (Fig. **3**).

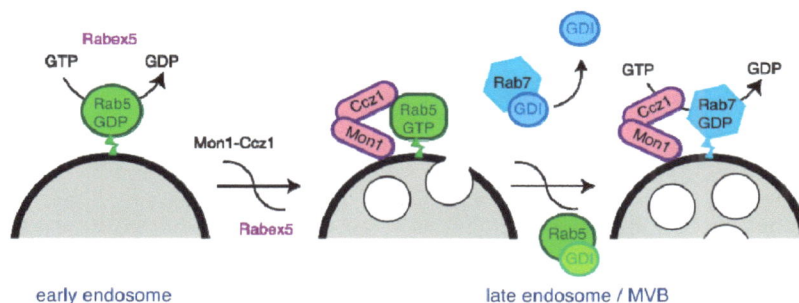

Figure 3: Model of endosomal maturation. The Mon1-Ccz1 complex is recruited to early endosomes *via* interaction with Rab5-GTP [77]. Mon1 thereby displaces Rabex5 from the membrane and together with Ccz1 recruits Rab7 by displacing it from GDI [77, 83]. The complex then promotes Rab7 activation [19]. The interaction with Vps39/HOPS may regulate this process. For details see text.

3.5. Role of the Rab7 Effector HOPS/Vps41 in late Endosome Fusion

Membrane fusion requires Rab-mediated tethering. For Rab7, the fusion relevant tethering complex is the HOPS complex (Fig. **4**). Its interplay with Rab7/Ypt7 has been mainly characterized in yeast. HOPS is a hetero-hexamer consisting of the Class C proteins Vps11, Vps16, Vps18, and Vps33 [67], which are also present in the homologous endosomal CORVET complex [84], and the HOPS-specific subunits Vps39 and Vps41 [16, 85]. HOPS binds to Ypt7 [85, 86], and both Vps39 and Vps41 recognize Ypt7 [16, 20, 21, 26]. Previous data indicated that Vps39 does not seem to discriminate between the nucleotide-state of Ypt7 [16, 20], and may control the activity of the Mon1-Ccz1 GEF *in vivo* [19]. Recently, we showed that both Vps39 [106] and Vps41 interact specifically with Ypt7-GTP [20, 26].

The analysis of Ypt7 in membrane fusion has been greatly facilitated by the vacuole fusion assay [87, 88] and its recent reconstitution using proteoliposomes [89, 90]. To tether membranes, HOPS seems to take advantage of its two binding sites: it does not only interact with Ypt7-GTP (*via* Vps41 and Vps39 [106]), but also with SNAREs [91], the membrane embedded fusion factors that mediate bilayer mixing [92]. Vacuole fusion requires Ypt7 on both membranes [10, 93]. As for mammalian Rab7, GTP-locked Ypt7 supports fusion, whereas GDP-locked Ypt7 is an inhibitor of vacuole fusion [94]. Our recent structural analyses demonstrate that HOPS has Vps41 and Vps39 at opposite ends of its tadpole-like structure [106]. We thus propose that HOPS tethers membranes by binding Ypt7-GTP on both, which precedes the SNARE-mediated fusion [95]. Interestingly, Ypt7 becomes dispensable for the vacuole fusion reaction if SNAREs are in excess [96], suggesting that SNAREs at high concentration can also tether membranes. When reconstituted into liposomes, Ypt7 is required for SNARE-mediated fusion if the lipid composition precludes the association of the HOPS complex with membranes [90], and promotes tethering [97], in agreement with our structural analyses [106]. This suggests that Ypt7 recruits HOPS prior SNARE-driven bilayer mixing [97, 98].

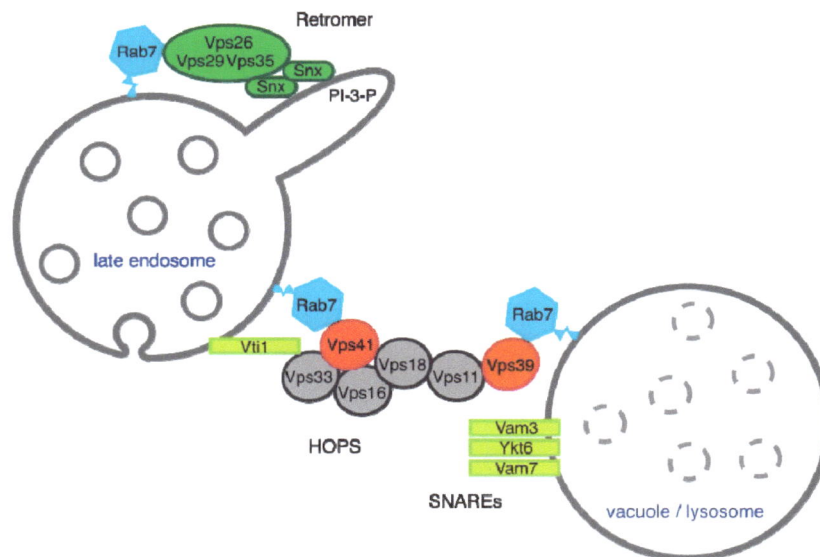

Figure 4: Coordination of receptor retrieval with fusion of late endosomes. Retromer interacts with Rab7-GTP [73, 74] to generate tubules and finally vesicles containing cargo receptors. The HOPS complex binds *via* Vps41 to Rab7-GTP [26] (our unpublished observations) and to SNAREs [91], presumably *via* Vps33 [105].

In vivo, Ypt7 may encounter Vps41 and the HOPS complex not only at the yeast vacuole, but also at the late endosome. Vps41 is a substrate of the casein kinase Yck3 [99]. In its absence, Vps41 accumulates at endosome-vacuole junctions [99], presumably in complex with Ypt7-GTP. Indeed, a phosphomimetic mutant in Vps41, which localizes mainly to the cytoplasm, is retrieved to vacuoles by Ypt7 overexpression [25]. Thus, Ypt7 may recruit Vps41/HOPS also to late endosomes to facilitate fusion of late endosomes with the vacuole.

4. ROLE OF RAB7 IN DISEASE

Mutations in Rab7 have been linked to some forms of Charcot Marie Tooth type 2B (CMT2B) disease, a neuropathy characterized by sensory loss and distal muscle weakness, often complicated by infections and amputations. CMT2B is caused by mutations in highly conserved residues in the GTP binding and hydrolysis domains of Rab7: L129F, K157N, N161T and V162M [100-102]. These four Rab7 mutants exhibit lower affinity for nucleotides, especially for GDP and reduced GTPase activity similar to the Rab7 Q67L mutant. Thus, they are mainly in a GTP-bound form and are able to bind the effector RILP and rescue EGF degradation in Rab7-depleted cells [102, 103]. In neurons, Rab7 regulates retrograde transport of axonal carriers. Thus, mutations associated to CMT2B may affect the interaction of Rab7 with specific effector proteins that are only present in peripheral neurons [104]. However, the Rab7 L129F mutant has the same reduced affinity for nucleotides, but no defect in the GTPase activity [28]. Thus, Rab7 mutants linked to this disease may undergo GEF and GAP-independent nucleotide cycling due to the decreased affinity for GDP and GTP [28].

5. SUMMARY AND OUTLOOK

Rab7 is a key regulator of late endosomal biogenesis. It is responsible for late endosomal membrane remodeling, fusion and cargo retrieval, as well as the positioning of late endosomes within cells. While many factors that bind Rab7 have been identified, their coordination remains to be resolved. For instance, the timing of Rab7-mediated activation during endosomal maturation, which involves the activities of

Mon1-Ccz1 and Vps39, is far from being understood. Likewise, the role of Rab7 in the cellular positioning of late endosomes in response to cholesterol levels or nutrient availability needs to be linked to endosome/autophagosome-lysosome fusion. Moreover, it would be expected that the Rab7-dependent recruitment of retromer and the retrieval of cargo would be coordinated with the fusion with lysosomes. How this is mediated by a single Rab7 isoform remains an open issue. At least in fission yeast, a division of labor seems to have occurred. Potentially, posttranslational modifications of Rab7 by ubiquitination may distinguish different states of Rab7. As Rab7 is a central factor in late endosomal biogenesis, involved in endosomal trafficking in neurons and linked to diseases, it will be important to clarify these key issues.

CONFLICT OF INTEREST

There is no conflict of interest from any of the authors.

REFERENCES

[1]   Chavrier P, Parton RG, Hauri HP, Simons K, Zerial M. Localization of low molecular weight GTP binding proteins to exocytic and endocytic compartments. Cell 1990;62(2):317-29.

[2]   Gorvel JP, Chavrier P, Zerial M, Gruenberg J. Rab5 controls early endosome fusion *in vitro*. Cell 1991;64(5):915-25.

[3]   Soldati T, Rancano C, Geissler H, Pfeffer SR. Rab7 and Rab9 are recruited onto late endosomes by biochemically distinguishable processes. J Biol Chem 1995;270(43):25541-8.

[4]   Bottger G, Nagelkerken B, van der Sluijs P. Rab4 and Rab7 define distinct nonoverlapping endosomal compartments. J Biol Chem 1996;271(46):29191-7.

[5]   Bucci C, Thomsen P, Nicoziani P, McCarthy J, van Deurs B. Rab7: a key to lysosome biogenesis. Mol Biol Cell 2000;11(2):467-80.

[6]   Peralta ER, Martin BC, Edinger AL. Differential effects of TBC1D15 and mammalian Vps39 on Rab7 activation state, lysosomal morphology, and growth factor dependence. J Biol Chem 2010;285(22):16814-21.

[7]   Wichmann H, Hengst L, Gallwitz D. Endocytosis in yeast: evidence for the involvement of a small GTP- binding protein (Ypt7p). Cell 1992;71(7):1131-42.

[8]   Raymond CK, Howald-Stevenson I, Vater CA, Stevens TH. Morphological classification of the yeast vacuolar protein sorting mutants: evidence for a prevacuolar compartment in class E vps mutants. Mol Biol Cell 1992;3(12):1389-402.

[9]   Schimmoller F, Riezman H. Involvement of Ypt7p, a small GTPase, in traffic from late endosome to the vacuole in yeast. J Cell Sci 1993;106(Pt 3):823-30.

[10]   Haas A, Scheglmann D, Lazar T, Gallwitz D, Wickner W. The GTPase Ypt7p of Saccharomyces cerevisiae is required on both partner vacuoles for the homotypic fusion step of vacuole inheritance. EMBO J 1995;14(21):5258-70.

[11]   Kashiwazaki J, Nakamura T, Iwaki T, Takegawa K, Shimoda C. A role for fission yeast Rab GTPase Ypt7p in sporulation. Cell Struct Funct 2005;30(2):43-9.

[12]   Kashiwazaki J, Iwaki T, Takegawa K, Shimoda C, Nakamura T. Two fission yeast rab7 homologs, Ypt7 and Ypt71, play antagonistic roles in the regulation of vacuolar morphology. Traffic 2009;10(7):912-24.

[13]   Poteryaev D, Fares H, Bowerman B, Spang A. Caenorhabditis elegans SAND-1 is essential for RAB-7 function in endosomal traffic. EMBO J 2007;26(2):301-12.

[14]   Sivars U, Aivazian D, Pfeffer SR. Yip3 catalyses the dissociation of endosomal Rab-GDI complexes. Nature 2003;425(6960):856-9.

[15]   Schoebel S, Oesterlin LK, Blankenfeldt W, Goody RS, Itzen A. RabGDI displacement by DrrA from Legionella is a consequence of its guanine nucleotide exchange activity. Mol Cell 2009;36(6):1060-72.

[16]   Wurmser AE, Sato TK, Emr SD. New component of the vacuolar class C-Vps complex couples nucleotide exchange on the ypt7 GTPase to SNARE-dependent docking and fusion. J Cell Biol 2000;151(3):551-62.

[17]   Rink J, Ghigo E, Kalaidzidis Y, Zerial M. Rab conversion as a mechanism of progression from early to late endosomes. Cell 2005;122(5):735-49.

[18]   Caplan S, Hartnell LM, Aguilar RC, Naslavsky N, Bonifacino JS. Human Vam6p promotes lysosome clustering and fusion *in vivo*. J Cell Biol 2001;154(1):109-22.

[19]   Nordmann M, Cabrera M, Perz A, *et al*. The Mon1-Ccz1 complex is the GEF for the Rab7 homolog Ypt7. Curr Biol 2010 Sept 28;20(18):1654-9.

[20] Ostrowicz CW, Brocker C, Ahnert F, *et al.* Defined subunit arrangement and Rab interactions are required for functionality of the HOPS tethering complex. Traffic 2010 Oct;11(10):1334-46.

[21] Binda M, Peli-Gulli MP, Bonfils G, *et al.* The Vam6 GEF controls TORC1 by activating the EGO complex. Mol Cell 2009;35(5):563-73.

[22] Zhang XM, Walsh B, Mitchell CA, Rowe T. TBC domain family, member 15 is a novel mammalian Rab GTPase-activating protein with substrate preference for Rab7. Biochem Biophys Res Commun 2005;335(1):154-61.

[23] Albert S, Will E, Gallwitz D. Identification of the catalytic domains and their functionally critical arginine residues of two yeast GTPase-activating proteins specific for Ypt/Rab transport GTPases. EMBO J 1999;18(19):5216-25.

[24] Eitzen G, Will E, Gallwitz D, Haas A, Wickner W. Sequential action of two GTPases to promote vacuole docking and fusion. EMBO J 2000;19(24):6713-6720.

[25] Cabrera M, Ostrowicz CW, Mari M, LaGrassa TJ, Reggiori F, Ungermann C. Vps41 phosphorylation and the Rab Ypt7 control the targeting of the HOPS complex to endosome-vacuole fusion sites. Mol Biol Cell 2009;20(7):1937-48.

[26] Brett CL, Plemel RL, Lobinger BT, Vignali M, Fields S, Merz AJ. Efficient termination of vacuolar Rab GTPase signaling requires coordinated action by a GAP and a protein kinase. J Cell Biol 2008;182(6):1141-51.

[27] Constantinescu AT, Rak A, Alexandrov K, Esters H, Goody RS, Scheidig AJ. Rab-Subfamily-Specific Regions of Ypt7p Are Structurally Different from Other RabGTPases. Structure (Camb) 2002;10(4):569-79.

[28] McCray BA, Skordalakes E, Taylor JP. Disease mutations in Rab7 result in unregulated nucleotide exchange and inappropriate activation. Hum Mol Genet 2010;19(6):1033-47.

[29] Feng Y, Press B, Wandinger-Ness A. Rab 7: an important regulator of late endocytic membrane traffic. J Cell Biol 1995;131(6 Pt 1):1435-52.

[30] Vitelli R, Santillo M, Lattero D, *et al.* Role of the small GTPase Rab7 in the late endocytic pathway. J Biol Chem 1997;272(7):4391-7.

[31] Press B, Feng Y, Hoflack B, Wandinger-Ness A. Mutant Rab7 causes the accumulation of cathepsin D and cation-independent mannose 6-phosphate receptor in an early endocytic compartment. J Cell Biol 1998;140(5):1075-89.

[32] Ceresa BP, Bahr SJ. rab7 activity affects epidermal growth factor:epidermal growth factor receptor degradation by regulating endocytic trafficking from the late endosome. J Biol Chem 2006;281(2):1099-106.

[33] Vanlandingham PA, Ceresa BP. Rab7 regulates late endocytic trafficking downstream of multivesicular body biogenesis and cargo sequestration. J Biol Chem 2009;284(18):12110-24.

[34] Cantalupo G, Alifano P, Roberti V, Bruni CB, Bucci C. Rab-interacting lysosomal protein (RILP): the Rab7 effector required for transport to lysosomes. EMBO J 2001;20(4):683-93.

[35] Jordens I, Fernandez-Borja M, Marsman M, *et al.* The Rab7 effector protein RILP controls lysosomal transport by inducing the recruitment of dynein-dynactin motors. Curr Biol 2001;11(21):1680-5.

[36] Wu M, Wang T, Loh E, Hong W, Song H. Structural basis for recruitment of RILP by small GTPase Rab7. EMBO J 2005;24(8):1491-501.

[37] Johansson M, Rocha N, Zwart W, *et al.* Activation of endosomal dynein motors by stepwise assembly of Rab7-RILP-p150Glued, ORP1L, and the receptor betaIII spectrin. J Cell Biol 2007;176(4):459-71.

[38] Harrison RE, Brumell JH, Khandani A, *et al.* Salmonella impairs RILP recruitment to Rab7 during maturation of invasion vacuoles. Mol Biol Cell 2004;15(7):3146-54.

[39] Jordens I, Westbroek W, Marsman M, *et al.* Rab7 and Rab27a control two motor protein activities involved in melanosomal transport. Pigment Cell Res 2006;19(5):412-23.

[40] Stinchcombe JC, Majorovits E, Bossi G, Fuller S, Griffiths GM. Centrosome polarization delivers secretory granules to the immunological synapse. Nature 2006;443(7110):462-5.

[41] Progida C, Malerod L, Stuffers S, Brech A, Bucci C, Stenmark H. RILP is required for the proper morphology and function of late endosomes. J Cell Sci 2007;120(Pt 21):3729-37.

[42] Progida C, Spinosa MR, De Luca A, Bucci C. RILP interacts with the VPS22 component of the ESCRT-II complex. Biochem Biophys Res Commun 2006;347(4):1074-9.

[43] Wang T, Hong W. RILP interacts with VPS22 and VPS36 of ESCRT-II and regulates their membrane recruitment. Biochem Biophys Res Commun 2006;350(2):413-23.

[44] Johansson M, Lehto M, Tanhuanpaa K, Cover TL, Olkkonen VM. The oxysterol-binding protein homologue ORP1L interacts with Rab7 and alters functional properties of late endocytic compartments. Mol Biol Cell 2005;16(12):5480-92.

[45] Rocha N, Kuijl C, van der Kant R, *et al.* Cholesterol sensor ORP1L contacts the ER protein VAP to control Rab7-RILP-p150 Glued and late endosome positioning. J Cell Biol 2009;185(7):1209-25.

[46]   Lebrand C, Corti M, Goodson H, *et al.* Late endosome motility depends on lipids *via* the small GTPase Rab7. EMBO J 2002;21(6):1289-1300.

[47]   Maxfield FR, Tabas I. Role of cholesterol and lipid organization in disease. Nature 2005;438(7068):612-21.

[48]   Choudhury A, Dominguez M, Puri V, *et al.* Rab proteins mediate Golgi transport of caveola-internalized glycosphingolipids and correct lipid trafficking in Niemann-Pick C cells. J Clin Invest 2002;109(12):1541-50.

[49]   Pagano RE. Endocytic trafficking of glycosphingolipids in sphingolipid storage diseases. Philos Trans R Soc Lond B Biol Sci 2003;358(1433):885-91.

[50]   Shah AH, Cianciola NL, Mills JL, Sonnichsen FD, Carlin C. Adenovirus RIDalpha regulates endosome maturation by mimicking GTP-Rab7. J Cell Biol 2007;179(5):965-80.

[51]   Pan X, Roberts P, Chen Y, *et al.* Nucleus-vacuole junctions in saccharomyces cerevisiae are formed through the direct interaction of Vac8p with Nvj1p . Mol Biol Cell 2000;11(7):2445-57.

[52]   Kvam E, Gable K, Dunn TM, Goldfarb DS. Targeting of Tsc13p to nucleus-vacuole junctions: a role for very-long-chain fatty acids in the biogenesis of microautophagic vesicles. Mol Biol Cell 2005;16(9):3987-98.

[53]   Roberts P, Moshitch-Moshkovitz S, Kvam E, O'Toole E, Winey M, Goldfarb DS. Piecemeal Microautophagy of Nucleus in Saccharomyces cerevisiae. Mol Biol Cell 2003;14(1):129-41.

[54]   Pankiv S, Alemu EA, Brech A, *et al.* FYCO1 is a Rab7 effector that binds to LC3 and PI3P to mediate microtubule plus end-directed vesicle transport. J Cell Biol 2010;188(2):253-69.

[55]   Pankiv S, Johansen T. FYCO1: Linking autophagosomes to microtubule plus end-directing molecular motors. Autophagy 2010;6(4).

[56]   Stenmark H, Aasland R, Toh BH, D'Arrigo A. Endosomal localization of the autoantigen EEA1 is mediated by a zinc- binding FYVE finger. J Biol Chem 1996;271(39):24048-54.

[57]   Rose A, Schraegle SJ, Stahlberg EA, Meier I. Coiled-coil protein composition of 22 proteomes--differences and common themes in subcellular infrastructure and traffic control. BMC Evol Biol 2005;5:66.

[58]   Schu PV, Takegawa K, Fry MJ, Stack JH, Waterfield MD, Emr SD. Phosphatidylinositol 3-kinase encoded by yeast VPS34 gene essential for protein sorting. Science 1993;260(5104):88-91.

[59]   Christoforidis S, Miaczynska M, Ashman K, *et al.* Phosphatidylinositol-3-OH kinases are Rab5 effectors. Nat Cell Biol 1999;1(4):249-52.

[60]   Stein MP, Feng Y, Cooper KL, Welford AM, Wandinger-Ness A. Human VPS34 and p150 are Rab7 interacting partners. Traffic 2003;4(11):754-71.

[61]   Simonsen A, Wurmser AE, Emr SD, Stenmark H. The role of phosphoinositides in membrane transport. Curr Opin Cell Biol 2001;13(4):485-92.

[62]   Lindmo K, Stenmark H. Regulation of membrane traffic by phosphoinositide 3-kinases. J Cell Sci 2006;119(Pt 4):605-14.

[63]   Dong J, Chen W, Welford A, Wandinger-Ness A. The proteasome alpha-subunit XAPC7 interacts specifically with Rab7 and late endosomes. J Biol Chem 2004;279(20):21334-42.

[64]   Kleijnen MF, Kirkpatrick DS, Gygi SP. The ubiquitin-proteasome system regulates membrane fusion of yeast vacuoles. EMBO J 2007;26(2):275-87.

[65]   Mizuno K, Kitamura A, Sasaki T. Rabring7, a novel Rab7 target protein with a RING finger motif. Mol Biol Cell 2003;14(9):3741-52.

[66]   Sakane A, Hatakeyama S, Sasaki T. Involvement of Rabring7 in EGF receptor degradation as an E3 ligase. Biochem Biophys Res Commun 2007;357(4):1058-64.

[67]   Rieder SE, Emr SD. A novel RING finger protein complex essential for a late step in protein transport to the yeast vacuole. Mol Biol Cell 1997;8(11):2307-27.

[68]   Yogosawa S, Hatakeyama S, Nakayama KI, Miyoshi H, Kohsaka S, Akazawa C. Ubiquitylation and degradation of serum-inducible kinase by hVPS18, a RING-H2 type ubiquitin ligase. J Biol Chem 2005;280(50):41619-27.

[69]   Bonifacino JS, Hurley JH. Retromer. Curr Opin Cell Biol 2008;20(4):427-36.

[70]   Seaman MN. Recycle your receptors with retromer. Trends Cell Biol 2005;15(2):68-75.

[71]   Seaman MN, McCaffery JM, Emr SD. A membrane coat complex essential for endosome-to-Golgi retrograde transport in yeast. J Cell Biol 1998;142(3):665-81.

[72]   Burda P, Padilla SM, Sarkar S, Emr SD. Retromer function in endosome-to-Golgi retrograde transport is regulated by the yeast Vps34 PtdIns 3-kinase. J Cell Sci 2002;115(Pt 20):3889-900.

[73]   Seaman MN, Harbour ME, Tattersall D, Read E, Bright N. Membrane recruitment of the cargo-selective retromer subcomplex is catalysed by the small GTPase Rab7 and inhibited by the Rab-GAP TBC1D5. J Cell Sci 2009;122(Pt 14):2371-82.

[74] Rojas R, van Vlijmen T, Mardones GA, *et al.* Regulation of retromer recruitment to endosomes by sequential action of Rab5 and Rab7. J Cell Biol 2008;183(3):513-26.

[75] Cabrera M, Ungermann C. Guiding endosomal maturation. Cell 2010;141(3):404-6.

[76] Kucharczyk R, Dupre S, Avaro S, Haguenauer-Tsapis R, Slonimski PP, Rytka J. The novel protein Ccz1p required for vacuolar assembly in saccharomyces cerevisiae functions in the same transport pathway as Ypt7p. J Cell Sci 2000;113(Pt 23):4301-11.

[77] Kinchen JM, Ravichandran KS. Identification of two evolutionarily conserved genes regulating processing of engulfed apoptotic cells. Nature 2010;464(7289):778-82.

[78] Wang CW, Stromhaug PE, Shima J, Klionsky DJ. The Ccz1-Mon1 protein complex is required for the late step of multiple vacuole delivery pathways. J Biol Chem 2002;277(49):47917-27.

[79] Kucharczyk R, Kierzek AM, Slonimski PP, Rytka J. The Ccz1 protein interacts with Ypt7 GTPase during fusion of multiple transport intermediates with the vacuole in S. cerevisiae. J Cell Sci 2001;114(Pt 17):3137-45.

[80] Wang CW, Stromhaug PE, Kauffman EJ, Weisman LS, Klionsky DJ. Yeast homotypic vacuole fusion requires the Ccz1-Mon1 complex during the tethering/docking stage. J Cell Biol 2003;163(5):973-85.

[81] Hoffman-Sommer M, Migdalski A, Rytka J, Kucharczyk R. Multiple functions of the vacuolar sorting protein Ccz1p in Saccharomyces cerevisiae. Biochem Biophys Res Commun 2005;329(1):197-204.

[82] Hoffman-Sommer M, Kucharczyk R, Piekarska I, Kozlowska E, Rytka J. Mutations in the Saccharomyces cerevisiae vacuolar fusion proteins Ccz1, Mon1 and Ypt7 cause defects in cell cycle progression in a num1Delta background. Eur J Cell Biol 2009;88(11):639-52.

[83] Poteryaev D, Datta S, Ackema K, Zerial M, Spang A. Identification of the switch in early-to-late endosome transition. Cell 2010;141(3):497-508.

[84] Peplowska K, Markgraf DF, Ostrowicz CW, Bange G, Ungermann C. The CORVET Tethering Complex Interacts with the Yeast Rab5 Homolog Vps21 and Is Involved in Endo-Lysosomal Biogenesis. Dev Cell 2007;12(5):739-50.

[85] Seals DF, Eitzen G, Margolis N, Wickner WT, Price A. A Ypt/Rab effector complex containing the Sec1 homolog Vps33p is required for homotypic vacuole fusion. Proc Natl Acad Sci U S A 2000;97(17):9402-7.

[86] Price A, Seals D, Wickner W, Ungermann C. The docking stage of yeast vacuole fusion requires the transfer of proteins from a cis-SNARE complex to a Rab/Ypt protein. J Cell Biol 2000;148(6):1231-8.

[87] Haas A. A quantitative assay to measure homotypic vacuole fusion *in vitro*. Methods Cell Sci. 1995;17:283-294.

[88] Cabrera M, Ungermann C. Purification and *in vitro* analysis of yeast vacuoles. Methods Enzymol 2008;451:177-96.

[89] Mima J, Hickey CM, Xu H, Jun Y, Wickner W. Reconstituted membrane fusion requires regulatory lipids, SNAREs and synergistic SNARE chaperones. EMBO J 2008;27(15):2031-42.

[90] Hickey CM, Stroupe C, Wickner W. The major role of the Rab Ypt7p in vacuole fusion is supporting HOPS membrane association. J Biol Chem 2009;284(24):16118-25.

[91] Stroupe C, Hickey CM, Mima J, Burfeind AS, Wickner W. Minimal membrane docking requirements revealed by reconstitution of Rab GTPase-dependent membrane fusion from purified components. Proc Natl Acad Sci U S A 2009;106(42):17626-33.

[92] Jahn R, Scheller RH. SNAREs - engines for membrane fusion. Nat Rev Mol Cell Biol 2006;7(9):631-43.

[93] Mayer A, Wickner W. Docking of yeast vacuoles is catalyzed by the Ras-like GTPase Ypt7p after symmetric priming by Sec18p (NSF). J Cell Biol 1997;136(2):307-17.

[94] Eitzen G, Thorngren N, Wickner W. Rho1p and Cdc42p act after Ypt7p to regulate vacuole docking. EMBO J 2001;20(20):5650-6.

[95] Ungermann C, Sato K, Wickner W. Defining the functions of trans-SNARE pairs. Nature 1998;396(6711):543-8.

[96] Starai VJ, Jun Y, Wickner W. Excess vacuolar SNAREs drive lysis and Rab bypass fusion. Proc Natl Acad Sci U S A 2007;104(34):13551-8.

[97] Hickey CM, Wickner W. HOPS initiates vacuole docking by tethering membranes before trans-SNARE complex assembly. Mol Biol Cell 2010;21(13):2297-305.

[98] Xu H, Jun Y, Thompson J, Yates J, Wickner W. HOPS prevents the disassembly of trans-SNARE complexes by Sec17p/Sec18p during membrane fusion. EMBO J. 2010; 29(12):1948-60.

[99] LaGrassa TJ, Ungermann C. The vacuolar kinase Yck3 maintains organelle fragmentation by regulating the HOPS tethering complex. J Cell Biol 2005;168(3):401-14.

[100] Verhoeven K, De Jonghe P, Coen K, *et al.* Mutations in the small GTP-ase late endosomal protein RAB7 cause Charcot-Marie-Tooth type 2B neuropathy. Am J Hum Genet 2003;72(3):722-7.

[101] Houlden H, King RH, Muddle JR, *et al.* A novel RAB7 mutation associated with ulcero-mutilating neuropathy. Ann Neurol 2004;56(4):586-90.

[102] De Luca A, Progida C, Spinosa MR, Alifano P, Bucci C. Characterization of the Rab7K157N mutant protein associated with Charcot-Marie-Tooth type 2B. Biochem Biophys Res Commun 2008;372(2):283-7.

[103] Spinosa MR, Progida C, De Luca A, Colucci AM, Alifano P, Bucci C. Functional characterization of Rab7 mutant proteins associated with Charcot-Marie-Tooth type 2B disease. J Neurosci 2008;28(7):1640-8.

[104] Cogli L, Piro F, Bucci C. Rab7 and the CMT2B disease. Biochem Soc Trans 2009;37(Pt 5):1027-31.

[105] Pieren M, Schmidt A, Mayer A. The SM protein Vps33 and the t-SNARE H(abc) domain promote fusion pore opening. Nat Struct Mol Biol 2010;17(6):710-7.

[106] Bröcker C, Kuhlee A, Gatsogiannis C, *et al*. Molecular architecture of the homotypic fusion and protein sorting (HOPS) tethering complex. Proc Natl Acad Sci USA 2012;109:1991-6.

CHAPTER 11

Transport from Late Endosomes to the Golgi: Rab9 GTPase

Eric J. Espinosa and Suzanne R. Pfeffer[*]

Stanford University School of Medicine, USA

Abstract: Rab9 functions in the retrieval of mannose 6-phosphate receptors (MPRs) from late endosomes and their subsequent delivery to the trans Golgi network (TGN). In this chapter, we will discuss how Rab9 is recruited onto membranes, how Rab9 functions to select and segregate MPRs into a specific microdomain depleted of Rab7, and ultimately, how Rab9-containing transport vesicles dock and fuse at the TGN.

Keywords: Rab9, Late Endosomes, Mannose 6-phosphate receptor, Lysosome, Retrograde Transport.

1. LATE ENDOSOME TO GOLGI TRANSPORT

Mannose 6-phosphate receptors bind newly synthesized lysosomal hydrolases in the trans-Golgi network and deliver them to pre-lysosomal compartments. There are two different MPRs, one of ~300kDa and a smaller, dimeric receptor of ~46kDa (reviewed in [1, 2]). The larger receptor is known as the cation-independent mannose 6-phosphate receptor (CI-MPR), while the smaller receptor requires divalent cations to efficiently recognize lysosomal hydrolases [2]. Both of these receptors bind terminal mannose 6-phosphate with similar affinity [3] and have similar signals in their cytoplasmic domains for intracellular trafficking [1, 4].

Early in the Golgi, the oligosaccharide chains of lysosomal hydrolases are tagged with terminal phosphomannosyl moieties [5, 6]. The lysosomal hydrolases encounter MPRs at the TGN [7]. There, MPR-ligand complexes interact with the heterotetrameric complex of the AP1 clathrin adaptor complex [8] and members of the GGA family [9-11] of clathrin adapters. Binding concentrates receptor-ligand complexes into tubular structures at the TGN [12] that later fuse with early endosomes containing internalized transferrin [12]. MPRs are then segregated from transferrin within early endosomes, yielding two distinct domains [12].

The return route taken by MPRs to the TGN has been somewhat controversial. Because depletion of AP1 [13] or so-called retromer proteins [14, 15] results in MPR sequestration in early endosomes, many have concluded that the MPR traffics directly from early endosomes to the TGN. The bacterial Shiga [16] and cholera toxins [17] use such a pathway while TGN46 [18] recycles back to the Golgi *via* recycling endosomes. However, MPRs bind their ligands in a pH dependent manner and need a pH <6 to dissociate the ligand from the MPR [19]. The pH of early endosomes is ~6.2-6.3, while that of late endosomes is 5.2-5.8 [20]. Rab9 is a late endosomal Rab protein that is absolutely required for MPR retrieval [21]. In addition, Press *et al.* [22] showed that the late endosome Rab7 was required for proper MPR trafficking. When they expressed inactive Rab7 in cells, the MPR was trapped in an early endosomes and could not return to the TGN [22]. More recent work has also confirmed the role that Rab7 plays in proper retromer localization and function [23, 24]. Thus, MPRs most likely passes through late endosomes to release lysosomal hydrolases before return to the TGN.

2. RAB9 IS THE RAB GTPASE NEEDED FOR MPR RECYCLING TO THE GOLGI

Kornfeld and colleagues showed that MPRs are localized primarily to late endosomes at steady-state [25]. When Rab7 and Rab9 were also found to be localized to late endosomes, this dramatically narrowed the

---

*Address correspondence to Suzanne R. Pfeffer: Department of Biochemistry, Stanford University School of Medicine, Stanford, CA 94305-5307, USA; Tel: 650-723-6169; Fax: 650-723-6783; E-mail: pfeffer@stanford.edu

search for a Rab GTPase needed for MPR recycling [26, 27]. An *in vitro* transport assay [28] showed that anti-Rab9 antibodies inhibited transport reactions up to 50% under conditions in which anti-Rab7 antibodies had no effect [27]. Furthermore, C-terminal prenylation of Rab9 was essential for transport, and cytosol enriched for Rab9 had higher activity than control cytosol when tested using the *in vitro* assay [27, 29]. These data showed that Rab9 enhanced MPR transport *in vitro*.

Analogous to Ras, Rab9 can be mutated into constitutively active and inactive forms. A serine at position 21 in Rab9 coordinates with a magnesium ion to stabilize Rab9:GTP [30]. Upon mutation of this residue to asparagine, the coordination is lost, yielding a Rab protein that prefers GDP fifty fold over GTP [21]. Similarly, a glutamine residue is nearly universally present in the so-called G3 motif of Rabs. In Rab9, this glutamine coordinates a water molecule that is responsible for the hydrolysis of the gamma phosphate of GTP. Rab9 has a slow, steady-state hydrolysis rate of 0.0052 min^{-1} [31]. Upon mutation of this glutamine into a hydrophobic residue (usually to a similarly sized leucine), the intrinsic rate of GTP hydrolysis is substantially lowered, effectively generating a constitutively active Rab. Because Rabs bind effectors and are considered active in their GTP bound state, these mutants can facilitate analysis of the physiological function of Rab9.

MPRs are transported from the Golgi to early endosomes, from early endosomes to either the cell surface or late endosomes, and also from late endosomes back to the TGN. Although most of the function of MPRs is to deliver newly synthesized lysosomal enzymes to the lysosome, the CI-MPR also clears IGF-II from the plasma, and can also internalize secreted lysosomal enzymes that are circulating in the bloodstream [1]. Once delivered to the cell surface, MPRs are endocytosed by clathrin-mediated endocytosis.

The consequence of improperly trafficked MPRs could lead to fewer receptors at the TGN to retrieve newly made hydrolases for delivery to lysosomes. These cells would therefore be predicted to display an increase in bulk-flow secretion of lysosomal enzymes, thus providing a cellular phenotype for defect in MPR trafficking. The lysosomal enzyme, cathepsin D, is synthesized as a 50kDa precursor enzyme that matures to 31kDa by the time it arrives in lysosomes [32]. Its mis-sorting to the plasma membrane, as judged by the appearance of the larger, 50kDa form in the medium, can be used to test for an MPR trafficking defect in cells.

When the constitutively inactive Rab9 S21N is expressed in cells, it acts as a dominant negative inhibitor of MPR trafficking [21]. Cathepsin D matures much more slowly and lysosomal enzymes are hyper-secreted by these cells [21]. In addition, depletion of Rab9 using siRNA causes MPRs to become up-regulated at the transcriptional level and at the same time, more rapidly degraded due to mislocalization to the lysosome [33] (Other blocks of endosome to Golgi transport have a similar consequence of upregulating MPRs). These data provide compelling evidence for Rab9's role in recycling of MPRs from late endosomes to the TGN in live cells.

When CFP-Rab7 and YFP-Rab9 are co-expressed in cells, they localize to distinct domains on a given late endosome compartment: for structures staining positive for both CFP-Rab7 and YFP-Rab9, only 15% of the two markers colocalize [34]. Forty percent of YFP-Rab9-containing domains are also positive for CI-MPRs, compared with 16% of CFP-Rab7-containing domains [34]. Finally, live cell video microscopy showed that Rab9 is present on transport vesicles moving from late endosomes to the TGN [34], consistent with Rab9's important role in this process.

3. LOCALIZATION OF RAB9 TO LATE ENDOSOMES

In the cytosol, prenylated, GDP-bearing Rabs exist in tight complex with a protein named GDI (GDP Dissociation Inhibitor [35]). While in complex with GDI, the exchange rate of GDP for GTP is minimal [35]. Shapiro and Pfeffer [36] were first to measure the affinity of Rab9-GDI association, and found it to be ≤20nM. Later, Soldati *et al.* [37] showed that purified complexes of prenylated Rab9 bound to GDI contain all of the information needed for selective delivery of Rab9 to endosome-enriched membranes [37]. GDI was released, and after a short lag, Rab9 on the membranes began to exchange GDP for GTP [37].

To distinguish if the machinery that brings Rab9 to the late endosome was specific for Rab9, competition assays were performed. Because Rab7 is on late endosomes with Rab9 [26], Rab7 would be expected to interfere with Rab9 recruitment if it is a general process. Like Rab9, Rab7 in complex with GDI was also recruited to endosome enriched membranes with a Km of 22nM [38]. Myc-Rab9 inhibited Rab9 recruitment with a K_i of 9nM (identical to the Km value of wild-type Rab9 recruitment), indicating that there is no difference between myc-Rab9 and wild-type Rab9 recruitment [38]. However, upon the addition of Rab7, the K_i increased over 10 fold to 112nM. This implies that Rab7 and Rab9 are recruited and stabilized on membranes by different machinery. Rab1 competed even more poorly, with a K_i of 405nM. Thus, it was concluded that Rab9 is recruited and stabilized on endosomes using a different set of proteins than Rab7. Whether this is due to proteins that load a Rab protein onto membranes or instead, to subsequent effector binding is not yet known.

Because complexes of prenylated Rabs bound to GDI had all of the information needed to accomplish Rab delivery to specific membranes [37] and because of the tight binding of Rab9 to GDI [36], a new class of enzyme was proposed that would catalyze the release of prenylated Rabs from GDI protein. These were termed GDI displacement factors (or GDFs), and such an activity was detected in crude membranes [39]. GDF was abolished with proteinase-K treatment, but not released by high salt, demonstrating that the GDF activity was, in fact, due to an integral membrane protein [39].

Since binding to GDI interferes with Rab nucleotide exchange, release of Rabs could be measured by their acquired ability to now exchange bound nucleotide. In this type of experiment, crude GDF was able to release GDI from endosomal Rabs 5, 7, and 9, but not from the ER and Golgi associated, Rab 1 or 2 proteins.

Purification of the membrane-associated GDF activity was incredibly difficult, so an informed approach was instead pursued. Because GDI is conserved from yeast to humans, all yeast proteins showing genetic interaction with Rabs or GDI were considered as candidates. A protein family was identified that bound prenylated Rab3A *via* yeast-two hybrid; these proteins were known as prenylated Rab acceptor (PRA) [40]. These proteins were also identified in yeast as interactors with Ypt1p and Ypt31p and called Ypt1 interacting proteins, or YIPs [41]. When expressed in cells, PRA1/Yip3 localized to endosomes and the Golgi while PRA2 localized to the ER [42]. In addition, PRA1/Yip3 was shown to have weak affinity for GDI [43]. Thus, Yip3 was tested in terms of its possible role as a GDI displacement factor for Rabs within the endocytic pathway.

Recombinant Yip3 protein was purified from membranes after expression in E. coli, and the purified protein catalyzed the dissociation of Rab9 from GDI [44]. When prenyl Rab9:GDI complexes were mixed with recombinant Yip3 and GTPγS, GDI was displaced and GTPγS was able to access Rab9's nucleotide binding site [44]. This reaction was catalytic because nearly 2 pmol of Rab protein was released from GDI, in reactions containing less than 15 fmol of Yip3 protein [44]. In addition, Yip3 was sufficient to recruit Rab9 onto purified liposomes from Rab9:GDI complexes [44]. This study was the first to identify a GDF for any Rab protein. This work showed that Yip3 could also work on Rab7 and Rab5 but not on Rab1. To date, one other protein has been shown to have GDF activity: the DrrA/SidM protein of Legionella [45, 46]. This protein shows both GDF and GEF activity. Two labs have since shown that this bacterial protein is such a potent a GEF for Rab1 that GEF action drives forward the GDF reaction [47, 48]. Whether any endogenous GEFs are as active as DrrA/SidM protein remains to be determined.

4. TIP47: A CARGO SELECTION PROTEIN FOR THE MANNOSE 6-PHOSPHATE RECEPTOR

TIP47 (Tail Interacting Protein of 47kDa) was initially discovered *via* a yeast-two hybrid screen looking for proteins that bind both the cation-independent (CI-) and cation-dependent (CD-) MPR cytoplasmic domains [49]. Like Rab9, TIP47 is essential for MPR trafficking *in vitro* and for MPR stability in living cells [49]. TIP47 recognizes a Phe-Trp signal in the CD-MPR [49] and a diproline hydrophobic interaction motif in the CI-MPR [50]. TIP47 binds to Rab9 tightly (K_d=95nM) and depletion of TIP47 from cells increases the cytosolic pool of Rab9 [33]. Thus, TIP47 stabilizes Rab9 on membranes and allows for efficient MPR recycling.

At this point it was clear that Rab9 was essential for MPR recycling, but its precise role was unclear. A very satisfying discovery was the ability of Rab9 to enhance the interaction of TIP47 with MPR cytoplasmic domains [51]. The K_D for TIP47 binding to the CI-MPR is 1μM [50]. When active Rab9 was added to the reaction, the affinity of TIP47 for the CI-MPR increased to 300nM, near the cytosolic concentration of TIP47 protein [51]. This increase in binding affinity would enhance TIP47 binding to MPRs when they are present in Rab9-containing compartments and not elsewhere. Since TIP47 is oligomeric [52], selective recruitment of TIP47 to Rab9 and MPR containing compartments would build a microdomain that can be detected by light microscopy [34].

To precisely define the residues of TIP47 involved in Rab9 binding, alanine scanning mutagenesis was employed. Mutation of TIP47 residues 167-169, SVV→AAA, decreased Rab9 binding by 75% *in vitro* [53]. When this construct was expressed in cells, Rab9-positive compartments became enlarged and MPRs were no longer trafficked effectively, consistent with TIP47 having a key function in MPR recycling [34]. It is likely that the dominant inhibitory effect of the mutant is due to its binding of MPRs in a non-productive manner.

5. TETHERING OF RAB9 VESICLES AT THE TGN

Transport vesicles interact with their targets by binding to so-called tethering factors. These proteinaceous factors are either large multi-subunit complexes (*i.e.* the TRAPP complex [54]) or long coil-coiled proteins (such as p115) [55, 56]. In addition to binding to transport vesicles, tethers also participate in bringing the correct SNARE proteins together for productive vesicle fusion [57]. This section will discuss the substantial progress that has been made in identifying proteins needed to tether and fuse Rab9-positive vesicles at the TGN.

With the goal of finding novel Rab9 effectors, a two-tiered, yeast-two hybrid screen was used with continued success. In this system, a human cDNA library is screened for positive interaction with a GTP-locked (Q66L) Rab9 protein; positive cDNAs are rescreened for lack of interaction with a GDP-locked (S21N) Rab9 protein. One protein identified in this manner is a long coiled-coil predicted protein called GCC185. GCC185 is one of four members of the so-called "GRIP" domain-containing Golgin proteins [58]. The GRIP domain is an approximately 50 amino acid sequence located at the extreme C-terminus of these proteins and this domain is sufficient to localize at least two GRIP Golgins to the TGN [58]. Two Rab9 binding sites were identified in GCC185: one is near the C-terminus, immediately upstream of the GRIP domain [59] and a second is found in the middle of the protein [60].

When GCC185 is depleted from cells using siRNA, MPRs are destabilized, the lysosomal enzyme, hexosaminidase, is mis-sorted to the cell surface, and MPR recycling to the Golgi is severely inhibited [59]. These phenotypes are consistent with an MPR trafficking defect [21]. Depletion of GCC185 also causes the Golgi to fragment into ministacks and disperses MPRs into peripheral, Rab9-positive compartments [59]. An intact Golgi requires the full length of GCC185; Golgi morphology could not be restored in GCC185-depleted cells with N-terminal deletions of GCC185 protein [60].

One of the more surprising aspects of GCC185 was the discovery that 14 different Rab proteins interact with the protein, across its entire length, as determined by yeast-two hybrid [60]. These Rabs are mostly Golgi-associated Rabs and may contribute low affinity interactions that hold the Golgi together laterally. Two other Drosophila GRIP domain proteins, Golgin245 and Golgin84, also display numerous Rab binding sites along their lengths [61].

Another tethering factor identified by Rab9-interaction yeast-two hybrid screen is RhoBTB3. RhoBTB3 is an unusual member of the Rho family of GTPases. First, it is 69kDa, nearly three times the size of other Rho family members [62]. It contains an N-terminal Rho domain and a C-terminal BTB (Bric-a-brac, Tramtrack, Broad-complex) domain [63]. The N-terminal Rho domain is not a GTPase like other Rho family members, but is instead an ATPase [64]. Rab9 binds to RhoBTB3 downstream of the BTB domain [64]. Like GCC185, RhoBTB3 is localized to the Golgi and depletion of the protein disperses MPRs into peripheral, Rab9-positive vesicles. Unlike GCC185 depletion, however, the Golgi appears swollen [64]. It is not known why loss of this protein yields a distinct phenotype. Mutations were identified that prevent RhoBTB3 from hydrolyzing ATP or from binding Rab9. Both functional regions were required to restore the proper MPR phenotype in cultured, RhoBTB3-depleted cells [64].

In membrane trafficking events, ATPases are used to disassemble SNARE proteins (NSF), ESCRT assemblies (Vps4) or clathrin coats (Hsc70). We therefore tested if RhoBTB3 might interact with the cargo adaptor protein, TIP47, that might be part of a vesicle coat. Native blue gel electrophoresis revealed that RhoBTB3 exists in a stable complex of ~475kDa with TIP47 on dodecyl maltoside-solubilized membranes [64]. Upon the addition of an ATP regeneration system and Rab9, TIP47 was released from this complex [64]. This supports a model where TIP47 remains on the Rab9 positive vesicle as a coat and is removed upon arrival at the Golgi. After TIP47 removal, SNARE pairing can occur and the vesicle fuses with the Golgi.

6. SNARE PROTEINS IN RAB9-VESICLE TRANSPORT

The last step of vesicle trafficking is membrane fusion catalyzed by SNARE proteins. In most cases, there are three Q-SNARES (or t-SNARES) on the target membrane and one R-SNARE (or v-SNARE) on an incoming vesicle [65, 66]. SNAREs contain regulatory, N-terminal helices, one or two SNARE domains that are used to provide the energy for membrane fusion, and a transmembrane anchor [67]. When the four SNARE domains come together, they form a very stable four-helix bundle that provides the energy to bring opposing membranes in very close proximity and ultimately fusion ensues [67]. Recently, it was shown that a single SNARE complex is sufficient to fuse liposome membranes [68]. Early experiments to analyze SNARE protein function were severely limited by the fact that fusion required hours to go to completion [69]. This contrasts with synaptic vesicle fusion which occurs on a millisecond timescale [70]. Thus, much work has been carried out to elucidate how SNAREs are regulated *in vivo*.

Each SNARE has a set of cognate SNARE partners that it can pair with to form the stable four-helix bundle. The N-terminal helices of the syntaxin family of SNAREs (Qa-SNAREs) interact with the SNARE domain effectively closing the syntaxin from productive pairing [71]. In addition, the Qa-SNAREs are clustered in bundles of ~75 molecules on membranes [72], weakly bound through the SNARE domain [73]. Early work with recombinant yeast SNAREs showed a high degree of promiscuity [74]. The N-terminal helices were thought to provide some specificity for SNARE pairing. Therefore, it was surprising that upon removal of these helices, liposomes were still able to only fuse with their correct cognate partners [75]. Thus, SNARE domains themselves encode the information that determines which SNAREs are acceptable partners [75].

Another way that SNARE pairing is regulated *in vivo* is through Sec1/Munc18-like (SM) proteins. Munc18 was identified as a protein that binds to the N-terminal region of syntaxin-1a and promotes the open conformation of the syntaxin [71]. Syntaxin 1 forms a t-SNARE complex with SNAP-25 (in this complex SNAP-25 contributes two SNARE motifs) and Munc18 has been shown to bind this complex with high affinity [76, 77]. Further evidence shows that Munc18 needs to bind the N-terminal region of syntaxin-1 in order to form the Syntaxin-1: SNAP-25 complex [77]. Thus, Munc18 regulates cognate SNARE interactions.

Vesicle tethering also enhances the fidelity of SNARE interactions and fusion of the correct membranes. The Golgi protein p115 tethers COPI vesicles en route from the ER to the Golgi [78], participates in intra-Golgi transport [55], and maintains the Golgi stacks [79]. p115 contains a domain that is weakly homologous to a SNARE domain and this domain can bind directly to either the Gos28 or Syntaxin 5 SNAREs [57]. Furthermore, addition of p115 to a reaction of Gos28 and Syntaxin 5 greatly increased the amount of Gos28 associated with Syntaxin 5 [57]. Thus, the Golgi tether p115 can catalyze SNARE pairing of specific SNARE complexes.

Because SNAREs are localized to specific compartments, the identification of the SNARE proteins involved in fusing a Rab9 transport vesicle with the Golgi was greatly simplified. The cytosolic domains of candidate SNARE proteins were purified and used as dominant negative inhibitors for MPR trafficking *in vitro* [80]. Out of twelve SNAREs that localize to either the TGN or to endosomes, four SNAREs were identified that inhibit MPR transport *in vitro*: STX10, STX16, Vti1a, and VAMP3 [80]. These SNAREs were shown earlier to form a SNARE complex *in vivo* [81]. This complex is distinct from that used in early endosome to Golgi trafficking; early endosome to TGN trafficking replaces STX10 with STX6 [82]. The SNARE STX16 (the Qa-SNARE) also binds the tethering protein, GCC185, directly, consistent with the roles of both proteins in Rab9 vesicle trafficking [80]. STX16 is also binds the SM protein, VPS45 (analogous to Munc18 binding Syntaxin 1), hinting at a further means to regulate the two STX16 containing SNARE complexes at the TGN [76].

The SNARE complex used by Rab9 transport vesicles is likely regulated indirectly by the Rab in that Rab9 may recruit a tethering factor, GCC185, to the vesicles, and GCC185 may participate in reactions needed for proper membrane fusion. Indeed, GCC185 binds multiple Rabs (including Rab9) along its length [60].

7. OTHER RAB9 EFFECTORS

There are several other Rab9 effectors with unknown functions. The Rab9 effector, p40, is a 40kDa protein that contains a Kelch motif that is predicted to fold into a β-barrel [83]. p40 decreases the intrinsic rate of Rab9 GTP hydrolysis and stimulates the transport of MPRs when added in purified form to *in vitro* transport reactions [83]. It has also been reported to bind directly to the phosphatidylinositol kinase, PIKfyve [84] although other labs have not been able to detect this interaction.

Recently, the BLOC-3 complex was identified as a Rab9 effector [85]. BLOC-3 is comprised of two proteins, HPS1 and HPS4 [86]. These proteins are important in the biogenesis of lysosome-related organelles and mutations in BLOC-3 proteins lead to Hermansky-Pudlak syndrome, a rare disorder that results in decreased pigmentation, bleeding problems and storage of ceroid lipofuscin [87]. What role Rab9 plays in binding BLOC-3 is unknown; it may recruit the protein to the surface of late endosome membranes.

Rab9 has recently been implicated in Atg5- and Atg7-independent autophagy. In Atg5 null mouse embryo fibroblasts, autophagosomes can be induced with the drug, etoposide *via* an alternative autophagy pathway [88]. These autophagosomes are positive for Rab9, and depletion of Rab9 decreased the number of autophagosome vacuoles in Atg5-null cells [88]. Depletion of Rab9 had no effect on classic autophagosome production, leading the authors to conclude that there are two distinct pathways for autophagy [88].

8. RAB9 IN DISEASE

HIV, Ebola, and measles virus are known to take advantage of Rab9 and its effectors to promote virulence [89]. Upon siRNA depletion of Rab9, p40, TIP47, or PIKfyve, HIV replication was drastically reduced and depletion of Rab9 prevented Ebola or measles from secreting new virus [89]. Salmonella also uses the Rab9 pathway to successfully replicate in eukaryotic cells. After endocytosis, Salmonella migrates to the perinuclear region in what is called the Salmonella containing vacuole (SCV) [90]. SCV formation requires Rab7 and the SCV is known to be decorated with LAMP1 but lacking MPRs [91]. Salmonella express the virulence protein, SifA that interacts with the host protein, SKIP [90]. SKIP contains an N-terminal RUN domain and a C-terminal pleckstrin homology (PH) domain [90]. SifA interacts with SKIP *via* its PH domain, and this interaction is essential for Salmonella replication in cells [90]. With the discovery of a role for Rab7 in SCV formation, Jackson *et al.* [92] wondered if Rab9 could also bind to SKIP and assist with Salmonella replication. They found that Rab9 does indeed bind to SKIP's PH domain and can compete with SifA for binding [92]. The physiological relevance of Rab9 binding to SKIP is not yet known in normal cells, but the fact that SKIP can bind kinesin suggests that SKIP may link late endosomes (*via* Rab9) to this microtubule-based motor [90].

Niemann-Pick type C (NPC) disease is an autosomal recessive disease with mutations affecting either the proteins NPC1 or NPC2 [93]. NPC disease cells are characterized by the accumulation of cholesterol and sphingolipids in late endosomes [93]. Over-expression of Rab9 appears to alleviate cholesterol accumulation, suggesting that Rab9 may facilitate egress of cholesterol from late endosomes [94]. However, upon siRNA depletion of Rab9 from HeLa cells, cholesterol did not accumulate, as would be expected if Rab9 normally traffics a significant amount of cholesterol out of late endosomes [95]. Loss of NPC1 leads to accumulation of MPRs in late endosomes, but overexpression of Rab9 can rescue the MPR defect in NPC1-depleted cells [95]. Rab5 and Rab9 also resist GDI-mediated removal from NPC1-cell membranes, suggesting that excess cholesterol might stabilize membrane-bound Rab proteins [95]. This was tested directly by GDI extraction; in liposomes containing high amounts of cholesterol, GDI extraction efficiency was decreased ~2 fold [95]. Thus, cholesterol alone can sequester Rab proteins on membranes, but importantly, the Rabs are trapped there in a non-functional form.

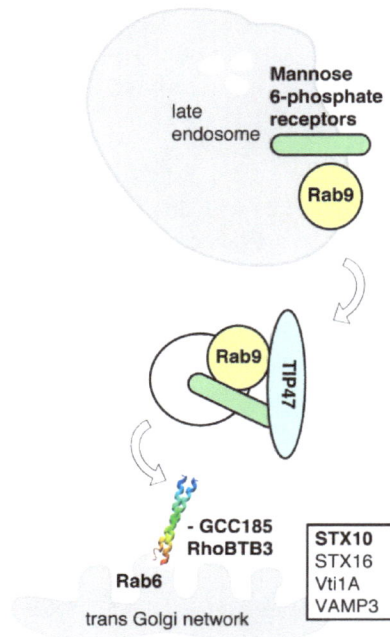

Figure 1: Rab9 mediated transport from late endosomes to the Golgi complex. Rab9 is localized predominantly on late endosomes. Cytosolic TIP47 (blue) is recruited onto late endosomes by Rab9 binding, and a Rab9-TIP47 complex binds mannose 6-phosphate receptors (green). TIP47 is a hexamer in cytosol and likely larger on membranes. A Rab9-containing vesicle then forms, and is tethered at the Golgi by the long coiled-coil protein GCC185. RhoBTB3 is also at the Golgi, and may uncoat TIP47 from the docked vesicle to permit fusion by a SNARE complex comprised of Syntaxin 10, Syntaxin 16, Vti1A and VAMP3.

9. CONCLUSION

Rab9 is a key constituent in the transport of MPRs from late endosomes to the TGN. A large body of work enables us to consider the following model for Rab9 action in MPR transport to the Golgi (Fig. **1**). Rab9:GDI is first recruited from the cytosol onto late endosome membranes, likely by action of the Yip3 GDF. TIP47 is then recruited onto the membrane by Rab9, where it stabilizes Rab9 on the membrane. TIP47:Rab9 complexes next bind the cytoplasmic domains of MPRs and a stable microdomain is formed. These interactions drive the formation of nascent transport vesicles that pinch off the late endosome and traffic along microtubule tracks [96] until they reach the TGN. Next, the vesicle encounters GCC185. GCC185 binds to the CLASP protein [97] that binds directly to microtubules [98]. In this manner, a vesicle moving along a microtubule will be brought directly to GCC185. RhoBTB3 is also nearby and can also bind Rab9. This activates RhoBTB3's ATPase activity [64], and TIP47 is removed from the vesicle. Next, GCC185 in some way brings the vesicle close to the TGN membrane, where the STX10:STX16:Vti1a SNARE complex is ready to bind the cognate, VAMP3 SNARE on the vesicle. Fusion with the TGN would then ensue. We expect that a Rab9 GAP enhances the hydrolysis of Rab9-GTP, thereby inactivating the Rab, and generating Rab9-GDP, a substrate for GDI membrane retrieval. GDI would then deliver Rab9 back to the late endosome.

While there is a great deal of information available in support of this model, there are still many gaps in our understanding. How is it that Rab9 is specifically recruited to late endosomes and not to other endocytic compartments? Is it simply a matter of microdomain formation by interaction with Rab9 specific effector proteins? What is the identity of the guanine nucleotide exchange factor that activates Rab9? What is the coat on Rab9-containing transport vesicles? How does GCC185 function to tether Rab9 vesicles? Certainly there are other Rab9 effectors that contribute to these processes; indeed, Rab5 has been reported to have perhaps as many as 30 distinct effector proteins. Nevertheless, much has been learned about this late endosome Rab that catalyzes vesicle formation at late endosomes and vesicle tethering at the TGN.

CONFLICT OF INTEREST

There is no conflict of interest from any of the authors.

REFERENCES

[1]   Ghosh P, Dahms NM, Kornfeld, S. Mannose 6-phosphate receptors: new twists in the tale. Nat Rev Mol Cell Biol 2003;4:202-12. Also see http://medlibrary.org/medwiki/Mannose_6-phosphate_receptor.

[2]   Hoflack B, Kornfeld S. Purification and characterization of a cation-dependent mannose 6-phosphate receptor from murine P388D1 macrophages and bovine liver. J Biol Chem 1985; 260:12008-12014.

[3]   Tong PY, Kornfeld S. Ligand interactions of the cation-dependent mannose 6-phosphate receptor Comparison with the cation-independent mannose 6-phosphate receptor. J Biol Chem 1989; 264:7970-7975.

[4]   Johnson KF, Chan W, Kornfeld S. Cation-dependent mannose 6-phosphate receptor contains two internalization signals in its cytoplasmic domain. Proc Natl Acad Sci USA 1990 87:10010-14.

[5]   Reitman ML, Kornfeld S. Lysosomal enzyme targeting N-Acetylglucosaminylphosphotransferase selectively phosphorylates native lysosomal enzymes. J Biol Chem 1981; 256:11977-80.

[6]   Waheed A, Hasilik A, von Figura K. UDP-N-acetylglucosamine:lysosomal enzyme precursor N-acetylglucosamine-1-phosphotransferase Partial purification and characterization of the rat liver Golgi enzyme. J Biol Chem 1982; 257:12322-31.

[7]   Duncan JR, Kornfeld S. Intracellular movement of two mannose 6-phosphate receptors: return to the Golgi apparatus. J Cell Biol 1988; 106:617-28.

[8]   Höning S, Sosa M, Hille-Rehfeld A, von Figura K. The 46-kDa mannose 6-phosphate receptor contains multiple binding sites for clathrin adaptors. J Biol Chem 1997; 272:19884–90.

[9]   Puertollano R, Aguilar RC, Gorshkova I, Crouch RJ, Bonifacino JS. Sorting of mannose 6-phosphate receptors mediated by the GGAs. Science 2001; 292:1712–16.

[10]   Zhu Y, Doray B, Poussu A, Lehto VP, Kornfeld S. Binding of GGA2 to the lysosomal enzyme sorting motif of the mannose 6- phosphate receptor. Science 2001; 292:1716–1718.

[11]   Takatsu H, Katoh Y, Shiba Y, Nakayama K. Golgi-localizing, gamma-adaptin ear homology domain, ADP-ribosylation factor-binding (GGA) proteins interact with acidin dileucine sequences within the cytoplasmic domains of sorting receptors through their Vps27p/Hrs/STAM (VHS) domains. J Biol Chem. 2001 Jul 27; 276 (30): 28541-5.

[12]   Waguri S, Dewitte F, Le Borgne R, *et al.* Visualization of TGN to endosome trafficking through fluorescently labeled MPR and AP-1 in living cells. Mol Biol Cell 2003; 14:142-55.

[13]   Meyer C, Zizioli D, Lausmann S, *et al.* Mu1A-adaptin-deficient mice: lethality loss of AP-1 binding and rerouting of mannose 6-phosphate receptors. EMBO J 2000;19:2193-2203.

[14]   Seaman MN. Cargo-selective endosomal sorting for retrieval to the Golgi requires Retromer. J Cell Biol 2004;165:111-22.

[15]   Arighi CN, Hartnell LM, Aguilar RC, Haft CR, Bonifacino JS. Role of the mammalian Retromer in sorting of the cation-independent mannose 6-phosphate receptor. J Cell Biol 2004;165:123-33.

[16]   Mallard F, Tenza D, Antony C, Salamero J, Goud B, Johannes L. Direct pathway from early/recycling endosomes to the Golgi apparatus revealed through the study of Shiga toxin B-fragment transport. J Cell Biol 1998;143:973-90.

[17]   Nichols BJ, Kenworthy AK, Polishchuk RS, *et al.* Rapid cycling of lipid raft markers between the cell surface and Golgi complex. J Cell Biol 2001;153:529-41.

[18]   Ghosh RN, Mallet WG, Soe TT, McGraw TE, Maxfield FR. An endocytosed TGN38 chimeric protein is delivered to the TGN after trafficking through the endocytic recycling compartment in CHO cells. J Cell Biol 1998; 142:923-36.

[19]   Borden LA, Einstein R, Gabel CA, Maxfield FR. Acidification-dependent dissociation of endocytosed insulin precedes that of endocytosed proteins bearing the mannose 6-phosphate recognition marker. J Biol Chem 1990; 265:8497-8504.

[20]   Yamashiro DJ, Maxfield FR. Acidification of morphologically distinct endosomes in mutant and wild-type Chinese hamster ovary cells. J Cell Biol 1987; 105:2723-33.

[21]   Riederer MA, Soldati T, Shapiro AD, Lin J, Pfeffer SR. Lysosome biogenesis requires Rab9 function and receptor recycling from endosomes to the trans-Golgi network. J Cell Biol 1994; 125:573-82.

[22]   Press B, Feng Y, Hoflack B, Wandinger-Ness A. Mutant Rab7 causes the accumulation of cathepsin D and cation-independent mannose 6-phosphate receptor in an early endocytic compartment. J Cell Biol 1998; 140:1075-89.

[23]   Rojas R, van Vlijmen T, Mardones GA, et al. Regulation of retromer recruitment to endosomes by sequential action of Rab5 and Rab7. J Cell Biol 2008; 183:513-26.

[24]   Seaman MN, Harbour ME, Tattersall D, Read E, Bright N. Membrane recruitment of the cargo-selective retromer subcomplex is catalysed by the small GTPase Rab7 and inhibited by the Rab-GAP TBC1D5. J Cell Sci 2009; 122:2371-82.

[25]   Griffiths G, Hoflack B, Simons K, Mellman I, Kornfeld S. The mannose 6-phosphate receptor and the biogenesis of lysosomes. Cell 1988 52:329-41.

[26]   Chavrier P, Parton RG, Hauri HP, Simons K, Zerial M. Localization of low molecular weight GTP binding proteins to exocytic and endocytic compartments Cell 1990; 62:317-29.

[27]   Lombardi D, Soldati T, Riederer MA, Goda Y, Zerial M, Pfeffer SR. Rab9 functions in transport between late endosomes and the trans Golgi network. EMBO J 1993; 12:677-82.

[28]   Goda Y, Pfeffer SR. Selective recycling of the mannose 6-phosphate/IGF-II receptor to the trans Golgi network *in vitro*. Cell 1988; 55:309-20.

[29]   Soldati T, Riederer MA, Pfeffer SR. Rab GDI: a solubilizing and recycling factor for rab9 protein. Mol Biol Cell 1993; 4:425-34.

[30]   Chen L, DiGiammarino E, Zhou XE, et al. High resolution crystal structure of human Rab9 GTPase: a novel antiviral drug target. J Biol Chem 2004; 279:40204-08.

[31]   Shapiro AD, Riederer MA, Pfeffer SR. Biochemical analysis of rab9 a ras-like GTPase involved in protein transport from late endosomes to the trans Golgi network. J Biol Chem 1993 268:6925-6931.

[32]   Rosenfeld MG, Kreibich G, Popov D, Kato K, Sabatini DD. Biosynthesis of lysosomal hydrolases: their synthesis in bound polysomes and the role of co- and post-translational processing in determining their subcellular distribution. J Cell Biol 1983; 93:135-43.

[33]   Ganley IG, Carroll K, Bittova L, Pfeffer S. Rab9 GTPase regulates late endosome size and requires effector interaction for its stability. Mol Biol Cell 2004; 15:5420-30.

[34]   Barbero P, Bittova L, Pfeffer SR. Visualization of Rab9-mediated vesicle transport from endosomes to the trans-Golgi in living cells. J Cell Biol 2002; 156:511-18.

[35]   Sasaki T, Kikuchi A, Araki S, et al. Purification and characterization from bovine brain cytosol of a protein that inhibits the dissociation of GDP from and the subsequent binding of GTP to smg p25A a ras p21-like GTP-binding protein. J Biol Chem 1990; 265:2333-7.

[36]   Shapiro AD, Pfeffer SR. Quantitative analysis of the interactions between prenyl Rab9 GDP dissociation inhibitor-alpha and guanine nucleotides. J Biol Chem 1995; 270:11085-90.

[37]   Soldati T, Shapiro AD, Svejstrup AB, Pfeffer SR. Membrane targeting of the small GTPase Rab9 is accompanied by nucleotide exchange. Nature 1994; 369:76-8.

[38]   Soldati T, Rancaño C, Geissler H, Pfeffer SR. Rab7 and Rab9 are recruited onto late endosomes by biochemically distinguishable processes. J Biol Chem 1995; 270:25541-48.

[39]   Dirac-Svejstrup AB, Sumizawa T, Pfeffer SR. Identification of a GDI displacement factor that releases endosomal Rab GTPases from Rab-GDI. EMBO J 1997; 16:465-72.

[40]   Martincic I, Peralta ME, Ngsee JK. Isolation and characterization of a dual prenylated Rab and VAMP2 receptor. J Biol Chem 1997; 272:26991-98.

[41]   Yang X, Matern HT, Gallwitz D. Specific binding to a novel and essential Golgi membrane protein (Yip1p) functionally links the transport GTPases Ypt1p and Ypt31p. EMBO J 1998; 17:4954-63.

[42]   Abdul-Ghani M, Gougeon PY, Prosser DC, Da-Silva LF, Ngsee JK. PRA isoforms are targeted to distinct membrane compartments. J Biol Chem 2001; 279:6225-33.

[43]   Hutt DM, Da-Silva LF, Chang LH, Prosser DC, Ngsee JK. PRA1 inhibits the extraction of membrane-bound rab GTPase by GDI1. J Biol Chem 2000; 275:18511-19.

[44]   Sivars U, Aivazian D, Pfeffer SR. Yip3 catalyses the dissociation of endosomal Rab-GDI complexes. Nature 2003; 425:856-59.

[45]   Ingmundson A, Delprato A, Lambright DG, Roy RC. Legionella pneumophila proteins that regulate Rab1 membrane cycling. Nature 2007; 450:365-69.

[46]   Machner MP, Isberg RR. A bifunctional bacterial protein links GDI displacement to Rab1 activation. Science 2007 318:974-977.

[47]   Schoebel S, Oesterlin LK, Blankenfeldt W, Goody RS, Itzen A. RabGDI displacement by DrrA from Legionella is a consequence of its guanine nucleotide exchange activity. Mol Cell 2009; 36:1060-72.

[48]   Suh HY, Lee DW, Lee KH, *et al*. Structural insights into the dual nucleotide exchange and GDI displacement activity of SidM/DrrA. EMBO J 2010; 29:496-504 .

[49]   Díaz E, Pfeffer SR. TIP47: a cargo selection device for mannose 6-phosphate receptor trafficking. Cell 1998; 93:433-43.

[50]   Krise JP, Sincock PM, Orsel JG, Pfeffer SR. Quantitative analysis of TIP47-receptor cytoplasmic domain interactions: implications for endosome-to-trans Golgi network trafficking. J Biol Chem 2000; 275:25188-93.

[51]   Carroll KS, Hanna J, Simon I, Krise J, Barbero P, Pfeffer SR. Role of Rab9 GTPase in facilitating receptor recruitment by TIP47. Science 2001; 292:1373-76.

[52]   Sincock PM, Ganley IG, Krise JP, *et al*. Self-assembly is important for TIP47 function in mannose 6-phosphate receptor transport. Traffic 2003; 4:18-25.

[53]   Hanna J, Carroll K, Pfeffer SR. Identification of residues in TIP47 essential for Rab9 binding. Proc Natl Acad Sci USA 2002; 99:7450-54.

[54]   Yu S, Satoh A, Pypaert M, Mullen K, Hay JC, Ferro-Novick S. mBet3p is required for homotypic COPII vesicle tethering in mammalian cells. J Cell Biol 2006; 174:359-68.

[55]   55 Waters MG, Clary DO, Rothman JE. A novel 115-kD peripheral membrane protein is required for intercisternal transport in the Golgi stack. J Cell Biol 1992; 118:1015–26.

[56]   Barroso M, Nelson DS, Sztul E. Transcytosis-associated protein (TAP)/p115 is a general fusion factor required for binding of vesicles to acceptor membranes. Proc Natl Acad Sci USA 1995; 92:527-31.

[57]   Shorter J, Beard MB, Seemann J, Dirac-Svejstrup AB, Warren G. Sequential tethering of Golgins and catalysis of SNAREpin assembly by the vesicle-tethering protein p115. J Cell Biol 2002; 157:45-62.

[58]   Munro S, Nichols BJ. The GRIP domain - a novel Golgi-targeting domain found in several coiled-coil proteins. Curr Biol 1999; 9:377-80.

[59]   Reddy JV, Burguete AS, Sridevi K, Ganley IG, Nottingham RM, Pfeffer SR. A functional role for the GCC185 golgin in mannose 6-phosphate receptor recycling. Mol Biol Cell 2006; 17:4353-63.

[60]   Hayes GL, Brown FC, Haas AK, Nottingham RM, Barr FA, Pfeffer SR. Multiple Rab GTPase binding sites in GCC185 suggest a model for vesicle tethering at the trans-Golgi. Mol Biol Cell 2009; 20:209-17.

[61]   Sinka R, Gillingham AK, Kondylis V, Munro S. Golgi coiled-coil proteins contain multiple binding sites for Rab family G proteins. J Cell Biol 2008; 183:607-15.

[62]   Ramos S, Khademi F, Somesh BP, Rivero F. Genomic organization and expression profile of the small GTPases of the RhoBTB family in human and mouse. Gene 2002; 298:147-57.

[63]   Salas-Vidal E, Meijer AH, Cheng X, Spaink HP. Genomic annotation and expression analysis of the zebrafish Rho small GTPase family during development and bacterial infection. Genomics 2005; 86:25-37.

[64]   Espinosa EJ, Calero M, Sridevi K, Pfeffer SR. RhoBTB3: a Rho GTPase-family ATPase required for endosome to Golgi transport. Cell 2009; 137:938-48.

[65]   Hong W. SNAREs and traffic. Biochim Biophys Acta 2005; 1744:493-517.

[66]   Südhof TC, Rothman JE. Membrane fusion: grappling with SNARE and SM proteins. Science 2009; 323:474-7.

[67]   Brunger AT, Weninger K, Bowen M, Chu S. Single-molecule studies of the neuronal SNARE fusion machinery. Ann Rev Biochem 2009; 78:903-28.

[68]   van den Bogaart G, Holt MG, Bunt G, Riedel D, Wouters FS, Jahn R. One SNARE complex is sufficient for membrane fusion. Nat Struc Mol Biol 2010; 17:358-64.

[69]   Weber T, Zemelman BV, McNew JA, *et al*. SNAREpins: minimal machinery for membrane fusion. Cell 1998; 92:759-72.

[70]   Sabatini BL, Regehr WG. Timing of neurotransmission at fast synapses in the mammalian brain. Nature 1996; 384:170-72.

[71]   Dulubova I, Sugita S, Hill S, *et al*. A conformational switch in syntaxin during exocytosis: role of munc18. EMBO J 1999; 18:4372-82.

[72] Sieber JJ, Willig KI, Kutzner C, *et al.* Anatomy and dynamics of a supramolecular membrane protein cluster. Science 2007; 317:1072-76.

[73] Lerman JC, Robblee J, Fairman R, Hughson FM. Structural analysis of the neuronal SNARE protein syntaxin-1A. Biochemistry 2000; 39:8470-79.

[74] Tsai MM, Banfield DK. Yeast Golgi SNARE interactions are promiscuous. J Cell Sci 2000; 113:145-52.

[75] Paumet F, Rahimian V, Rothman JE. The specificity of SNARE-dependent fusion is encoded in the SNARE motif. Proc Natl Acad Sci USA 2004; 101:3376-80.

[76] Dulubova I, Yamaguchi T, Gao Y, *et al.* How Tlg2p/syntaxin 16 'snares' Vps45. EMBO J 2002; 21:3620-31.

[77] Burkhardt P, Hattendorf DA, Weis WI, Fasshauer D. Munc18a controls SNARE assembly through its interaction with the syntaxin N-peptide. EMBO J 2008; 27:923-33.

[78] Sönnichsen B, Lowe M, Levine T, Jamsa E, Dirac-Svejstrup B, Warren G. A role for giantin in docking COPI vesicles to Golgi membranes. J Cell Biol 1998; 140:1013–21.

[79] Shorter J, Warren G. A role for the vesicle tethering protein p115 in the post-mitotic stacking of reassembling Golgi cisternae in a cell-free system. J Cell Biol 1999; 146:57-70.

[80] Ganley IG, Espinosa E, Pfeffer SR. A syntaxin 10-SNARE complex distinguishes two distinct transport routes from endosomes to the trans-Golgi in human cells. J Cell Biol 2008; 180:159-72.

[81] Wang Y, Tai G, Lu L, Johannes L, Hong W, Tang BL. Trans-Golgi network syntaxin 10 functions distinctly from syntaxins 6 and 16. Mol Membr Biol 2005; 22:313-25.

[82] Mallard F, Tang BL, Galli T, *et al.* Early/recycling endosomes-to-TGN transport involves two SNARE complexes and a Rab6 isoform. J Cell Biol 2002; 156:653-64.

[83] Díaz E, Schimmöller F, Pfeffer SR. A novel Rab9 effector required for endosome-to-TGN transport. J Cell Biol 1997; 138:283-90.

[84] Ikonomov OC, Sbrissa D, Mlak K, *et al.* Active PIKfyve associates with and promotes the membrane attachment of the late endosome-to-trans-Golgi network transport factor Rab9 effector p40. J Biol Chem 2003; 278:50863-71.

[85] Kloer DP, Rojas R, Ivan V, *et al.* Assembly of the biogenesis of lysosome-related organelles complex-3 (BLOC-3) and its interaction with Rab9. J Biol Chem 2010; 285:7794-7804.

[86] Nazarian R, Falcón-Pérez JM, Dell'Angelica EC. Biogenesis of lysosome-related organelles complex 3 (BLOC-3): a complex containing the Hermansky-Pudlak syndrome (HPS) proteins HPS1 and HPS4. Proc Natl Acad Sci USA 2003; 100:8770-75.

[87] Raposo G, Marks MS, Cutler DF. Lysosome-related organelles: driving post-Golgi compartments into specialization. Curr Opin Cell Biol 2007; 19:394-401.

[88] Nishida Y, Arakawa S, Fujitani K, *et al.* Discovery of Atg5/Atg7-independent alternative macroautophagy. Nature 2009; 461:654-58.

[89] Murray JL, Mavrakis M, McDonald NJ, *et al.* Rab9 GTPase is required for replication of human immunodeficiency virus type 1 filoviruses and measles virus. J Virol 2005; 79:11742-51.

[90] Boucrot E, Henry T, Borg JP, Gorvel JP, Méresse S. The intracellular fate of Salmonella depends on the recruitment of kinesin. Science 2005; 308:1174-78.

[91] Méresse S, Steele-Mortimer O, Finlay BB, Gorvel JP. The rab7 GTPase controls the maturation of Salmonella typhimurium-containing vacuoles in HeLa cells. EMBO J 1999 18:4394-4403.

[92] Jackson LK, Nawabi P, Hentea C, Roark EA, Haldar K. The Salmonella virulence protein SifA is a G protein antagonist. Proc Natl Acad Sci USA 2008; 105:14141-46.

[93] Sturley SL, Patterson MC, Balch W, Liscum L. The pathophysiology and mechanisms of NP-C disease. Biochim Biophys Acta 2004; 1685:83-87

[94] Choudhury A, Dominguez M, Puri V, *et al.* Rab proteins mediate Golgi transport of caveola-internalized glycosphingolipids and correct lipid trafficking in Niemann-Pick C cells. J Clin Invest 2002; 109:1541-50.

[95] Ganley IG, Pfeffer SR. Cholesterol accumulation sequesters Rab9 and disrupts late endosome function in NPC1-deficient cells. J Biol Chem 2006; 281:17890-99.

[96] Itin C, Ulitzur N, Mühlbauer B, Pfeffer SR. Mapmodulin cytoplasmic dynein and microtubules enhance the transport of mannose 6-phosphate receptors from endosomes to the trans-golgi network. Mol Biol Cell 1999; 10:2191-97.

[97] Efimov A, Kharitonov A, Efimova N, *et al.* Asymmetric CLASP-dependent nucleation of noncentrosomal microtubules at the trans-Golgi network. Dev Cell 2007; 12:917-30.

[98] Akhmanova A, Hoogenraad CC, Drabek K, *et al.* Clasps are CLIP-115 and -170 associating proteins involved in the regional regulation of microtubule dynamics in motile fibroblasts. Cell 2001; 104:923-35.

CHAPTER 12

Novel Rab GTPases

Maria Luisa Rodrigues and José B. Pereira-Leal[*]

Instituto Gulbenkian de Ciência, Portugal

Abstract: Rab GTPases have been identified more than 20 years ago, and their central role as regulators of protein trafficking has attracted considerable attention to this protein family. Hundreds of Rabs have been described in the literature based on bioinformatics analysis, but only a small proportion has been experimentally characterized. Using the human Rab family as an example, we discuss here how our knowledge of the Rab universe is biased towards evolutionarily older, more highly and widely expressed proteins. Newly described Rab proteins have thus received little attention. We explore the types of functional and structural novelty that newly characterized Rabs are unveiling, and discuss the importance of these poorly characterized proteins by exploring their participation in human disease.

Keywords: Rab, Evolution, Novelty, Conservation, Expression.

1. INTRODUCTION

Rabs are well known to be regulators of membrane trafficking, but the particular organelle, pathway and trafficking step in which each Rab participates is unknown for many Rab proteins. Several taxonomical groups expanded their Rab families independently, giving rise to taxon- and sometimes species-specific Rabs whose functions are completely unknown. Extreme examples are *Entamoeba histolytica and E. invadens,* each having more than 100 Rabs [1, 2], and *Trichomonas vaginalis* which has nearly 300 [3]. In contrast, other organisms display extremely reduced, but not fully overlapping, Rab repertoires - *Plasmodium falciparum* has 11 Rabs [4], *Trypanosomas brucei* has 16 [5] and many fungi have very small Rab complements as a result of what was likely a streamlining of an ancestral larger family - *e.g., Schizosaccharomyces pombe* has just 8 Rabs [6].

Comparative analysis of the Rab family in multiple organisms revealed a small common core of Rab proteins basal to eukaryotes [6, 7], which have been extensively studied. The organisms with the smallest Rab families include most or all of these ancestral Rabs, but even those include specific ones, typically with poorly characterized functions.

The rapid pace of genome sequencing has resulted, as of May 2010, in nearly 250 sequenced eukaryotic genomes (obtained from the Superfamily database [8]), and more than 1,300 genome projects in progress (GOLD database [9]). The advent of new, fast sequencing technologies will increase this number even more. Our preliminary survey revealed that the 250 sequenced eukaryotic genomes encode in excess of 7,500 Rab genes (unpublished results). The overwhelming majority of these sequences have no associated functional information. Although we can infer a function for some by orthology, for many we can at best predict a role in protein trafficking. However, a recent study by the Field lab, showing that two *Trypanosoma brucei* divergent Rab-like proteins seem to be involved in prevention of infection of its vector organism, suggests that taxon-specific Rabs may play unexpected roles, beyond their canonical function in organizing the endomembrane system [10].

In this chapter we focus on the human Rab family to discuss the notion of novelty in the Rab family, contrasting evolutionary novelty with recent discovery. We discuss the relative importance that novel *vs.* old Rabs may have and explore what types of functional and structural novelty newly characterized Rabs are unveiling.

Address correspondence to José B. Pereira-Leal: Instituto Gulbenkian de Ciência, Rua da Quinta Grande 6, P-2781-901, Portugal; Tel: +351-21-446-4528; +351-21-440-7973; E-mail: jleal@igc.gulbenkian.pt

2. WHAT IS A NOVEL RAB?

The word novelty applied to the Rab family can have different readings. It can describe those Rabs that were recently discovered - Rab35 would be more recent than Rab1, as human Rabs have been numbered in order of discovery. It may also represent those whose function has been more recently elucidated. In that regard, Rab27 is 'older' than Rab12, as the first started to be characterized as early as 1995 due to its involvement in several human diseases [11, 12], whereas the second has received little attention so far. Moreover, novelty can mean evolutionary novelty, *i.e.* those Rabs that appear in restricted branches of the taxonomic tree. We now discuss these two major types of novelty - historical *vs.* evolutionary.

2.1. Newly Described Rabs are Poorly Characterized

Rab GTPases were first described in the 1980's [13-15] (see Chapter 1 in this eBook for a more detailed historical account). Since then, numerous studies have been published on this protein family, and as of May 2010 there are in excess of 3700 hits in PubMed (query: Rab GTPases OR Ypt). However, a limited number of Rabs (5, 11, 7, 3, 1, 4 and 6, in this order) account for the great majority of these publications (Fig. **1A**). These include the "housekeeping" Rabs (Rabs 1, 4, 5, 6, 7 and 11), which perform generic functions in the endocytic and secretory pathways [16, 17]. There is a trend for the first Rabs to be discovered to be more studied, and as shown in Fig. **1B**, this trend has been accentuated over time.

2.2. Newly Described Rabs Tend to be Evolutionary Innovations

Comparative genomics studies suggested that the last common eukaryotic ancestor (LCEA) already possessed a relatively large set of Rab proteins, comprising at least Rabs 1, 2, 4, 5, 6, 7, 8, 11 and 18 [7, 18, 19]. Some of these are absent in specific eukaryotic groups as a result of secondary losses - *e.g.,* Rabs 2 and 18 were lost in fungi [6] and the secondary loss of Rab4 was a frequent event in the evolution of Rabs [20]. These ancestral Rabs include the earliest to be discovered, as indicated by their low numbering.

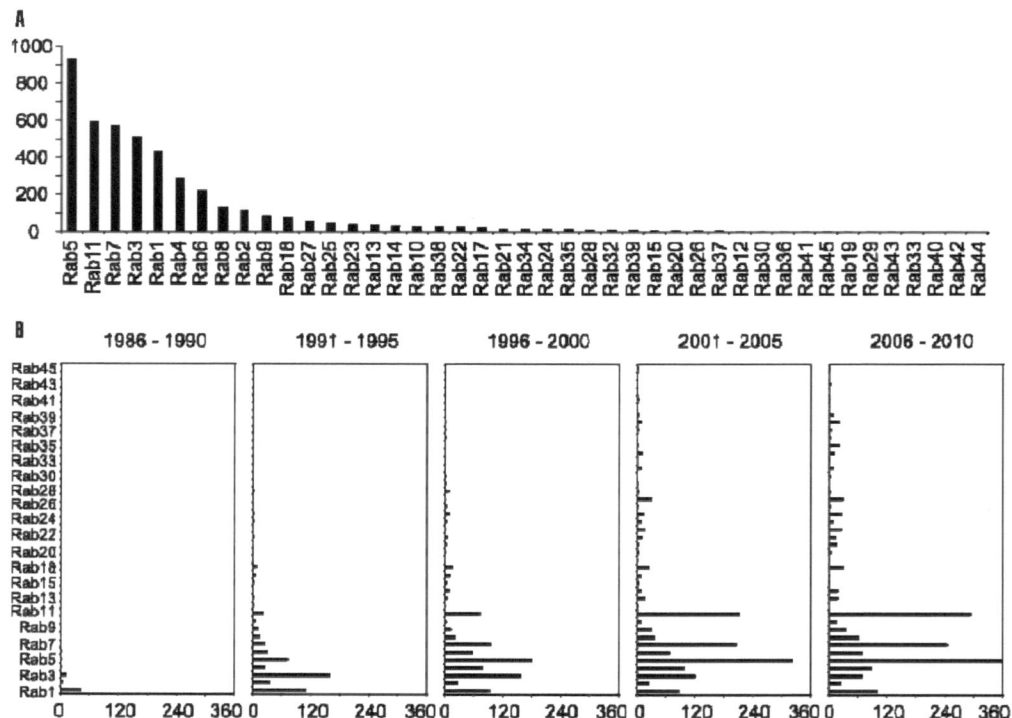

Figure 1: Publications referring to Rab GTPases. The number of Rab articles is given by the total number of PubMed hits (**A**) and by the number of hits obtained for different time intervals (**B**), for each Rab, as of May 2010.

We manually computed a phylogenetic profile for all the human Rab subfamilies (Fig. **2**) in order to test whether this trend held, *i.e.* the order of discovery is somehow related to their order of appearance in evolution. We did not consider isoforms (*e.g.,* Rab1a, Rab1b), as subfamilies have been shown to expand in a taxon-specific manner [6, 21]. The small number of organisms we considered here illustrates that the lowest numbered Rabs are amongst the ancestral set, even though this rule can break, for example for Rab18 which is present in higher plants [22]. By transversing the evolutionary tree in Fig. **2**, we can see that the emergence of Opisthokont/Amoebozoa branch parallels the emergence of Rab 21 [2] and that the major expansions in terms of new subfamilies happened at the emergence of the Metazoa/Choanoflagellate branch (Rabs 9, 14, 22, 23, 28 and 32), then of Metazoa (Rabs 3, 10, 27, 30, 33, 35, 37, 39, 43, 45), and finally of Vertebrates (Rabs 12, 13, 15, 17, 19, 20, 24, 34, 36, 38 and 44). The emergence of mammals is solely accompanied by the appearance of three novel Rab subfamilies (Rabs 25, 41 and 42). Thus, there is a weak tendency for "older" Rabs to be evolutionarily ancestral, *i.e.* predicted to be present in the Last Eukaryotic Common Ancestor. Conversely, most 'new' Rabs appear at the earliest in the Metazoa/Choanoflagellates.

2.3. Recent Rabs are Tissue-Specific and Lowly Expressed

Ancestral Rabs are likely to play house-keeping functions, and as such to be present in every tissue in multicellular organisms, whereas it is plausible that those Rabs that emerged concomitantly with the origin of animals may play tissue-specific functions. We investigated this hypothesis by analyzing tissue specificity and average mRNA (Figs. **3** and **4**), for 79 human tissues.

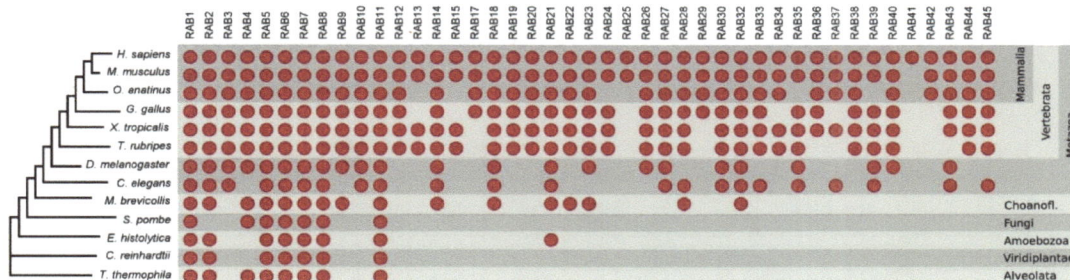

Figure 2: Phylogenetic profile of human Rabs. The presence of human Rab orthologues in different species is indicated as red dots. The organisms selected as representatives of major eukaryotic groups were: *Homo sapiens, Mus musculus, Ornithorhynchus anatinus, Gallus gallus, Xenopus tropicalis, Takifugu rubripes, Drosophila melanogaster, Caenorhabditis elegans, Monosiga brevicollis, Schizosaccharomyces pombe, Entamoeba histolytica, Chlamydomonas reinhardtii* and *Tetrahymena thermophila*. Orthologues were manually searched using BLAST [92].

The older a Rab is, and also the earliest it was discovered, the more likely it is to be widely expressed across human tissues (Fig. **3**). Conversely, those Rabs more recently discovered and evolutionarily more recent tend to be expressed in a narrower range of tissues. Thus the evolutionary conserved Rab1A, which regulates the early steps of the secretory pathway, and Rab11A, which mediates endocytic recycling (see Chapter 2 in this eBook), can be found in all tissues. In contrast, animal specific innovations like Rab3A or Rab25 have more restricted expression profiles, indicating tissue-specific functions. However, this is not always the case, as exemplified by Rab27A and Rab35, which although being animal innovations, display very wide-spread expression profiles. As noted before, the expression patterns of members of each subfamily can vary considerably [23], suggesting tissue-specificity and functional specialization in each sub-family. This has been shown, for example, for Rab5 isoforms in mammals [24-26]. Another recent example is that of Rab27A and B, which not only have different patterns of tissue expression, but even when expressed in the same tissue and operating in the same exosome secretion pathway, function at different points of the pathway and display distinct sub-cellular localizations [27].

The total level of expression of different Rabs varies in orders of magnitude (Fig. **4** - see Fig. legend for details of expression level). The Rabs with highest expression levels are Rabs 1A/B, 2A, 5A, 6A/B/C, 7A,

8A, 10, 11A, 13 and 22B (with average > 200, corresponding to a percentile of about 80%). These overlap with the most widely expressed Rabs (Fig. **3**). The most highly expressed Rabs are also part of the ancestral set, with the exception of Rabs 10, 13, 22B, 23 and 35 - we discuss some of these particular Rabs below. It is worth remarking that besides differences in expression patterns, there are striking differences between the levels of expression within subfamilies, both for the ancestral widely-expressed Rabs, as well as for more recent ones [23].

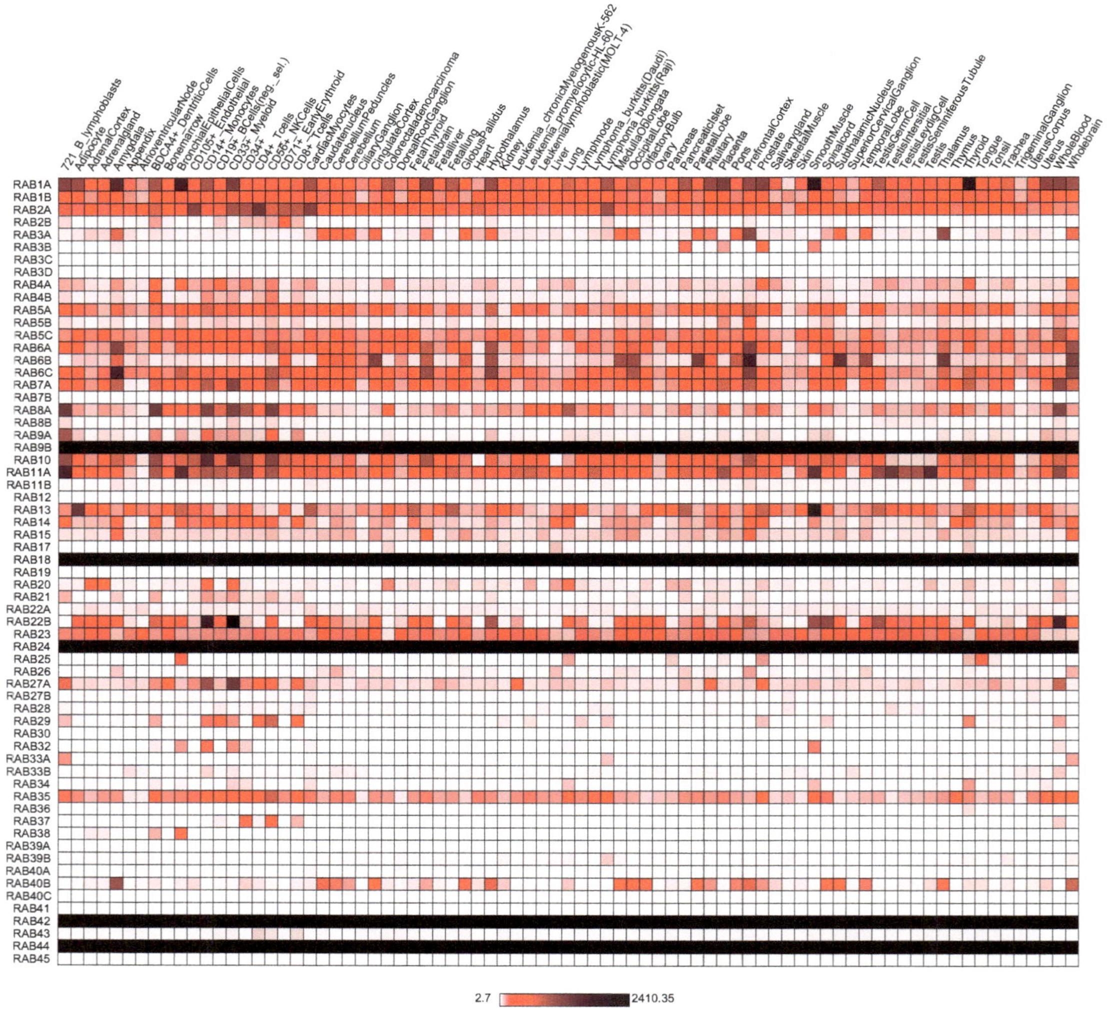

Figure 3: Tissue-specific gene expression of human Rabs. The expression of Rabs in 79 different human tissues is displayed as a colored map, where each square (expression of a given Rab in a given tissue) is colored according to its expression level. A non-linear color ramp is defined from white (at minimum expression value) to red (at 10% of the maximum value) and then to black (maximum expression value) to highlight differences at different expression levels. Profiles of Rabs for which no expression data is available are shown in grey. Data was obtained from the averaged Human U133A/GNF1H Gene Atlas dataset [28] from BioGPS database [93] (gnf1h-gcrma.zip file, downloaded from http://biogps.gnf.org/downloads/). When more than one probeset was available, the one with highest average expression was used in the analysis. Heatmaps for all Rab probesets, as well as for the log2 transformed data is available in the supplementary materials section of our web site (http://www.evocell.org). The figure was created using Cairo (http://www.cairographics.org/).

2.4. Novelty - A Summary

The term "novel Rab", as we have stated above, can have a variety of different interpretations. It is however apparent that these are not independent. The evolutionarily older Rabs, *i.e.* those that emerged first in evolution, are also those that are expressed in most tissues and at higher levels. Conversely, there is a trend for evolutionarily recent Rabs to display both lower levels of expression as well as restricted expression patterns. The study of Rabs has to some extent paralleled this - we first discovered the older Rabs, and possibly because of their widely required function, we have invested more in dissecting their functions. It is also possible that their higher levels of expression may have rendered them easier to identify and study in the pre-human genome days. In fact, of the 10 Rabs with highest average expression, 7 are in the top 10 of most studied Rabs (Rabs 1, 2, 5, 6, 7, 8 and 11). An historical bias towards the study of abundant proteins was previously highlighted by the analysis of the Human and Mouse Gene Atlases, which showed that transcripts corresponding to the known genes (from the RefSeq database) were expressed at a higher level than those that were predicted [28]. Note however that the association between time of discovery and number of publications cannot be ignored (Fig. **1B**), as Rabs 3, 4 and 9 have a moderate average expression but are among those that received most attention.

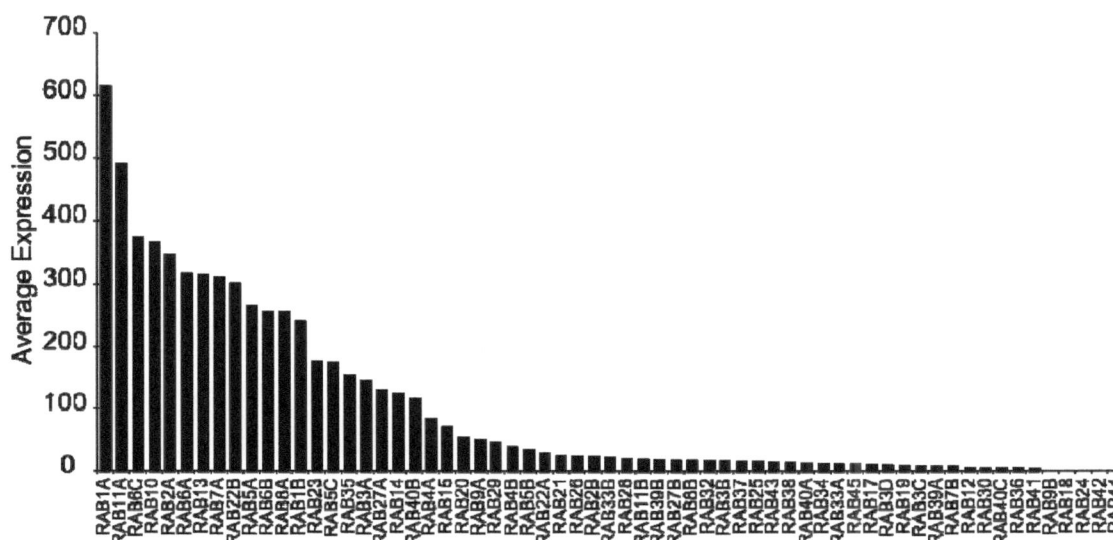

Figure 4: Average expression of Rabs in human tissues. Rabs are ordered by average expression in the 79 different human tissues (same data as in Fig. **3**). The five Rabs that are not present in the Human U133A/GNF1H Gene Atlas dataset (Rabs 9B, 18, 24, 42 and 44, shown in grey in Fig. **3**) are displayed at the tail of the bar plot, with zero average expression.

3. WHAT DO WE LEARN FROM NOVEL RABS?

From Fig. **1** it is evident that we have concentrated on understanding a small number of Rab GTPases. These are discussed in the other chapters of this eBook. We now consider three poorly characterized proteins to ask what can we expect to learn from these "novel Rabs".

3.1. A New Function for an 'Old' Rab

Rab10 is among the first to be described [29] and is one of the most widely and highly expressed human Rabs (Figs. **2** and **3**). However, it has received comparably little attention. This animal-specific protein belongs to the same functional group as Rab8 and Rab13 [19], proteins involved in polarized transport [30, 31]. Similarly, Rab10 also appears to play a role in polarized intracellular trafficking. It localizes to the Golgi in MDCK cells and its activated mutants retain biosynthetic cargo at the TGN [32]. These mutants also mistarget cargo to the apical surface instead of the normal basolateral destination, consistent with a role in basolateral sorting in the early stages of MDCK cells polarization [32]. In contrast, in fully polarized MDCK cells, Rab10 is found mostly in an endocytic compartment and regulates basolateral recycling to the endosome [33]. In adipocytes, Rab10 plays a role in regulating the glucose transporter GLUT4 levels at the plasma membrane [34]. Activated mutants increase the levels of GLUT4 at the plasma membrane, which suggests that Rab10 regulates the exocytic step of GLUT4 homeostasis (see [35] for review). Interestingly, Rab8 and Rab10 appear to be partially redundant in their basolateral sorting activity [32], but not in GLUT4 translocation [36].

Rab10 was recently found to also play a role in phagosome maturation. Phagocytosis is a specialized form of endocytosis, in which large particles (0.5 - 25 μm) are internalized by membrane invagination, forming the phagosome (see [37] for a recent review on the cell biology of the phagosome). After phagosome formation, these "mature" by recruitment of components that permit fusion to the lysosome, forming the phagolysosome, and followed by degradation of the cargo. Rab5 and Rab7 which are known to be involved in endosomal maturation and fusion to lysosomes also play the same role in phagosome maturation (reviewed in [38]). Rab10 is a newcomer to this pathway. It was first associated to the phagosome when it was found by large-scale proteomic identification of the components of phagosomes in fruit fly's haemocyte-derived S2 cells [39] and mouse macrophage-like cell line [40]. However, its specific role was only recently elucidated. Cardoso and colleagues showed that it transiently associates with the phagosomes containing latex beads early in its formation, even prior to Rab5 acquisition, however it associated poorly with *Mycobacterium tuberculosis*-containing phagosomes [41]. This bacteria induces the arrest of phagosome maturation thus preventing the formation of the phagolysosome. Overexpression of activated Rab10 mutant was able to partly overcome the bacterial-induced block, but was not sufficient for lysosomal fusion [41]. Rab10 was also found to localize to the vacuole formed by the obligate intracellular parasite *Anaplasma phagocytophilum* [42], an organelle that is similar to the 'endocytic-like' phagosomes formed by *M. tuberculosis* [37]. *M. Tuberculosis* and A. *phagocytophilum* appear to have followed different strategies. While the former prevents Rab10 association to the phagolysosome, in the the later Rab10 association to the vacuole is dependent on protein synthesis and is independent of the nucleotide-state of Rab10, leading the authors to suggest that the bacteria actively regulates Rab10 association to the vacuolar membrane [42]. The precise role that Rab10 plays in phagosome maturation, and how it is related to its polarized trafficking roles is still unclear. However, this involvement in phagocytosis of pathogenic organisms is likely to attract renewed interest by the community.

3.2. An Unexpected Pathway for a 'New' Rab

Rab35, initially known as Ray or Rab1c, was first identified in 1994 from a human fetal skeletal muscle cDNA library and shown to be very related to Rab1A, Rab1B and Ypt1 [43]. Rab35 and Rab1B sequences are rather similar (sequence identity and similarity of 64% and 81%, respectively), but they diverge significantly at the C-terminus, where Rab35 displays a glutamine repeat and a polybasic sequence. The high similarity between Rab35 and Rab1 places these proteins in the same functional group [19], suggesting that they may function in related trafficking pathways. However, this is not the case, as Rab35 has been demonstrated to work in endocytic recycling, whereas Rab1 regulates trafficking between ER and Golgi (see Chapter 2). Also, unlike other cases where a recent duplicate becomes tissue-specific (*e.g.,* Rab25 *vs.* Rab11), Rab35 is widely expressed [43] (Fig. **3**).

The first clue about the function of Rab35 was obtained in a screen for cytokinesis regulators. Downregulation by RNAi and overexpression of a dominant negative mutant resulted in the accumulation of binucleated cells [44]. Rab35 localizes to plasma membrane and endocytic compartments and is present in clathrin-coated vesicles, regulating a fast recycling step that is essential for late stages of cytokinesis in both human cells and Drosophila [44]. Rab35's role in endocytic recycling is also important in the establishment of the immunological synapse [45]. Furthermore, expression of dominant negative mutant causes reduced exosome secretion in oligodendrocytes, with accumulation of LAMP-1 positive vesicles, suggestive of a transport route between late endomes and the plasma membrane [46]. In *C. elegans*, Rab35 is equally involved in clathrin-dependent recycling, as shown by its important role in yolk receptor recycling to the plasma membrane [47].

Rab35 appears to be involved in other trafficking events, such as control of phagosome-lysosome fusion [48] and neurite outgrowth and cell shape modulation, suggesting a role in actin cytoskeleton remodeling

[49]. This link to actin remodeling was established by the observation that Rab35 regulates actin assembly by recruiting the actin-bundling protein fascin as a downstream effector at the leading edge of cell protrusions [50]. The notion that closely related Rabs have related functions inspired a previous proposal of the existence of "Rab functional groups" [19], however the functional characterization of this novel Rab begs the question of whether such functional inferences can always be made.

3.3. A New Structure for a 'New' Rab

Rab45 is an animal-specific multi-domain protein containing a C-terminal Rab domain, an N-terminal EF-hand domain and a coiled-coil motif at the middle region (Fig. **5**). It is also known as RASEF (Ras and EF-hand domain-containing protein). Together with Rab44, they are the only human Rabs that are fused to other domains. Biochemical characterization of Rab45/RASEF proved the guanine-nucleotide binding activity of its Rab domain and showed that the protein is able to self-interact through its middle region, most likely as a result of coiled-coils interactions [51]. Moreover, the Rab domain and the middle region of Rab45/RASEF were responsible for the observed perinuclear localization of the protein in HeLa cells, which also required the nucleotide binding. The cellular function of Rab45/RASEF is still poorly understood.

Gene fusions are a common event in evolution, frequently indicating a functional association between the fused genes [52, 53]. The EF-hand is the most abundant and functionally diverse eukaryotic Ca^{2+}-binding domain (For review see for example [54, 55]). EF-hand domains are found in Rab effectors, such as in class II FIPS (reviewed in [56]), which suggests that this fusion may reflect a typical interaction between a Rab and an effector. An interesting avenue will be to determine whether Rabs can interact and fuse with any type of the multiplicity of EF-hand domains.

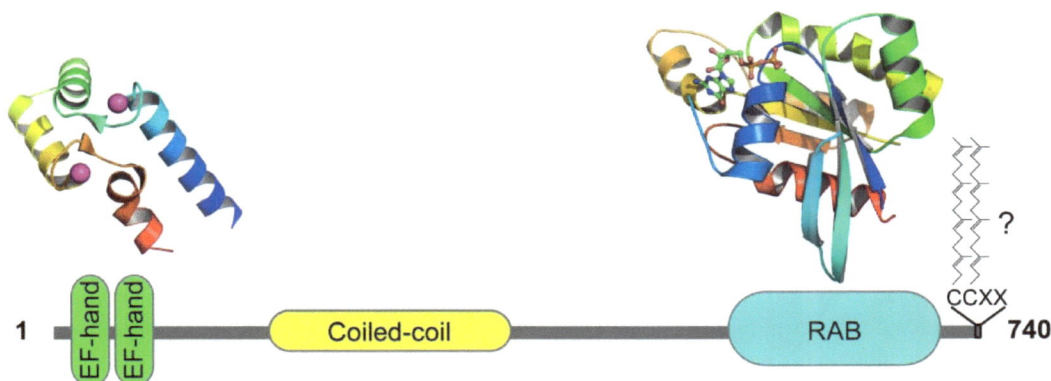

Figure 5: Rab45/RASEF. Schematic representation of the Rab45 domain arrangement, as identified by the SMART online research tool [94, 95]. The EF-hand domain (a pair of calcium binding motifs), the coiled-coil motif and the RAB domain extend from residues 12 to 74, 170 to 362 and 542 to 711, respectively. The three-dimensional structures of EF-hand (residues 3 to 75) and RAB (residues 533 to 712) domains of human Rab45 have been individually determined (2PMY and 2P5S PDB codes, respectively). The secondary structure representation is shown in a rainbow color scheme, ranging from blue, at the N-terminal, to red, at the C-terminal regions. The GDP molecule that is bound to the Rab domain is represented in ball & sticks and the EF-hand calcium ions are depicted as pink spheres. The Rab45 CCXX C-terminal motif is expected to be prenylated by the addition of two geranylgeranyl lipid groups. The three-dimensional structure figures were done with PyMOL (The PyMOL Molecular Graphics System, Version 1.2r3pre, Schrödinger, LLC., http://www.pymol.org/).

Rab45/RASEF was identified as one of the four novel genes in the commonly deleted region of del(9q) in acute myeloid leukemia [57] and one of the genes in 9q21.32 susceptibility locus in familial cutaneous malignant melanoma (CMM) [58]. Importantly, down-regulation of Rab45/RASEF was observed in 7 out of 10 nonfamilial metastatic CMM, in contrast to breast tumor samples, which exhibited expression levels that were similar to those of control cells. Interestingly, no deletion in 9q21 chromosomal region was found in sporadic CMM tumors with decreased Rab45/RASEF expression [58], prompting the search for mutations or epigenetic modifications that accounted for Rab45/RASEF inactivation [59]. No mutations, other than a known SNP (R262C, in the predicted coiled-coil domain), were detected in Rab45/RASEF from sporadic uveal melanoma samples and uveal melanoma cells. Instead, promoter methylation was observed in all cases with absent gene expression and found to correlate with loss of heterozygosity (LOH). Homozygous R262C tumors with a methylated promoter region were significantly associated with decreased survival, strongly suggesting a tumor suppressor role of Rab45/RASEF in uveal melanoma. Further studies are however required to support the putative importance of this novel Rab in cancer.

4. ARE NOVEL RABS IMPORTANT

We have seen above that our knowledge of the (human) Rab universe is severely biased for the older and most expressed Rabs. We have also discussed how there is still plenty of novelty at the molecular and cellular biology level to be dissected in most recently described and poorly studied Rabs. We now consider another level at which novel Rabs may be important - involvement in human disease. We consider their role in inherited diseases and cancer, but not in the the pathogenesis of infectious diseases (for this, see for example [60]).

4.1. Somatic and Germline Mutations

Small GTPases are (in)famous for their role in tumorigenesis. Somatic mutations that render members of the Ras family constitutively active, *i.e.* unable to switch OFF by hydrolyzing GTP [61] are found in a disproportionate amount of human cancers (*e.g.,* [62]). Somatic mutations in Rab genes are, in contrast to Ras-family members but similarly to Rho proteins [63], infrequent in human cancers. A survey of the COSMIC database (Catalogue of Somatic Mutations in Cancer) [64] does not reveal any hit in the "Cancer Genes" list (identified from the Cancer Gene Census), which comprises about 400 genes that have been causally implicated in cancer [65]. However, some Rabs are found in the "Other Genes with Mutations" list, such as Rab3C, Rab4B, Rab5C, Rab8A, Rab8B, Rab28, Rab31/22B, Rab36, Rab38 and Rab41, all of them with only one to three curated references and with mutations in no more than two unique samples. None of the identified missense mutations appears to affect the GTPase cycle [66-68], pointing to loss of function as the most likely mechanism of a putative involvement in tumorigenesis. Furthermore, a recent report suggests that loss of function of Rab21 by deletion may play a critical role in ovarian and prostate cancers [69]. As of May 2010, alterations of the GTPase cycle of Rabs, old or new, do not appear to play any role in tumorigenesis.

The same somatic inactivating mutations that in Ras render these GTPases oncogenic can be found in the germline. They cause a variety of human diseases - Noonan, Costello, Autoimmune lymphoproliferative and Cardio-facio-cutaneous syndromes ("RASopathies", reviewed in [70]. Rab germline mutations have been described and linked to human disease (Table **1**). The sample is still small, but of the five Rabs with a known mutation, four are animal-specific or also present in choanoflagellates. Interestingly, these four cause disease by loss of function mutations [11, 71-73]. Rab7A is the only ancestral gene that is known to cause human diseases. It is also the only Rab for which activating mutations have been described [74, 75]. The effect of all diseases caused by Rab mutations is tissue-specific, even for the widely expressed, activated Rab7 in Charcot-Marie-Tooth disease. It is likely that in most cases, the loss of function of house-keeping, ancestral Rabs is lethal, and as such these Rabs are not likely to be involved in human inherited diseases. In fact, despite the paucity of Rab knock-out data in model organisms, there is a trend for deletion of house-keeping Rabs to be lethal such as Ypt1 in yeast [13] or Rab5 in fruit fly [76]. Conversely, loss of function mutations in taxon-specific Rabs are more likely to result in viable organisms, and are thus better candidates for being involved in inherited diseases. One example is the mouse mutant for Rab38, which displays a tissue-restricted phenotype [77]. However, expansion of subfamilies by gene duplication may provide some partial redundancy, potentially allowing viability of loss of function in ancestral Rabs. This is true even in yeast, where the deletion of Ypt31 and 32 is viable, but the deletion of both is lethal [78].

4.2. Rab Expression Changes in Cancer

A growing number of studies is suggesting a role for changes in the level of expression of Rabs in tumorigenesis [79], and in particular of Rab25 in epithelial cancers [80], which can act as an oncogene [81] or as a tumor suppressor [82, 83], depending on tissue and tumor type. In order to assess if "novel" and "old" Rabs can have different impact on cancers we surveyed the Oncomine cancer microarray database, version 4.2 [84] (Fig. **6**). The noisy nature of microarray data and the incompleteness of this database render this analysis very preliminary. Nevertheless, some trends are obvious. The first is that there is frequent disregulation of Rab levels in most tumors - 42 out of 63 Rabs surveyed here were up or down regulated in at least one tumor type, and only one of the tumors surveyed (Myeloma) did not report changes in Rab levels.

Table 1: Rab genes that are causally mutated in human diseases and animal models of human diseases (*).

Rab	disease	References
Rab7	Charcot-Marie-Tooth disease type 2B	[91]
Rab23	Carpenter Syndrome	[72]
Rab27a	Grishelli syndrome type 2	[11]
Rab38	Hermansky-Pudlak syndrome*	[73, 77]
Rab39B	X-linked mental retardation associated with autism, epilepsy and macrocephaly	[71]

The second is that the same Rab is frequently upregulated in one tumor and down regulated on another. This raises questions about any causal role that these changes may have, but the example of Rab25 discussed above suggests that both may be relevant for disease. Independent studies have reported up-regulation of Rabs in cancers - Rab1A as an early event in tongue squamous cell carcinomas [85], Rab5A potentially playing a role in EGF trafficking in hepatocellular carcinomas [86], Rab20 in triple-negative breast cancer [87], Rab22B/Rab31 as an independent prognostic marker in breast cancer [88] or Rab23 in diffuse-type gastric carcinomas [89]. Conversely, down regulation of Rabs by epigenetic mechanisms in tumorigenesis has also been described. Promoter/exon1 hypermethylation of Rab37 was found in about half (41/71) of non-small cell lung cancer (NSCLC) patients and was strongly associated with the occurrence of metastasis [90]. The epigenetic silencing of Rab45/RASEF in uveal melanoma, discussed above, was linked to decreased survival, supporting a tumor-suppressing role for Rab45/RASEF [59].

The third major trend observable in Fig. **6** is that both old and new, as well as high and lowly expressed Rabs are up/down regulated in multiple tumors. Thus gene expression changes of both house-keeping and tissue-specific Rabs may play a role in tumor progression.

5. FUTURE DIRECTIONS

Our knowledge of the (human) Rab universe is very biased towards the most ancestral, widely- and highly-expressed Rab proteins. However disease and evolutionary data suggests that there is much to surprise us in the most recently discovered Rabs, both at the functional and structural levels.

Rab proteins have expanded independently in a variety of lineages. Different organisms evolved different "Rab strategies" - for example, while fungi reduced their repertoires relative to the predicted ancestral Rab set without adding many new Rabs [6], *Entamoeba histolytica* also lost several ancestral Rabs, but "invented" many new ones [2]. Plants followed yet a distinct route of massive expansions of existing classes [22]. Our preliminary data suggests that the currently sequenced genomes code in excess of 7,500 Rab GTPases, for which we have almost no functional information. Very few of these will be experimentally studied, so we need to define very clearly how much information we can predict from sequence alone, and which proteins are likely to yield significant functional novelty upon investigation - those are the ones where one should invest in functional characterization. Bioinformaticians and cell biologists need to work together to make sense of this ever-growing Rab universe.

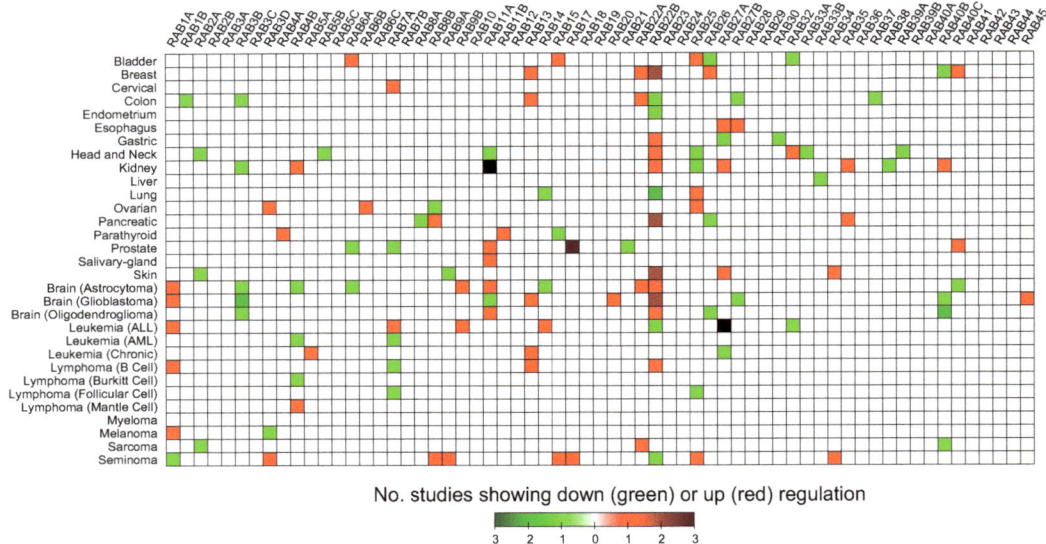

Figure 6: Rabs differentially expressed in cancer. Rabs reported in the Oncomine 3.0 cancer database [84] to be up- (red) or down- (green) regulated, in different cancer types. The search for differentially expressed Rabs was done using the following criteria: presence in the top 1% list of up- or down- differentially expressed genes, with a p-value $< 10^{-4}$. Color intensity is proportional to the number of studies reporting Rabs differentially expression in different types of cancer (1, 2 or 3 studies). Cancer types were defined as in the Oncomine database, except brain, leukemia and lymphoma tumors, which were subdivided in different cancer subypes. Leukemia (ALL) includes acute adult T-cell, B-cell acute and T-cell acute lymphoblastic leukemias; Leukemia (AML) corresponds to acute myeloid leukemia; Leukemia (Chronic) includes chronic adult T-cell, chronic lymphocytic and hairy cell leukemias and Lymphoma (B Cell) includes centroblastic, diffuse large B-cell, primary effusion anf marginal zone B-cell lymphomas. Cancer types are listed according to the following order: carcinomas (bladder to skin), brain tumors, hematopoietic tissue tumors (leukemia, lymphoma, and myeloma), melanoma, sarcoma and seminoma. Cases with conflicting evidence are shown in grey. The figure was created using Matplotlib (http://matplotlib.sourceforge.net/) from Python.

ACKNOWLEDGEMENTS

We wish to thank Marc Gouw and Yoan Diekmann for helpful discussions and for sharing of preliminary data. We thank Monica Bettencourt-Dias, Marc Gouw and Yoan Diekmann for critical reading of the manuscript. MLR is a recipient of a post doctoral fellowship supported by Fundação para a Ciência e a Tecnologia (SFRH/BPD/46395/2008).

CONFLICT OF INTEREST

There is no conflict of interest from any of the authors.

REFERENCES

[1]   Nakada-Tsukui K, Saito-Nakano Y, Husain A, Nozaki T. Conservation and function of Rab small GTPases in Entamoeba: Annotation of E. invadens Rab and its use for the understanding of Entamoeba biology. Exp Parasitol 2010 Nov;126(3):337-47.

[2]   Saito-Nakano Y, Loftus BJ, Hall N, Nozaki T. The diversity of Rab GTPases in Entamoeba histolytica. Exp Parasitol 2005;110:244-52.

[3]   Carlton JM, Hirt RP, Silva JC, *et al.* Draft genome sequence of the sexually transmitted pathogen Trichomonas vaginalis. Science 2007;315:207-12.

[4]   Quevillon E, Spielmann T, Brahimi K, Chattopadhyay D, Yeramian E, Langsley G. The Plasmodium falciparum family of Rab GTPases. Gene 2003;306:13-25.

[5]   Ackers JP, Dhir V, Field MC. A bioinformatic analysis of the RAB genes of Trypanosoma brucei. Mol Biochem Parasitol 2005;141:89-97.

[6]   Pereira-Leal JB. The Ypt/Rab family and the evolution of trafficking in fungi. Traffic 2008;9:27-38.

[7]   Dacks JB, Field MC. Evolution of the eukaryotic membrane-trafficking system: origin, tempo and mode. J Cell Sci 2007;120:2977-85.

[8]   Wilson D, Pethica R, Zhou Y, *et al.* SUPERFAMILY--sophisticated comparative genomics, data mining, visualization and phylogeny. Nucleic Acids Res 2009;37:D380-6.

[9]   Liolios K, Chen IM, Mavromatis K, *et al.* The Genomes On Line Database (GOLD) in 2009: status of genomic and metagenomic projects and their associated metadata. Nucleic Acids Res 2010;38:D346-54.

[10]   Natesan SK, Peacock L, Leung KF, Matthews KR, Gibson W, Field MC. The trypanosome Rab-related proteins RabX1 and RabX2 play no role in intracellular trafficking but may be involved in fly infectivity. PLoS One 2009;4:e7217.

[11]   Menasche G, Pastural E, Feldmann J, *et al.* Mutations in RAB27A cause Griscelli syndrome associated with haemophagocytic syndrome. Nat Genet 2000;25:173-76.

[12]   Seabra MC, Ho YK, Anant JS. Deficient geranylgeranylation of Ram/Rab27 in choroideremia. J Biol Chem 1995;270:24420-27.

[13]   Segev N, Mulholland J, Botstein D. The yeast GTP-binding YPT1 protein and a mammalian counterpart are associated with the secretion machinery. Cell 1988; 52: 915-24

[14]   Touchot N, Chardin P, Tavitian A. Four additional members of the ras gene superfamily isolated by an oligonucleotide strategy: molecular cloning of YPT-related cDNAs from a rat brain library. Proc Natl Acad Sci U S A 1987;84:8210-14.

[15]   Salminen A, Novick PJ. A ras-like protein is required for a post-Golgi event in yeast secretion. Cell 1987;49:527-38.

[16]   Zerial M, McBride H. Rab proteins as membrane organizers. Nat Rev Mol Cell Biol 2001;2:107-17.

[17]   Stenmark H. Rab GTPases as coordinators of vesicle traffic. Nat Rev Mol Cell Biol 2009;10:513-25.

[18]   Jekely G. Small GTPases and the evolution of the eukaryotic cell. Bioessays 2003;25:1129-38.

[19]   Pereira-Leal JB, Seabra MC. Evolution of the Rab family of small GTP-binding proteins. J Mol Biol 2001;313:889-901.

[20]   Dacks JB, Peden AA, Field MC. Evolution of specificity in the eukaryotic endomembrane system. Int J Biochem Cell Biol 2009;41:330-40.

[21]   Field H, Farjah M, Pal A, Gull K, Field MC. Complexity of trypanosomatid endocytosis pathways revealed by Rab4 and Rab5 isoforms in Trypanosoma brucei. J Biol Chem 1998;273:32102-10.

[22]   Rutherford S, Moore I. The Arabidopsis Rab GTPase family: another enigma variation. Curr Opin Plant Biol 2002;5:518-28.

[23]   Gurkan C, Lapp H, Alory C, Su AI, Hogenesch JB, Balch WE. Large-scale profiling of Rab GTPase trafficking networks: the membrome. Mol Biol Cell. 2005;16:3847-64.

[24]   Alvarez-Dominguez C, Stahl PD. Interferon-gamma selectively induces Rab5a synthesis and processing in mononuclear cells. J Biol Chem 1998;273:33901-04.

[25]   Alvarez-Dominguez C, Stahl PD. Increased expression of Rab5a correlates directly with accelerated maturation of Listeria monocytogenes phagosomes. J Biol Chem 1999;274:11459-62.

[26]   Barbieri MA, Roberts RL, Gumusboga A, *et al.* Epidermal growth factor and membrane trafficking. EGF receptor activation of endocytosis requires Rab5a. J Cell Biol 2000;151:539-50.

[27]   Ostrowski M, Carmo NB, Krumeich S, *et al.* Rab27a and Rab27b control different steps of the exosome secretion pathway. Nat Cell Biol. 2010;12:19-30; sup pp 1-13.

[28]   Su AI, Wiltshire T, Batalov S, *et al.* A gene atlas of the mouse and human protein-encoding transcriptomes. Proc Natl Acad Sci U S A 2004;101:6062-67.

[29]   Chavrier P, Vingron M, Sander C, Simons K, Zerial M. Molecular cloning of YPT1/SEC4-related cDNAs from an epithelial cell line. Mol Cell Biol 1990;10:6578-85.

[30]   Huber LA, Pimplikar S, Parton RG, Virta H, Zerial M, Simons K. Rab8, a small GTPase involved in vesicular traffic between the TGN and the basolateral plasma membrane. J Cell Biol 1993;123:35-45.

[31]   Zahraoui A, Joberty G, Arpin M, *et al.* A small rab GTPase is distributed in cytoplasmic vesicles in non polarized cells but colocalizes with the tight junction marker ZO-1 in polarized epithelial cells. J Cell Biol 1994;124:101-15.

[32]   Schuck S, Gerl MJ, Ang A, *et al.* Rab10 is involved in basolateral transport in polarized Madin-Darby canine kidney cells. Traffic 2007;8:47-60.

[33]   Babbey CM, Ahktar N, Wang E, Chen CC, Grant BD, Dunn KW. Rab10 regulates membrane transport through early endosomes of polarized Madin-Darby canine kidney cells. Mol Biol Cell 2006;17:3156-75.

[34]   Sano H, Eguez L, Teruel MN, *et al.* Rab10, a target of the AS160 Rab GAP, is required for insulin-stimulated translocation of GLUT4 to the adipocyte plasma membrane. Cell Metab 2007;5:293-303.

[35]   Watson RT, Pessin JE. Intracellular organization of insulin signaling and GLUT4 translocation. Recent Prog Horm Res 2001;56:175-93.

[36]   Sano H, Roach WG, Peck GR, Fukuda M, Lienhard GE. Rab10 in insulin-stimulated GLUT4 translocation. Biochem J 2008;411:89-95.

[37]   Haas A. The phagosome: compartment with a license to kill. Traffic 2007;8:311-30.

[38]   Kinchen JM, Ravichandran KS. Phagosome maturation: going through the acid test. Nat Rev Mol Cell Biol 2008;9:781-95.

[39]   Stuart LM, Boulais J, Charriere GM, *et al*. A systems biology analysis of the Drosophila phagosome. Nature 2007;445:95-101.

[40]   Garin J, Diez R, Kieffer S, *et al*. The phagosome proteome: insight into phagosome functions. J Cell Biol 2001;152:165-180.

[41]   Cardoso CM, Jordao L, Vieira OV. Rab10 regulates phagosome maturation and its overexpression rescues Mycobacterium-containing phagosomes maturation. Traffic 2010;11:221-35.

[42]   Huang B, Hubber A, McDonough JA, Roy CR, Scidmore MA, Carlyon JA. The Anaplasma phagocytophilum-occupied vacuole selectively recruits Rab-GTPases that are predominantly associated with recycling endosomes. Cell Microbiol 2010 Sep 1;12(9):1292-307.

[43]   Zhu AX, Zhao Y, Flier JS. Molecular cloning of two small GTP-binding proteins from human skeletal muscle. Biochem Biophys Res Commun 1994;205:1875-82.

[44]   Kouranti I, Sachse M, Arouche N, Goud B, Echard A. Rab35 regulates an endocytic recycling pathway essential for the terminal steps of cytokinesis. Curr Biol 2006;16:1719-25.

[45]   Patino-Lopez G, Dong X, Ben-Aissa K, *et al*. Rab35 and its GAP EPI64C in T cells regulate receptor recycling and immunological synapse formation. J Biol Chem 2008;283:18323-30.

[46]   Hsu C, Morohashi Y, Yoshimura S, *et al*. Regulation of exosome secretion by Rab35 and its GTPase-activating proteins TBC1D10A-C. J Cell Biol 2010;189:223-32.

[47]   Sato M, Sato K, Liou W, Pant S, Harada A, Grant BD. Regulation of endocytic recycling by C. elegans Rab35 and its regulator RME-4, a coated-pit protein. EMBO J 2008;27:1183-96.

[48]   Smith AC, Heo WD, Braun V, *et al*. A network of Rab GTPases controls phagosome maturation and is modulated by Salmonella enterica serovar Typhimurium. J Cell Biol 2007;176:263-68.

[49]   Chevallier J, Koop C, Srivastava A, Petrie RJ, Lamarche-Vane N, Presley JF. Rab35 regulates neurite outgrowth and cell shape. FEBS Lett 2009;583:1096-1101.

[50]   Zhang J, Fonovic M, Suyama K, Bogyo M, Scott MP. Rab35 controls actin bundling by recruiting fascin as an effector protein. Science 2009;325:1250-4.

[51]   Shintani M, Tada M, Kobayashi T, Kajiho H, Kontani K, Katada T. Characterization of Rab45/RASEF containing EF-hand domain and a coiled-coil motif as a self-associating GTPase. Biochem Biophys Res Commun. 2007;357:661-67.

[52]   Enright AJ, Iliopoulos I, Kyrpides NC, Ouzounis CA. Protein interaction maps for complete genomes based on gene fusion events. Nature 1999;402:86-90.

[53]   Marcotte EM, Pellegrini M, Ng HL, Rice DW, Yeates TO, Eisenberg D. Detecting protein function and protein-protein interactions from genome sequences. Science 1999;285:751-3.

[54]   Grabarek Z. Structural basis for diversity of the EF-hand calcium-binding proteins. J Mol Biol 2006;359:509-25.

[55]   Haiech J, Moulhaye SB, Kilhoffer MC. The EF-Handome: combining comparative genomic study using FamDBtool, a new bioinformatics tool, and the network of expertise of the European Calcium Society. Biochim Biophys Acta.2004;1742:179-83.

[56]   Horgan CP, McCaffrey MW. The dynamic Rab11-FIPs. Biochem Soc Trans 2009;37:1032-36.

[57]   Sweetser DA, Peniket AJ, Haaland C, *et al*. Delineation of the minimal commonly deleted segment and identification of candidate tumor-suppressor genes in del(9q) acute myeloid leukemia. Genes Chromosomes Cancer 2005;44:279-91.

[58]   Jonsson G, Bendahl PO, Sandberg T, *et al*. Mapping of a novel ocular and cutaneous malignant melanoma susceptibility locus to chromosome 9q21.32. J Natl Cancer Inst 2005;97:1377-82.

[59]   Maat W, Beiboer SH, Jager MJ, Luyten GP, Gruis NA, van der Velden PA. Epigenetic regulation identifies RASEF as a tumor-suppressor gene in uveal melanoma. Invest Ophthalmol Vis Sci 2008;49:1291-98.

[60]   Gruenberg J, van der Goot FG. Mechanisms of pathogen entry through the endosomal compartments. Nat Rev Mol Cell Biol 2006;7:495-504.

[61]   Gibbs JB, Sigal IS, Poe M, Scolnick EM. Intrinsic GTPase activity distinguishes normal and oncogenic ras p21 molecules. Proc Natl Acad Sci U S A 1984;81:5704-08.

[62]   Bos JL. ras oncogenes in human cancer: a review. Cancer Res 1989;49:4682-89.

[63] Ellenbroek SI, Collard JG. Rho GTPases: functions and association with cancer. Clin Exp Metastasis 2007;24:657-72.

[64] Forbes SA, Bhamra G, Bamford S, *et al.* The Catalogue of Somatic Mutations in Cancer (COSMIC). Curr Protoc Hum Genet 2008;Chapter 10:Unit 10.11.

[65] Futreal PA, Coin L, Marshall M, *et al.* A census of human cancer genes. Nat Rev Cancer 2004;4:177-183.

[66] Sjoblom T, Jones S, Wood LD, *et al.* The consensus coding sequences of human breast and colorectal cancers. Science 2006;314:268-74.

[67] Parsons DW, Jones S, Zhang X, *et al.* An integrated genomic analysis of human glioblastoma multiforme. Science. 2008;321:1807-12.

[68] Dalgliesh GL, Furge K, Greenman C, *et al.* Systematic sequencing of renal carcinoma reveals inactivation of histone modifying genes. Nature 2010;463:360-63.

[69] Pellinen T, Tuomi S, Arjonen A, *et al.* Integrin trafficking regulated by Rab21 is necessary for cytokinesis. Dev Cell 2008;15:371-85.

[70] Tidyman WE, Rauen KA. The RASopathies: developmental syndromes of Ras/MAPK pathway dysregulation. Curr Opin Genet Dev 2009;19:230-6.

[71] Giannandrea M, Bianchi V, Mignogna ML, *et al.* Mutations in the small GTPase gene RAB39B are responsible for X-linked mental retardation associated with autism, epilepsy, and macrocephaly. Am J Hum Genet 2010;86:185-95.

[72] Jenkins D, Seelow D, Jehee FS, *et al.* RAB23 mutations in Carpenter syndrome imply an unexpected role for hedgehog signaling in cranial-suture development and obesity. Am J Hum Genet 2007;80:1162-70.

[73] Oiso N, Riddle SR, Serikawa T, Kuramoto T, Spritz RA. The rat Ruby (R) locus is Rab38: identical mutations in Fawn-hooded and Tester-Moriyama rats derived from an ancestral Long Evans rat sub-strain. Mamm Genome 2004;15:307-14.

[74] Spinosa MR, Progida C, De Luca A, Colucci AM, Alifano P, Bucci C. Functional characterization of Rab7 mutant proteins associated with Charcot-Marie-Tooth type 2B disease. J Neurosci 2008;28:1640-48.

[75] McCray BA, Skordalakes E, Taylor JP. Disease mutations in Rab7 result in unregulated nucleotide exchange and inappropriate activation. Hum Mol Genet 2010;19:1033-47.

[76] Dobie KW, Kennedy CD, Velasco VM, *et al.* Identification of chromosome inheritance modifiers in Drosophila melanogaster. Genetics 2001;157:1623-37.

[77] Loftus SK, Larson DM, Baxter LL, *et al.* Mutation of melanosome protein RAB38 in chocolate mice. Proc Natl Acad Sci U S A 2002;99:4471-76.

[78] Jedd G, Mulholland J, Segev N. Two new Ypt GTPases are required for exit from the yeast trans-Golgi compartment. J Cell Biol 1997;137:563-80.

[79] Chia WJ, Tang BL. Emerging roles for Rab family GTPases in human cancer. Biochim Biophys Acta 2009;1795:110-16.

[80] Agarwal R, Jurisica I, Mills GB, Cheng KW. The emerging role of the RAB25 small GTPase in cancer. Traffic 2009;10:1561-68.

[81] Cheng KW, Lahad JP, Kuo WL, *et al.* The RAB25 small GTPase determines aggressiveness of ovarian and breast cancers. Nat Med 2004;10:1251-56.

[82] Cheng JM, Volk L, Janaki DK, Vyakaranam S, Ran S, Rao KA. Tumor suppressor function of Rab25 in triple-negative breast cancer. Int J Cancer 2009 Jun 15;126(12):2799-812.

[83] Nam KT, Lee HJ, Smith JJ, *et al.* Loss of Rab25 promotes the development of intestinal neoplasia in mice and is associated with human colorectal adenocarcinomas. J Clin Invest 2010;120:840-49.

[84] Rhodes DR, Kalyana-Sundaram S, Mahavisno V, *et al.* Oncomine 3.0: genes, pathways, and networks in a collection of 18,000 cancer gene expression profiles. Neoplasia 2007;9:166-80.

[85] Shimada K, Uzawa K, Kato M, *et al.* Aberrant expression of RAB1A in human tongue cancer. Br J Cancer 2005;92:1915-21.

[86] Fukui K, Tamura S, Wada A, *et al.* Expression of Rab5a in hepatocellular carcinoma: Possible involvement in epidermal growth factor signaling. Hepatol Res 2007;37:957-65.

[87] Turner N, Lambros MB, Horlings HM, *et al.* Integrative molecular profiling of triple negative breast cancers identifies amplicon drivers and potential therapeutic targets. Oncogene 2010;29:2013-23.

[88] Kotzsch M, Sieuwerts AM, Grosser M, *et al.* Urokinase receptor splice variant uPAR-del4/5-associated gene expression in breast cancer: identification of rab31 as an independent prognostic factor. Breast Cancer Res Treat 2008;111:229-40.

[89] Hou Q, Wu YH, Grabsch H, *et al.* Integrative genomics identifies RAB23 as an invasion mediator gene in diffuse-type gastric cancer. Cancer Res 2008;68:4623-30.

[90] Wu CY, Tseng RC, Hsu HS, Wang YC, Hsu MT. Frequent down-regulation of hRAB37 in metastatic tumor by genetic and epigenetic mechanisms in lung cancer. Lung Cancer 2009;63:360-367.

[91] Verhoeven K, De Jonghe P, Coen K, *et al.* Mutations in the small GTP-ase late endosomal protein RAB7 cause Charcot-Marie-Tooth type 2B neuropathy. Am J Hum Genet 2003;72:722-27.

[92] Altschul SF, Madden TL, Schaffer AA, *et al.* Gapped BLAST and PSI-BLAST: a new generation of protein database search programs. Nucleic Acids Res 1997;25:3389-3402.

[93] Wu C, Orozco C, Boyer J, *et al.* BioGPS: an extensible and customizable portal for querying and organizing gene annotation resources. Genome Biol 2009;10:R130.

[94] Letunic I, Doerks T, Bork P. SMART 6: recent updates and new developments. Nucleic Acids Res. 2009;37:D229-32.

[95] Schultz J, Milpetz F, Bork P, Ponting CP. SMART, a simple modular architecture research tool: identification of signaling domains. Proc Natl Acad Sci U S A 1998;95:5857-64.

INDEX

A

Actin, 9, 25, 34, 41, 47, 50-51, 55-56, 60, 62-64, 77-83, 88, 99-100, 125, 160-161
ALS, 95-97
Alsin, 7, 95-97
Alzheimer's disease, 27
APPL1/2, 97, 100-101
ARF, 5, 7, 25-26, 108-109, 111-112. 116, 125
Autophagy, 6, 11, 18, 20-23, 26, 29, 133, 149

B

Bicaudal-D1/BICD1, 10, 36, 38, 40

C

Cancer, 3, 5, 27, 101, 103, 162-163
Carpenter syndrome, 163
Cascade, 7, 25-26, 48-49, 93, 96, 101-103, 117
Charcot Marie Tooth Type 2B/CMT2B, 138, 162-163
Clathrin, 8, 93, 96, 98, 101-102, 108, 111-113, 123-125, 128, 144-145, 148, 160
Coordination, 5-7, 25-26, 29, 88, 95, 123, 125, 138-139, 145
COG, 20, 23, 36, 38
Conservation, 3, 20, 24, 34, 155
Cytokinesis, 10, 36-37, 41, 108, 111, 117, 160
Cystic fibrosis, 28

D

DENN, 6-7, 22, 36, 58
Diabetes, 3, 5, 26
Docking, 10-11, 34, 36, 41, 47-49, 51, 55, 61-64, 66-67, 77, 83-88, 109, 115

E

Early endosome, 7, 10, 12, 19, 38, 49, 93-103, 108-112, 116, 123, 132-133, 135, 137, 144-145, 149
EEA1, 10, 97, 99-101, 103, 112, 135
Endocytosis, 3-4, 9, 26-28, 34, 55-56, 67, 93-97, 99-103, 123-124, 128, 136, 145, 149, 160
Endosome Positioning, 134
Evolution, 155-159, 161, 163
Exocyst, 10-11, 41, 47-51, 63, 128
Exocytosis, 3, 11-12, 24, 27, 34, 40, 47, 50-51, 55-59, 61, 64-68, 84-85, 88, 115, 128

F

Fusion, 3-11, 18, 25-26, 40-41, 47, 49-51, 59-60, 64-68, 77-78, 84-88, 93-103, 112-113, 116, 125, 128, 132-134, 136-139, 147-150, 160-161
FYVE, 10, 99, 112, 135-136

G

GAP, 4-8, 18, 22, 22-25, 27-29, 35-36, 39, 41, 48-49, 58-59, 79, 81, 94-97, 101, 116, 133, 138, 150
GAPCenA, 35-36, 41, 116
GAPex-5, 7, 95-97
GCC185, 36-38, 147-151
GDF, 5-6, 8, 47, 58, 132-133, 146, 150
GDI, 5-6, 8, 36, 47-48, 58, 81, 98, 133, 137, 145-146, 150-151
GEF, 4-8, 18, 20, 22, 24-29, 35, 39-41, 47-51, 58-59, 79, 81, 93-98, 100-103, 116, 132-133, 137-138, 146

www.ingramcontent.com/pod-product-compliance
Lightning Source LLC
Chambersburg PA
CBHW041705210326
41598CB00007B/543